Mathematics

for Caribbean Schools

Students' Book 1

Third Edition

Althea A Foster   Terry Tomlinson

HODDER
EDUCATION
AN HACHETTE UK COMPANY

Orders: please contact Hachette UK Distribution, Hely Hutchinson Centre, Milton Road, Didcot, Oxfordshire, OX11 7HH. Telephone: +44 (0)1235 827827. Email education@hachette.co.uk Lines are open from 9 a.m. to 5 p.m., Monday to Friday. You can also order through our website: www.hoddereducation.com

First published by Pearson Education Limited
This edition published 2007
Published from 2015 by Hodder Education,
An Hachette UK Company
Carmelite House
50 Victoria Embankment
London EC4Y 0DZ
www.hoddereducation.com

2024
IMP 10

ISBN: 978-1-4058-4777-3

Set in 9.5/12 pt Stone Serif

Printed in Great Britain by Ashford Colour Press Ltd

Hachette UK's policy is to use papers that are natural, renewable and recyclable products and made from wood grown in well-managed forests and other controlled sources. The logging and manufacturing processes are expected to conform to the environmental regulations of the country of origin.

Acknowledgements
The Publishers wish to acknowledge the work of J B Channon, A McLeish Smith, H C Head and M F Macrae which laid the foundation for this series.

The Publishers are grateful to the following for their permission to reproduce photographs:
J. Allan Cash for Fig. 7.33(a) and Nestlé Co. Ltd. for Fig. 7.15(a).

Preface

This series of four volumes, of which this is the first, is intended for use primarily by students who are preparing to sit the certificate examinations held by the Caribbean Examinations Council and by the individual countries in the Caribbean.

Each volume represents material which may be covered by the average student in one year approximately (although some students may need more time) so that there is ample time for the series to be completed over a four to five year period. Emphasis has been placed on detailed explanation of concepts, principles and methods of working out problems. In addition, many problems have been included, both as worked examples to illustrate particular approaches to solving problems in the teaching text, and also as exercises for practice and reinforcement of the concepts. This has been done in a deliberate attempt to provide a stimulus to teachers in developing their strategies for teaching different topics; but, especially, to provide guidance to students as they work or revise on their own.

At the end of each chapter a summary of the main points developed in the chapter has also been included. The summary is followed by practice exercises to give students an opportunity to consolidate the material in the chapter. Key reference words are printed in bold type throughout the text. The text has been arranged sequentially so that each chapter may use material covered in previous chapters. However, in order to use relevant examples at some points in the text, ideas/concepts, which may need a short explanation or reminder by the teacher, may be introduced, e.g. {counting numbers} in Example 2, Chapter 2 – a concept which is fully explained in Chapter 3. A few chapters are independent of the previous chapters and so may be omitted without loss of continuity at a first working. However, it is believed that the text will be used most efficiently by working through the chapters in the given order.

We are indebted to both the teachers and the students whose questions and responses over the years have undoubtedly influenced our thinking and our approach to the teaching of the subject as exemplified in this series.

This revised edition seeks to incorporate changes in keeping with amendments to the syllabuses in the various Caribbean countries and that of the Caribbean Examinations Council. Attention has also been given to the suggestions of Caribbean teachers whose positive reaction and responses to the series have been most encouraging.

To the teacher

This edition of the series of four texts, revised with respect to content and its sequencing and to pedagogy (to a lesser extent), will be found useful in providing help and guidance in how the topics are taught and the order in which they are taken. The texts do not attempt to prescribe specific approaches to the teaching of any topic. Teachers are free to adopt or modify the suggested approaches. It is the teacher who must decide on the methodology to be used to create the most suitable learning conditions in the classroom and to provide challenging activities which motivate the students to think and yet give them a chance to succeed in finding solutions.

It is vitally important that teachers use the 'Oral' exercises to initiate class discussion in the careful development of concepts. Whenever the opportunity arises, teachers are urged to use, and thus reinforce, concepts taught earlier, so that, for example, the use of estimation and approximation in the calculation of numerical values is practised throughout the course.

In addition, it is widely accepted that learning is aided by doing. Thus concrete/practical examples and real-life applications must be provided, whenever possible, as well as the use of pictures, flow charts and other diagrammatic representations to deepen the understanding of abstract/theoretical ideas. Some problems in the exercises require the use of diagrams which are tedious and/or time-consuming to produce. In

order to keep down costs to the student/school and reduce the tedium and time wastage, it is suggested that teachers should use a copying machine for producing the necessary material.

Group work

In some instances, students may improve their performance by working in a group where the insecurity and stigma of not knowing 'where to start' in solving a problem is not as manifest. Throughout the texts, therefore, in addition to the 'Oral' exercises, we have also identified relevant exercises in which the open-ended questions are applicable to 'Group Work', namely, in Chapters 4, 7, 12, 13, 18 and 19. After alternative strategies for tackling a problem have been discussed by the group, the actual solution(s) may be carried out in the initial group, in smaller groups, or by students working individually. It is important that adequate *time and thought* be given to the process of trying different approaches, and to considering the reasonableness of the 'answer'.

'Good' social behaviour, such as listening to and respecting another person's suggestions and working co-operatively in teams, is also a worth-while long-term benefit of such group activities.

In an attempt to assist the teacher and the student in quickly identifying and reviewing necessary background knowledge, we have included information to be referenced under the heading 'Pre-requisites' at the beginning of each chapter. It must be remembered that the main new ideas of each chapter are highlighted in the 'Summary' at the end of the chapter. The point at which the 'Revision Exercises and Tests' are used is a matter for the individual teacher's choice. If the series of related chapters has been taught as sequenced in the printed text, then the revision material may be used at that time, or may be omitted until the entire book has been completed. This material may also be used as supplemental problems for the quicker students. In order to comply with the requests of teachers, a 'Practice Examination (Papers 1 and 2)' has also been included at the end of the book.

Use of the electronic calculator

The widespread use of the electronic calculator in today's world demands that suitable attention to its potential and usefulness be given in the early years of secondary schooling. However, it is also very important to ensure that the calculator, by its premature introduction, is not misused and seen only as a 'number cruncher'.

In this series the use of the calculator is formally introduced in Book 2. In Book 1, the calculations to be performed have for the most part been maintained within limits which do not require the use of a calculator. It is believed that, in the first year of secondary schooling, the opportunity must be provided for the understanding and use of basic mathematical processes which utilise concepts such as place value in numbers, the sequencing of operations, approximation and estimation, factorisation and the application of short cuts in computation. Such understanding is fundamental to the efficient and effective utilisation of this potentially valuable tool in promoting and aiding the development of mathematical ideas and thought.

Finally, it is unfortunate that Mathematics is perceived by a large majority of students as a 'necessary evil', a subject that they have to 'get at CXC' in order to become employable. Teachers have a significant responsibility in helping to change this attitude, and in having students appreciate that, by acquiring the skills and techniques to solve problems in mathematics, they also acquire the tools and the ability to solve problems in the real world.

To the student

Before attempting the problems in the exercises, study and discuss the worked examples until you understand the concept. The oral exercises are intended to encourage discussion. This will help to clarify lingering misunderstandings. In particular, in solving word problems you have to get thoroughly familiar with the problem. The next step involves translating the problem into mathematical symbols and language, for example, into an equation or an inequality, or into a graph. The next steps are applying the required mathematical operations, and finally, checking the original word problem. Remember to check that the variables are in the same units. Another useful hint is to look for patterns in similar problems and in the methods of solution. Concrete materials such as cans, coins, balls, stones and boxes are very useful aids for clearing up doubts − not for wasting time!

A calculator is an excellent machine when used wisely, but you must bear in mind that it needs to be used by a clear-thinking human who fully comprehends the mathematical concepts. Dividing 4.0 by 3 and giving your answer as 1.3333333 indicates, among other things, a lack of appreciation of the idea of accuracy.

Nothing can replace the neat appearance of an answer that is well set out − the date, the page and exercise from which the problem has been taken, a statement of the facts given, the necessary calculations performed and the conclusions drawn which result in a final answer. This whole process helps you to think clearly.

Finally, the more a concept is applied, the clearer it becomes. Thus, PRACTICE and more practice in working out examples is an essential ingredient to success.

Althea A. Foster
E. M. Tomlinson

Contents

Contents

Number systems (1)
The place-value system and number bases

Counting

It is most likely that mathematics began when people started to count and measure. Counting and measuring are part of everyday life. Nearly every language in the world contains words for numbers and measures.

People have always used their fingers to help them when counting. This led them to collect numbers in groups: sometimes fives (fingers of one hand), sometimes tens (both hands) and even in groups of twenty (fingers and toes). For example, a herd boy with 23 cattle might say that he had 'four fives and three' cattle, or 'two tens and three' cattle, or 'one twenty and three' cattle. It would depend on the custom and language of his people. In every case, the number of cattle would be the same.

When people group numbers in fives we say that they are using a **base five** method of counting. Most people use **base ten** when counting. For this reason, base ten is used internationally.

Symbols for numbers

As civilizations developed, spoken languages were written down using **symbols**. Symbols are letters and marks which represent sounds and ideas. Numbers were also written down. We use the word **numerals** for number symbols. For example, $\mathcal{N}$ and ― are the numerals for *two* in Arabic and Chinese. We use the numeral 2 for *two*.

Tally system

Tally marks were probably the first numerals. Herdsmen made tally marks to represent the numbers of animals they had. The tally marks were scratched on stones or sometimes cut on sticks (Fig. 1.1).

Fig. 1.1

We still use the tally system. It is very useful when counting a large number of objects. We usually group tally marks in fives: hence ⊞ ⊞ ⊞ || means *three fives and two*, or *seventeen*. Notice that the fifth tally is marked across the other four: |||| = 4, ⊞ = 5.

Exercise 1a (Oral)

What numbers do the following tally marks represent?

1. ⊞⊞||||
2. ⊞|||
3. ⊞ ⊞ ⊞ |
4. ⊞⊞ ⊞⊞
5. ⊞ ⊞ ⊞ ⊞ ⊞ ⊞ ⊞ ||
6. ⊞ ⊞ || ⊞ ⊞
7. ⊞⊞ ⊞ ⊞ ⊞ ⊞⊞ ⊞ ⊞ ⊞ |||
8. ⊞ ⊞ ⊞ ⊞ ⊞ ⊞ ⊞
9. ⊞⊞⊞⊞⊞⊞⊞ ⊞⊞⊞⊞⊞⊞ ||||
10. ⊞ ⊞ ⊞ ⊞ ⊞ |
11. ⊞ ⊞ ⊞ ||| ⊞ ⊞ ⊞
12. ⊞ ⊞ ⊞ ⊞ |

The place-value system

The tally system is inconvenient when numbers are large or fractional. The most convenient way of representing numbers is to use a **place-value system**. The base ten place-value system uses ten symbols: 1, 2, 3, 4, 5, 6, 7, 8, 9 for one to nine and 0 for zero. These symbols are sometimes called **digits**.

In this system, the numbers of units, tens, hundreds; ..., are each represented by a single digit. However, each digit must be written in a special position or place. For any whole number, the **units place** is at the right-hand end of the number, the **tens place** is next to the units place on the left, and so on. For example, the number 3 902 means 3 thousands, 9 hundreds, 0 tens and 2 units (Fig. 1.2).

Fig. 1.2

As numbers grow we use more places. The symbol 0, zero, shows that a place is empty. Hence ten thousand is 10 000, using two digits and five places.

An advantage of the place-value system is that it can represent fractions. Just as there are places for whole numbers such as hundreds, tens, units, so we make places for decimal fractions such as tenths, hundredths, thousandths, and so on. However, it is necessary to show where the whole numbers end and the decimal fractions begin. We use a **decimal point** to separate the whole numbers from the decimal fractions (Fig. 1.3).

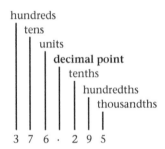

Fig. 1.2

In Fig. 1.3, the 2 in the first position after the decimal point shows that there are *two tenths*. 9 is in the second position after the point; this stands for *nine hundredths*.

Exercise 1b

① The value of the 5 in 6508 is five hundred. What is the value of each of the following?
   (a) the 4 in 6402
   (b) the 4 in 2984
   (c) the 2 in 6402
   (d) the 2 in 2984
   (e) the 2 in 10 269
   (f) the 9 in 2984
   (g) the 9 in 10 269
   (h) the 6 in 6402
   (i) the 6 in 10 269
   (j) the 0 in 6402
   (k) the 0 in 10 269
   (l) the 1 in 10 269

② What is the value of each of the following?
   (a) the 8 in 1.85
   (b) the 8 in 16.08
   (c) the 5 in 1.85
   (d) the 5 in 5.691
   (e) the 6 in 5.691
   (f) the 0 in 19.08
   (g) the 1 in 1.85
   (h) the 1 in 5.691
   (i) the 1 in 19.08
   (j) the 7 in 0.072

③ A student does the sum 352 + 79 and writes:

   352
   + 79
   ─────

   Why is it not correct to write the sum like this? How should the sum be written?

④ The following additions and subtractions have been set out badly.
   Set them out correctly. (Do not work out the answers.)

   (a)   3107
        + 26
         147
        ─────

   (b)   6203
        + 97
        ─────

   (c)   1429
        + 6580
        ─────

   (d)   6700
        + 34
        ─────

⑤ Rearrange the digits in the following numbers to make (i) the highest possible number, (ii) the lowest possible number.
   (a) 263
   (b) 728
   (c) 506
   (d) 2815
   (e) 2407

⑥ The following additions and subtractions have been set out badly.
   Set them out correctly. (Do not work out the answers.)

   (a)   60.91
        + 3.2
        ──────

   (b)   26.3
        − 1.7
        ──────

   (c) $4.49
        + $56.20
        ──────

   (d)   42.5
        − 9.65
        ──────

Number bases

For most purposes, numbers are written in the **base ten** or **denary** system. The placing of the digits shows their values. For example 2053 means **2** thousands, **0** hundreds, **5** tens, **3** units:

$$2053 = 2 \times 1000 + 0 \times 100 + 5 \times 10 + 3 \times 1$$

while

$$5302 = 5 \times 1000 + 3 \times 100 + 0 \times 10 + 2 \times 1$$

This may also be written:
$$5302 = 2 \times 1 + 0 \times 10 + 3 \times 100 + 5 \times 1000$$

It is possible to expand any denary number in powers of ten:
$$25 = 2 \times 10 + 5 \times 1$$
$$147 = 1 \times 100 + 4 \times 10 + 7 \times 1$$
$$3706 = 3 \times 1000 + 7 \times 100 + 0 \times 10 + 6 \times 1$$

Other number systems are sometimes used. For example, the **base eight** or **octal** system is based on powers of eight. In base eight, 25 means **2** eights, **5** units.

$$25_{eight} = 2 \times 8 + 5 \times 1$$
$$147_{eight} = 1 \text{ sixty-four, } 4 \text{ eights, } 7 \text{ units}$$
$$= 1 \times 64 + 4 \times 8 + 7 \times 1$$
$$= 7 \times 1 + 4 \times 8 + 1 \times 64$$

Fig. 1.4 shows the place values of the digits in the number 147_{eight}.

sixty-fours

eights

units

1 4 7

Fig. 1.4

Notice that 147_{eight} is short for 147 in base eight.

Exercise 1c (Oral)

Expand the following to show the place values of the digits.

① 2389_{ten} ② 647_{eight}

③ $35\,154_{eight}$ ④ 4102_{five}

⑤ 1011_{two} ⑥ $22\,010_{three}$

⑦ $26\,523_{eight}$ ⑧ 1100_{two}

⑨ $2\,102_{three}$ ⑩ $11\,001_{two}$

Other bases of counting: seven, sixty, twenty-four

There are seven days in a week. Suppose that a baby is 2 weeks and 5 days old. This means that it is 2 lots of 7 days and 5 days old, 19 days altogether.

Example 1

Find the total of 1 week 5 days, 6 days and 3 weeks 4 days. Give the answer (a) in weeks and days, (b) in days.

(a) *working*:

	wk	d
	1	5
		6
	3	4
	6	1

method: In the days column,
$$(5 + 6 + 4) \text{ days} = 15 \text{ days}$$
$$= 2 \times 7 \text{ days} + 1 \text{ day}$$
$$= 2 \text{ weeks} + 1 \text{ day}$$

Write down 1 day and carry 2 weeks.
answer: 6 weeks and 1 day

(b) 6 wk 1 d $= 6 \times 7$ days + 1 day
$$= 42 \text{ days} + 1 \text{ day}$$
$$= 43 \text{ days}$$

Example 2

Today is Wednesday. What day of the week will it be in 160 days' time?

Change 160 days to days and weeks.
160 d $= 160 \div 7$ wk $= 22$ wk 6 d
After 22 weeks it will be Wednesday again.
Count a further 6 days on from Wednesday:

  Thursday, Friday, Saturday,
  Sunday, Monday, *Tuesday*

It will be Tuesday in 160 days' time.

There are 60 seconds in a minute and 60 minutes in an hour.

Example 3

Find the number of seconds in 3 min 49 s.

Number of seconds in 3 min   = 3 × 60 s
                             = 180 s
Number of seconds in 3 min 49 s = 180 s + 49 s
                             = 229 s

Example 4

Add the following times together. Give the answer in hours and minutes.

3 h 40 min, 2 h 25 min, 28 min, 1 h 35 min

working: h   min
            3   40
            2   25
                28
            1   35
            8   8

method: In the minutes column:

(40 + 25 + 28 + 35) min
        = 128 min
        = 2 × 60 min + 8 min
        = 2 h + 8 min

Write down 8 min and carry 2 h.
answer: 8 h 8 min

Example 5

Find the difference between 6 days 3 hours and 2 days 13 hours.

working: d   h
            6   3
          − 2   13
            3   14

method: In the hours column, it is impossible to subtract 13 h from 3 h. Borrow 1 day, i.e. 24 h, from the days column then subtract in the usual way:

  d   h   d   h   d   h
  6   3   →   5   3   + 24 →   5   27
− 2 13   − 2 13   − 2   13
  3 14   3 14   3   14

answer: 3 days 14 hours

---

1 A baby is 3 weeks and 4 days old. What is its age in days?

2 During a dry season it did not rain for 128 days. How many weeks and days is this?

3 Add the following.

(a) wk d   (b) wk d
      4 4   1 5
      1 2   2 6
    _____   _____

(c) wk d   (d) wk d
      1 4   3 1
      1 4   2 6
    _____   _____

4 Suppose today is Thursday.
   What day of the week will it be after
   (a) 20 days   (b) 50 days
   (c) 70 days   (d) 100 days?

5 Find the number of seconds in the following times.
   (a) 2 min   (b) 10 min 54 s
   (c) 3 min 22 s   (d) 1 h

6 Find the number of minutes in the following.
   (a) 240 s   (b) 5 h
   (c) 2 h 34 min   (d) 1 day

7 Find the number of hours in the following.
   (a) 420 min   (b) 2 days
   (c) 3 d 11 h   (d) 1 week

8 Add the following.

(a) min   s   (b) min   s
      1   42   24   5
          37   5   55
    _____   _____

(c) h   min   (d) h   min   s
     2   12   1   34   28
     1   49   45   50
    _____   _____

(e) d   h   (f) d   h   min
     4   14   3   15   40
     1   22   2   8   40
    _____   _____

---

9 Do the following subtractions.

(a)
```
   min  s
   15   10
−   8   25
──────────
```

(b)
```
   wk   d
    5    3
−        6
──────────
```

(c)
```
    d    h
    6   11
−   4   17
──────────
```

(d)
```
   h   min   s
   3    24   8
− 1    50  28
──────────────
```

Converting base ten numbers to other bases

To convert from base ten to another base, we can find the number in the new base by inspection.

Example 6

Convert 37_{ten} (a) to base eight, (b) to base five.

(a) Since 37 is less than 64, there are no sixty-fours in 37. To find the number of eights in 37, divide by 8.

$$37 \div 8 = 4, \text{ remainder } 5$$
$$37 = 4 \text{ eights} + 5 \text{ units}$$
$$37_{ten} = 45_{eight}$$
$$Check: 45_{eight} = 4 \times 8 + 5 \times 1$$
$$= 32 + 5 = 37_{ten}$$

(b) Since 37 is greater than 25, there must be a 25 in 37.

$$25 \div 25 = 1, \text{ remainder } 12$$
$$37 = 1 \text{ twenty-five} + 12 \text{ units}$$

Consider the 12 units. Since $12 > 5$, there must be some fives in 12.

$$12 \div 5 = 2, \text{ remainder } 2$$
$$12 = 2 \text{ fives} + 2 \text{ units}$$
$$\text{Hence } 37 = 1 \text{ twenty-five} + 2 \text{ fives}$$
$$+ 2 \text{ units}$$
$$37_{ten} = 122_{five}$$
$$Check: 122_{five} = 1 \times 25 + 2 \times 5 + 2 \times 1$$
$$= 25 + 10 + 2 = 37_{ten}$$

In Example 6 part (b), first calculate the digit in the units place by dividing by the new base. The

remainder in this division gives that digit. If the quotient is greater than the base, then there is a digit (which may be zero) in the next place. When the quotient is zero, all the remainders including zero give the number in the new base.

Example 7

```
5 │37    rem
5 │ 7 + 2    (i.e. 7 × 5   + 2 × 1)
5 │ 1 + 2    (i.e. 1 × 25  + 2 × 5)
  │ 0 + 1    (i.e. 0 × 125 + 1 × 25)
        ↑                      ↑
```

Repeated division by 5 gives remainders. Reading the remainders *upwards* gives $37_{ten} = 122_{five}$ (see the arrows above.)

To change from base ten to another base:

1 Divide the base ten number by the new base number.

2 Continue dividing until the quotient is zero, writing down the remainder each time, even when it is zero.

3 Start at the last remainder and read the remainders upwards to get the answer.

Example 8

Convert 75_{ten} (a) to base three, (b) to base two.

(a)
```
3 │75    rem
3 │25 + 0 ↑
3 │ 8 + 1 │
3 │ 2 + 2 │
  │ 0 + 2 │
```
$$75_{ten} = 2210_{three}$$

(b)
```
2 │75    rem
2 │37 + 1 ↑
2 │18 + 1 │
2 │ 9 + 0 │
2 │ 4 + 1 │
2 │ 2 + 0 │
2 │ 1 + 0 │
  │ 0 + 1 │
```
$$75_{ten} = 1\ 001\ 011_{two}$$

Example 8 shows that if a remainder is 0 it must be written down.

Notice that when converting to base three, the remainders can only be 0, 1 or 2. Hence a base three number contains no digits other than 0, 1 and 2. In the same way, a base eight number contains no digits greater than 7. In base two, the only digits are 0 and 1.

Exercise 1e

Each of the given numbers is in base ten.
Convert the numbers to the bases shown.

1. 15 to base eight
2. 20 to base five
3. 11 to base two
4. 12 to base three
5. 64 to base three
6. 27 to base two
7. 76 to base five
8. 18 to base eight
9. 35 to base three
10. 31 to base eight
11. 49 to base two
12. 49 to base five
13. 99 to base five
14. 56 to base three
15. 88 to base eight
16. 98 to base two
17. 128 to base eight
18. 115 to base five
19. 725 to base three
20. 129 to base two
21. 100 to base five
22. 733 to base eight
23. 256 to base two
24. 543 to base three

Converting from other bases to base ten

Example 9

Convert 1264_{eight} to base ten.

1st method: By expanding the given number:

$$1264_{eight} = 4 \times 1 + 6 \times 8 + 2 \times 64 + 1 \times 512$$
$$= 4 + 48 + 128 + 512$$
$$= 692_{ten}$$

2nd method: By repeated multiplication:

```
  1   2      6        4
× 8   ↓      |        |
  8 + 2 =   10        |
        × 8 ↓         |
          80 + 6 =   86
              × 8 ↓
                688 + 4 = 692
```

$$1264_{eight} = 692_{ten}$$

Notice that the method of repeated multiplication is the reverse of the repeated division method for conversion from base ten. Hence start with the digit in the place of greatest

value, multiply by the base number and add the digit in the next place to the product. Then multiply that answer by the base.

Continue until the digit in the units place is added. This is the last step.

Example 10

Convert the following to base ten. (a) $11\,101_{two}$ (b) 432_{five}

(a) By expanding

$$11\,101_{two}$$
$$= 1 \times 1 + 0 \times 2 + 1 \times 4 + 1 \times 8 + 1 \times 16$$
$$= 1 + 4 + 8 + 16$$
$$= 29_{ten}$$

(b) By repeated multiplication:

```
    4   3        2
  × 5   ↓        |
    20 + 3 =    23
          × 5   ↓
            115 + 2 = 117_ten
```

Exercise 1f

Convert the following to base ten.

1. 17_{eight}
2. 31_{eight}
3. 231_{eight}
4. 472_{eight}
5. 631_{eight}
6. 43_{five}
7. 24_{five}
8. 102_{five}
9. 324_{five}
10. 212_{three}
11. 120_{three}
12. 111_{three}
13. 2102_{three}
14. 1202_{three}
15. 111_{two}
16. 1010_{two}
17. $11\,110_{two}$
18. $10\,100_{two}$
19. $110\,100_{two}$
20. $10\,011_{two}$
21. 505_{eight}
22. 2043_{five}
23. 1443_{five}
24. 2021_{three}

Summary

Symbols are used to represent ideas. A number is an idea. The symbols used to represent numbers are called **numerals**. When the numerals used are 0, 1, 2, 3, 4, 5, 6, 7, 8, 9, they are called **digits**.

In order that digits are not repeated a large number of times, and that very many digits do not have to be used, the **place-value system** was devised. In this system the place or the position of the digit in the number determines its value.

Numbers may be collected in different groups, such as in groups of five (fingers on a hand), or groups of seven (days in a week), or in twos (on/off switch, up/down, above/below). These groups are called **number bases**. The most commonly used number base is base ten. Nowadays, base two is widely used in computers.

In the base ten place-value number system, digits to the right of a decimal point represent different parts of a whole, that is decimal fractions.

Practice exercise P1.1

① Complete the table.

Base number	Greatest digit
10	9
8	
7	
5	
2	

② Write the value of the columns for a five-digit number in
(a) base eight   (b) base five.

③ State the value in **base ten** of each of the underlined digits.
(a) $2\underline{4}578_{ten}$ (b) $101\underline{1}001_{two}$
(c) $23\underline{2}7_{eight}$ (d) $3.0\underline{5}67_{ten}$
(e) $7.001\underline{2}_{ten}$ (f) $123.\underline{9}01_{ten}$

④ Write the following numbers in expanded form.
(a) 101_{ten} (b) 101_{eight} (c) 101_{two}

⑤ Find the value of the numbers in 4(b) and 3(c) in base ten.

⑥ Convert the following numbers in base ten to the given bases.
(a) to base eight
   (i)   23, 234, 2345
   (ii) 63, 173, 544
(b) to base two
   (i)   9, 13, 26, 39, 141
   (ii) 11, 17, 23, 89, 132

⑦ Convert the following to base ten (if possible).
(a) 327_{eight} (b) $100\,101_{two}$
(c) 244_{five} (d) 170_{two}
(e) $111\,101_{two}$

Chapter 2

Sets (1)
Elements of sets, subsets, Venn diagrams

Sets

A **set** is a collection of objects. The **members** of a set may have something in common, such as the set of mathematical instruments in Fig. 2.1. Each member of this set can be used for drawing and measuring shapes.

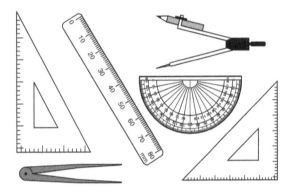

Fig. 2.1

Alternatively, there may be no obvious connection between the members of a group. Fig. 2.2 shows a group which does not appear to have anything in common.

Fig. 2.2

A set can be described, or defined, in two ways: either by making a list of its members, or by describing the property or rule that connects its members. For example, if M is the set shown in Fig. 2.1, then the set can be described by listing its members:

M = {60° set square, 45° set square, protractor, ruler, compasses, dividers}

or by using the defining property:

M = {mathematical instruments}

Notice the use of the capital letter to stand for the set and the curly brackets to contain the members of the set. Read M = {mathematical instruments} as 'M is the set of mathematical instruments'.

Another way of representing a set is to show its members within a boundary as in Fig. 2.3.

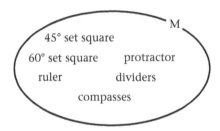

Fig. 2.3

A diagram like that in Fig. 2.3 is called a **Venn diagram**.

Example 1

Given the set F = {table, chair, mat},

(a) pick out a member that is out of place, giving a possible reason;

(b) state a common defining property for the members of F;

(c) hence give three more members of your set;

(d) draw a Venn diagram to show F.

(a)   *mat*: the only one with no legs
or *table*: people often sit on the other two
or *chair*: the only word that does not
            contain the letter *t*

(b)   F = {household furniture}
or F = {things which cost less than $100}
or F = {words containing the letter *a*}

(c) Assuming F = {household furniture}, then
other members of the set could be: bed, chest,
cupboard, bookshelf, cabinet, and so on.

(d)

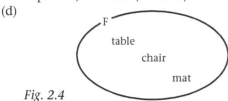

Fig. 2.4

Notice in parts (a) and (b) that there is more
than one possible correct answer. This shows
the importance of defining sets carefully.

Exercise 2a

Questions 1−4 are suitable for class discussion.

① In each of the following sets, pick out the
member that is out of place. Give reasons.
(a) {Kingstown, Barbados, Kingston, Castries}
(b) {bee, aeroplane, wasp, fly}
(c) {2, 4, 6, 7, 8}
(d) {table, cow, chair, chicken, dog}
(e) {(2 + 4), (11 − 5), (2 × 5), (24 ÷ 4)}

Use the following list for answering question 2:
*engine, cow, driver, petrol, passenger, tractor, soldier,
conductor, peace, wheel*

② Write down the members of the given list
which belong to the following sets.
(a) {human beings}
(b) {living things}
(c) {liquids}
(d) {things normally found on a bus}

③ Give a common defining property for each
of the following sets. There may be more
than one correct answer in each case.
(a) {tractor, cow}
(b) {petrol, passenger, peace}
(c) {engine, wheel, tractor}
(d) {peace}

④ List three members of each of the following
sets.
(a) {students in your class whose names
contain the letter *l*}
(b) {records in the Top 20}
(c) {even numbers}
(d) {Cabinet Ministers}
(e) {things on the teacher's desk}

⑤ If T is the set shown in Fig. 2.5,
(a) write down T in two different ways;
(b) draw a Venn diagram representing T.

Fig. 2.5

Elements of a set

The members of a set are often called **elements**.

Example 2

List the elements of the sets A = {even numbers
less than 50} and B = {counting numbers}.

A = {even numbers less than 50}
  = {2, 4, 6, 8, ..., 48}
B = {counting numbers}
  = {1, 2, 3, 4, ...}

Notice the following:

1 there is a comma between one element and
the next;

2 when it is impracticable or impossible to
write down all the elements we use dots, ...,
to show that we have left some out. The dots
stand for 'and so on'.

The following symbols are used to show
whether or not elements belong to a set:
    ∈ meaning 'is a member of' or 'belongs to'
    ∉ meaning 'is *not* a member of' or
            'does *not* belong to'.

For example, in Fig. 2.1
    protractor ∈ {mathematical instruments}
    cooking pot ∉ {mathematical instruments}

are short for 'a protractor is an element of the
set of mathematical instruments' and 'a cooking
pot does not belong to the set of mathematical
instruments' respectively.

The number of elements in a set

n(A) is short for the **number** of elements in set A.

Example 3

If W = {whole numbers between 2 and 10}
find n(W).

If W = {whole numbers between 2 and 10}
then W = {3, 4, 5, 6, 7, 8, 9}*

Set W contains 7 elements (by counting),
hence n(W) = 7.

*Note: It is assumed that the numbers 2 and 10
are *not* included in the set.

Equal sets

Two sets are **equal** if they contain the **same
elements**.

Example 4

If S = {b, a, d, c} which of the following sets are
equal to S?
(a) P = {b, a, d, c, a, d}
(b) Q = {d, a, c, e}
(c) R = {first four letters of the alphabet}

(a) P contains the same elements as S with
     elements a and d repeated.
     Hence P = S, since **any element which is
     repeated is counted** *once* **only**.
(b) e ∈ Q but e ∉ S,
     hence Q ≠ S (≠ means 'is not equal to')
(c) R = {first four letters of alphabet}
     R = {a, b, c, d}
     Hence R = S, since the elements of R are
     simply a rearrangement of those of S.

Exercise 2b

❶ List the elements of the following sets, using
   three dots, …, if necessary.
   (a) {letters of the alphabet}
   (b) {odd numbers less than 30}
   (c) {all the even numbers}
   (d) {months in a year}

❷ Write down the next three members of the
   following sets.
   (a) {3, 7, 11, 15, …}
   (b) {2, 4, 8, 16, …}
   (c) {7, 14, 21, 28, …}
   (d) {10, 21, 32, 43, …}
   (e) {1, 4, 9, 16, …}

❸ Make each of the following statements true
   by writing ∈ or ∉ in place of the *.
   (a) 15 * {3, 6, 9, 12, 15, 18, …}
   (b) 17 * {1, 2, 3, …, 7, 8, 9}
   (c) 2½ * {whole numbers}
   (d) If A = {Asian countries}, then *India* * A
   (e) If T = {four-sided shapes}, then *square* * T
   (f) 11 * {1, 3, 5, 7, …, 19}

❹ Rewrite the following using set notation.
   (a) A pencil is a member of the set of
        writing instruments.
   (b) A pencil is not a member of the set of
        animals.
   (c) Bread does not belong to the set of
        vehicles.
   (d) 3 does not belong to the set of letters
        of the alphabet.
   (e) 3 belongs to the set of digits.
   (f) 3 is a digit.
   (g) An eagle is not a member of the
        mammalian class.
   (h) An eagle is a bird.

❺ Complete the following.
   (a) If P = {a, e, g, i, l, n, s, t},   n(P) =
   (b) If F = {fingers on your hands},   n(F) =
   (c) If D = {days of the week},   n(D) =
   (d) If N = {whole numbers between 3
             and 13},   n(N) =
   (e) If X = {4, 3, 6, 1, 4, 5},   n(X) =
   (f) If B = {0},   n(B) =
   (g) If L = {12, 13, 14, 15, …, 26},   n(L) =
   (h) If K = {0, 3, 6, 9, …, 21},   n(K) =

❻ If A = {p, q, r, s},   B = {t, u, r, q},
   C = {p, q, r, q, s, q} and D = {t, r, q, u}
   decide whether each of the following is true
   or false.
   (a) A = B   (b) n(A) = n(B)
   (c) A = D   (d) n(A) = n(D)
   (e) u ∈ C   (f) t ∉ D
   (g) B = D   (h) B ≠ C
   (i) n(C) = 6   (j) A = C

Empty sets and infinite sets

Some sets have no elements in them. For example:

{squares with 5 sides}
{Antiguans who are 4 m tall}
{secondary students of age 3 years}

Since all these sets are empty, they are all the same. We call such a set the **empty set** or **null set**. We write { } or ∅ to represent the empty set.

If Y = {every third natural number}
then Y = {3, 6, 9, 12, 15, ...}

It is impossible to list all the elements of Y. This is an example of an **infinite set**. Question 2 of Exercise 2b contains more examples of infinite sets. An infinite set has a never-ending list of elements.

Exercise 2c (Oral)

Decide whether each of the following is an empty set or an infinite set.

1. {animals bigger than elephants in Africa}
2. {numbers which are divisible by 4}
3. {numbers divisible by 4 which are *not* divisible by 2}
4. {teachers with two heads}
5. {odd numbers greater than 7 but less than 9}
6. {fractions between 0 and 1}
7. {fractions less than $\frac{1}{100}$}
8. {cows that can fly}
9. {whole numbers divisible by 1}
10. {trees with $1 notes instead of leaves}

Subsets

Fig. 2.6 shows a bowl containing a mango, a banana and an orange.

Fig. 2.6

If a girl is allowed to choose any fruit she likes from the bowl, she may pick any of the following sets:

{mango}, {banana}, {orange}
{mango, banana}, {mango, orange},
{banana, orange}
{ }
{mango, banana, orange}

Each of these sets is a **subset** of the original set of fruit. Notice that the empty set and the whole set are also subsets. In symbols we write:

{mango, banana} ⊂ {mango, banana, orange}

where ⊂ means 'is a subset of'.

The original given set is often called the **universal set**. The universal set contains all the elements which can be used in a given problem. The symbol used to represent the universal set is U, or ℰ.

If A = {mango, banana},
then A ⊂ U (or A ⊂ ℰ).
U ⊃ A means U contains A
(or ℰ contains A).

Since {pawpaw} is *not* a subset of the given universal set, we can write {pawpaw} ⊄ U where ⊄ means 'is *not* a subset of'.

The relationship between subsets and universal sets can be shown in a Venn diagram.

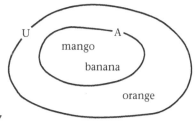

Fig. 2.7

In Fig. 2.7 the loop labelled U contains all the elements of the universal set. It also contains the subset A.

Example 5

Given U = {a, b, c, d, e, f, g, h, i, j},
V = {vowels},
C = {letters of the word cabbage},
H = {letters of the word hid}.
(a) List the members of V and C.
(b) Draw a Venn diagram showing U, V, C and H.

(a) V = {a, e, i}
    C = {c, a, b, g, e}
Notice that V contains only those vowels found in U, and that the elements, b and a, are *not* repeated in C.

(b)

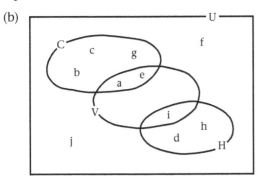

Fig. 2.8

Exercise 2d

① Write down all the subsets of the following sets.
   (a) {knife, fork, spoon}
   (b) {2, 3, 4}

② Given A = {a, b}
   (a) Which of the sets Ø, {a}, {a, b}, {a, c}, {a, b, c} are subsets of A?
   (b) How many subsets does A have?

③ If U = {earth, air, fire, water} write down all the subsets of U.

④ (a) Copy and complete Table 2.1.
   (b) Hence estimate the number of subsets of set Z if n(Z) = 5.

Table 2.1

Number of elements in a set	Number of subsets in that set
0	
1	
2	
3	
4	

⑤ If U = {mango, banana, orange} which of the following statements is (are) correct?
   (a) {banana} ⊂ U
   (b) banana ⊂ U
   (c) {banana} ∈ U
   (d) banana ∈ U

⑥ Write the following in symbols.
   (a) {dogs} is a subset of {mammals}
   (b) {cars} is not a subset of {cities}
   (c) A is a subset of B
   (d) Y contains X

⑦ Write the following in words.
   (a) F ⊂ G
   (b) A ⊄ B
   (c) M ⊅ N
   (d) D ⊃ E

⑧ Copy the Venn diagram given in Fig. 2.9.

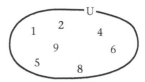

Fig 2.9

   (a) Draw the subsets
        A = {numbers less than 6},
        B = {odd numbers}.
   (b) Write down those elements which are members of both A and B.

⑨ If U = {days of the week} and
   T = {days beginning with T}, draw a Venn diagram showing the elements of U and T.

⑩ Given U = {1, 2, 3, ..., 9, 10},
   N = {even numbers},
   F = {numbers divisible by 5},
   G = {numbers greater than 3}.
   (a) Write down the following statement using symbols:
        5 and 10 are both greater than 3.
   (b) If *a* is a number such that a ∈ N, a ∈ F and a ∈ G, what is *a*?
   (c) If *b* is a number such that b ∈ N and b ∉ G, what is *b*?
   (d) Draw a Venn diagram showing U, N, F and G.

Summary

A set is usually named by a capital letter, e.g. V may be used to name the set of vowels. The members of the set are called its **elements**; the elements are separated by commas and written between braces, e.g.

$$V = \{a, e, i, o, u\}$$

The symbol $\in$ shows that an element is a member of a set, e.g. $a \in V$; similarly $d \notin V$ (*not* a member).

A set and its members may be shown graphically in a **Venn diagram**.

n(F) means the number of elements in the set F. Note that if an element is repeated, it is counted *only once*, e.g. $F = \{r, s, t, r\}$, then $n(F) = 3$, because r is a repeated element.

Sets are **equal** if they contain the same elements.

An **empty set** contains no elements and is written as $\{\,\}$ or $\emptyset$.

If there is an **infinite** (never-ending) number of elements, this is shown by three dots at the end of named elements, e.g. $\{1, 2, 3, 4, \ldots\}$.

Some members of one set may belong to another set, e.g. the set of vowels V and the set of consonants C belong to the alphabet A. Then V and C are **subsets** of A and this is written as $V \subset A$, $C \subset A$. Also, $A \supset V$, $A \supset C$. The set E of even numbers is *not* a subset of A and this is written as $E \not\subset A$. Also $A \not\supset E$.

The **universal set** contains all the elements used in a particular problem and is written as U (or $\mathscr{E}$).

Practice exercise P2.1

1. Complete the statements below by replacing ✿ with the correct symbol from the following list. $\in$, $\notin$, $\subset$, $\not\subset$, $\supset$ and $\not\supset$

    (a) boys ✿ {males}
    (b) {girls} ✿ {females}
    (c) keyboard ✿ {computer}
    (d) 7 ✿ {factors of 35}
    (e) {30, 45} ✿ {multiples of 15}
    (f) metre ✿ {units of weight}
    (g) {solids} ✿ {cube, prism, cone}
    (h) 5 ✿ {factors of 9}
    (i) $\{1.6, \frac{1}{3}\}$ ✿ {counting numbers}
    (j) {animals} ✿ {cat, dog, bird, tree}

2. State whether the following are *true* or *false*. Give your reason.

    (a) {Bob, Dan, Tom} $\subset$ {girls}
    (b) pyramid $\in$ {solids}
    (c) {vowels} $\subset$ {consonants} = {alphabet}
    (d) $n(M) = 7$, if $M = \{3, 4, 5, 6, 5, 4, 3\}$
    (e) $\{2, 3, 4\}$ and (x, y, z) are equal sets.

Practice exercise P2.2

1. In each of the following sets, select the member that does not belong to the set. Give a reason for your choice.

    (a) A = {pot, jug, plate, cup, glass, pan}
    (b) B = $\{(3 \times 3), 3^3, \sqrt{81}, (27 \div 3)\}$
    (c) J = {cone, cylinder, pyramid, triangle, prism}
    (d) C = {b, e, g, i, k}
    (e) F = $\{\frac{1}{4}, 0.25, \frac{2}{5}, 25\%, \frac{2}{8}\}$
    (f) S = {mean, range, mode, median}
    (g) M = {metre, kilogram, litre, tonne, mile}

2. List the members of each of the following sets.

    (a) J = {factors of 32}
    (b) K = {multiples of 7}
    (c) X = {odd numbers greater than 10 and less than 20}
    (d) D = {prime numbers between 5 and 20}

3. Write the next three terms of each of the following sets. Give reasons.

    (a) R = {Z, Y, X, W, V, __, __, __}
    (b) F = $\{\frac{1}{2}, \frac{2}{3}, \frac{3}{4}, \frac{4}{5},$ __, __, __$\}$
    (c) K = {32, 16, 8, 4, 2, __, __, __}

4. List the first five members of each of the following sets.

    (a) {even numbers}
    (b) {prime numbers}
    (c) {common multiples of 2 and 5}
    (d) {decimals correct to 1 decimal place between 0 and 1}

5 Use brackets and describe the following sets in words.

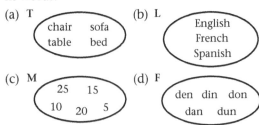

(a) T — chair sofa table bed

(b) L — English French Spanish

(c) M — 25 15 10 20 5

(d) F — den din don dan dun

Practice exercise P2.3

1 U = {a, b, c, d, e, ..., j}
F = {letters of the word *feed*}
H = {letters of the word *hid*}

(a) List the members of F and H.
(b) Draw a Venn diagram to show U, F and H.
(c) State n(F) and n(H).
(d) Are F and H equal sets? Give the reason for your answer.

2 F = {factors of 20}
O = {odd numbers less than 11}
Q = {quadrilateralss with all sides equal}

(a) List the members of each set.
(b) Write the number of members as n( ).

3 F = {fishes that walk}
P = {prime numbers that are less than 10}
W = {whole numbers}
E = {common factor of all even numbers}
C = {consonants}
N = {positive numbers that are less than 0}

(a) Decide whether each set is
m: an empty set , **i**: an infinite set
or **s**: a finite set.
(b) If **s**, state the number of members as n( ).

4 (a) Give three examples of an empty set.
(b) Give three examples of an infinite set.

Practice exercise P2.4

1 U = {scarf, jacket, skirt, helmet, bonnet, boots, sandals}
H = {headgear}
F = {footwear}
J = {jewellery}

(a) Write a phrase to describe U.
(b) Name and list the members of the subsets of U.
(c) Is there any set that is not a subset? Name this set.
(d) Show this by drawing a Venn Diagram.

2 30 students apply for a scholarship.
10 have college degrees
18 have school certificates
8 have both qualifications.

(a) Draw a Venn diagram to show this information.
(b) How many students have no qualification?

3 Use each of the following sets in turn as the universal set.
P = {polygons}
F = {fractions}
A = {animals}
H = {family}

(a) List the members of the universal set.
(b) Name two subsets and describe with a phrase what is common to the members of each subset.
(c) List the members of each subset.

Pre-requisites
■ the place-value system

⑪	27	⑫	36	⑬	54	⑭	56
⑮	60	⑯	90	⑰	120	⑱	144
⑲	180	⑳	210				

Natural numbers, whole numbers

$$\mathbb{N} = \{1, 2, 3, 4, ...\}$$

$\mathbb{N}$ is the set of natural numbers. These are the numbers used for counting. $\mathbb{N}$ is an infinite set.

$$W = \{0, 1, 2, 3, ...\}$$

W is the set of whole numbers. There are no fractions in the set of whole numbers. W is an infinite set.

Note $\mathbb{N}$ is a subset of W: $\mathbb{N} \subset W$

Factors

$$40 \div 8 = 5 \quad \text{and} \quad 40 \div 5 = 8$$

We say that 8 and 5 are **factors** of 40.
If a number can be divided by another number without a remainder, the second number is a factor of the first.

The numbers 1, 2, 4, 5, 8, 10, 20, 40 all divide into 40. They are the set of factors of 40.

Factors of 40 = {1, 2, 4, 5, 8, 10, 20, 40}

We can write 40 as a product of two factors in eight ways:
$$40 = 1 \times 40 = 2 \times 20 = 4 \times 10 = 5 \times 8$$
$$= 8 \times 5 = 10 \times 4 = 20 \times 2 = 40 \times 1$$

Exercise 3a

Write down the set of factors of the following numbers.

❶	12	❷	18	❸	20	❹	24
❺	28	❻	30	❼	32	❽	48
❾	63	❿	72				

Which members of {2, 3, 4, 5, 6, 7, 8, 9} are factors of the following numbers?

Even numbers and odd numbers

Let F_6 be the set of factors of 6, then
$$F_6 = \{1, 2, 3, 6\}$$
In the same way,
$$F_8 = \{1, 2, 4, 8\}$$
$$\text{and } F_{10} = \{1, 2, 5, 10\}$$

Notice that 2 is common to all three sets. The numbers 6, 8, 10 are all divisible by 2. Numbers that are divisible by 2 are **even numbers**. Numbers that are *not* divisible by 2 are called **odd numbers**.

Exercise 3b

❶ Which members of {3, 6, 9, 12, 15, 8} are
(a) odd numbers, (b) even numbers?

List the elements of the following sets.

❷ {even numbers less than 12}

❸ {factors of 24 that are odd numbers}

❹ {factors of 16 that are odd numbers}

❺ {odd numbers between 23 and 35}

❻ {even numbers greater than 123 and less than 135}

Prime numbers and composite numbers

A **prime number** has only two factors, itself and 1.

2, 3, 5, 7, 11, 13, ... are prime numbers. 1 has only one factor, itself. 1 is *not* a prime number.

$$P = \{2, 3, 5, 7, 11, 13, ...\}$$

where P is the set of prime numbers. P is an infinite set.

Example 1

List the elements of the following sets.
(a) {factors of 5}
(b) {factors of 11}
(c) {factors of 6}
(d) {factors of 15}

(a) {1, 5}   (b) {1, 11}
(c) {1, 2, 3, 6}   (d) {1, 3, 5, 15}

Notice that 6 and 15 each have four factors. These numbers are called **composite numbers**. A composite number is a number that has more than two factors. Numbers can be classified as being either prime numbers or composite numbers.

Exercise 3c

❶ Write down the set of prime numbers between 1 and 30.

❷ Write down the set of composite numbers between 12 and 40.

❸ Copy the 1−100 number square in Fig. 3.1 into your exercise book.

1	2	3	4	5	6	7	8	9	10
11	12	13	14	15	16	17	18	19	20
21	22	23	24	25	26	27	28	29	30
31	32	33	34	35	36	37	38	39	40
41	42	43	44	45	46	47	48	49	50
51	52	53	54	55	56	57	58	59	60
61	62	63	64	65	66	67	68	69	70
71	72	73	74	75	76	77	78	79	80
81	82	83	84	85	86	87	88	89	90
91	92	93	94	95	96	97	98	99	100

Fig. 3.1

You are to shade every number which is not a prime number.
(a) 1 is *not* a prime number. Shade the square with 1 in it.
(b) 2 is a prime number. Leave this square unshaded, then shade every second number following 2. This has been done in Fig. 3.1.
(c) 3 is a prime number. Leave this square unshaded, then shade every third number following 3.

(d) Carry on with the next unshaded number, 5, then shade every fifth number after 5.
(e) Carry on until you can shade no more numbers.

This method of finding prime numbers is called the **sieve of Eratosthenes**. Eratosthenes was a Greek who lived over 2000 years ago.

Prime factors

The **prime factors** of a number are the factors of the number which are prime. It is possible to write every non-prime number as a product of its prime factors. For example:

$$15 = 3 \times 5$$
$$24 = 2 \times 2 \times 2 \times 3$$
$$42 = 2 \times 3 \times 7$$

In set notation:

prime factors of 15 = {3, 5}
prime factors of 24 = {2, 3}
prime factors of 42 = {2, 3, 7}

Notice that although $24 = 2 \times 2 \times 2 \times 3$, the set of prime factors of 24 is simply {2, 3}. It is not necessary to repeat the factor 2.

To find the prime factors of a number:
(a) Start with the lowest prime number, 2. Find out if this will divide into the number. If it will not divide, try the next prime number, 3. And so on, trying 5, 7, 11, 13, ... in turn.
(b) If a prime number will divide, check if it will divide again before moving on to the next prime.

Example 2

(a) Express 15 288 as a product of its prime factors. (b) Hence write down the set of prime factors of 15 288.

(a) $15\,288 = 2 \times 7644$
$\qquad = 2 \times 2 \times 3822$
$\qquad = 2 \times 2 \times 2 \times 1911$
$\qquad = 2 \times 2 \times 2 \times 3 \times 637$
$\qquad = 2 \times 2 \times 2 \times 3 \times 7 \times 91$
$\qquad = 2 \times 2 \times 2 \times 3 \times 7 \times 7 \times 13$

The working can be set out as a continued division as follows:

```
 2 | 15 288
 2 |  7644
 2 |  3822
 3 |  1911
 7 |   637
 7 |    91
13 |    13
   |     1
```

Hence 15 288 = 2 × 2 × 2 × 3 × 7 × 7 × 13.

(b) Prime factors of 15 288 = {2, 3, 7, 13}

Exercise 3d

Use the above method

(a) to express the following as products of their prime factors,

(b) to write down their sets of prime factors.

① 12	② 18	③ 28	④ 30
⑤ 72	⑥ 84	⑦ 108	⑧ 105
⑨ 180	⑩ 216	⑪ 288	⑫ 875
⑬ 900	⑭ 880	⑮ 1512	⑯ 17 325

Index form

Remember that 10^3 is a short way of writing 10 × 10 × 10 or 1000. The digit 3 in 10^3 is called the **index** or **power** (Fig. 3.2).

10^3 —— index or power
—— base

Fig. 3.2   *10 to the power 3*

In Example 2 we saw that

15 288 = 2 × 2 × 2 × 3 × 7 × 7 × 13

In **index form** 15 288 = $2^3 × 3 × 7^2 × 13$. The use of index form saves space and can help to prevent errors in counting and copying.

Exercise 3e

① Write the following in index form.
   (a) 7 × 7 × 7
   (b) 3 × 3 × 3 × 3 × 3
   (c) 2 × 2
   (d) 10 × 10 × 10 × 10

(e) 6 × 6 × 6 × 6 × 6 × 6 × 6
(f) 8 × 8 × 8 × 8 × 8 × 8 × 8 × 8 × 8 × 8
(g) 4 × 4 × 4 × 5 × 5 × 5 × 5 × 5 × 5 × 5
(h) 11 × 11 × 17 × 17 × 17 × 17 × 17 × 17

② Rewrite your answers to Exercise 3d(a) in index form where possible.

③ Express the following as a product of primes in index form.

(a) 24	(b) 48	(c) 63
(d) 92	(e) 136	(f) 441
(g) 360	(h) 625	(i) 512
(j) 720	(k) 729	(l) 1000
(m) 1232	(n) 1404	(o) 1225
(p) 7290		

④ Write down the set of prime factors of each number given in question 3.

Squares and square roots

$7^2 = 7 × 7 = 49$

In words, 'the square of 7 is 49'. We can turn this statement round and say, 'the square root of 49 is 7'.

In symbols, 49 = 7.

The symbol $\sqrt{49}$ means *the square root of*.

To find the square root of a number, first find its factors.

Example 3

Find $\sqrt{11\,025}$.

method: Try the prime numbers in turn.

```
3 | 11 025
3 |  3675
5 |  1225
5 |   245
7 |    49
7 |     7
  |     1
```

$11\,025 = 3^2 × 5^2 × 7^2$
$= (3 × 5 × 7) × (3 × 5 × 7)$
$= 105 × 105$

Thus $\sqrt{11\,025} = 105$

It is not always necessary to write a number in its prime factors.

Example 4

Find $\sqrt{6400}$.

$$6400 = 64 \times 100$$
$$= 8^2 \times 10^2$$

Thus $\sqrt{6400} = 8 \times 10 = 80$

Exercise 3f

Find by factors the square roots of the following.

① 225	② 196	③ 324
④ 441	⑤ 484	⑥ 400
⑦ 900	⑧ 1600	⑨ 2500
⑩ 4900	⑪ 576	⑫ 784
⑬ 729	⑭ 625	⑮ 1225
⑯ 1936	⑰ 1764	⑱ 2025
⑲ 2304	⑳ 2916	㉑ 3025
㉒ 3600	㉓ 3969	㉔ 8100
㉕ 3136	㉖ 4356	㉗ 5625
㉘ 6561	㉒ 7056	㉚ 7744

Perfect squares

$9 = 3 \times 3$	$\sqrt{9} = 3$
$25 = 5 \times 5$	$\sqrt{25} = 5$
$225 = 15 \times 15$	$\sqrt{225} = 15$
$9216 = 96 \times 96$	$\sqrt{9216} = 96$

We say that 9, 25, 225, 9216 are **perfect squares** because their square roots are whole numbers. A perfect square is a whole number whose square root is also a whole number.

It is always possible to express a perfect square in factors with even indices. For example,

$$9216 = 96^2 = 3^2 \times 32^2$$
$$= 3^2 \times 4^2 \times 8^2$$
$$= 3^2 \times 2^{10}$$

Example 5

Find the smallest number by which 540 must be multiplied so that the product is a perfect square.

2	540
2	270
3	135
3	45
3	15
5	5
	1

$540 = 2^2 \times 3^3 \times 5$

The index of 2 is even.

The indices of 3 and 5 are odd.

One more 3 and one more 5 will make all the indices even. The product will then be a perfect square. The number required $= 3 \times 5 = 15$.

Exercise 3g

Find the smallest numbers by which the following must be multiplied so that their products are perfect squares.

① 24	② 54	③ 45	④ 99
⑤ 84	⑥ 162	⑦ 405	⑧ 240
⑨ 432	⑩ 147	⑪ 252	⑫ 504

Common factors

Let F_{12} be the set of factors of 12, then

$F_{12} = \{\mathbf{1}, 2, \mathbf{3}, 4, 6, 12\}$
and $F_{21} = \{\mathbf{1}, \mathbf{3}, 7, 21\}$
and $F_{33} = \{\mathbf{1}, \mathbf{3}, 11, 33\}$

Notice that 1 and 3 are common to all three sets. 1 and 3 are **common factors** of 12, 21 and 33. If C is the set of common factors, then

$C = \{1, 3\}$

There may be more than two common factors of a set of numbers. For example.

$F_{28} = \{\mathbf{1}, \mathbf{2}, 4, \mathbf{7}, \mathbf{14}, 28\}$
$F_{42} = \{\mathbf{1}, \mathbf{2}, 3, 6, \mathbf{7}, \mathbf{14}, 21, 42\}$
$F_{70} = \{\mathbf{1}, \mathbf{2}, 5, \mathbf{7}, 10, \mathbf{14}, 35, 70\}$

Common factors $= \{1, 2, 7, 14\}$

Notice that since 2 and 7 are common prime factors of 28, 42 and 70, then 14 (2×7) must also be a common factor of these numbers.

1 is a common factor of *all* numbers.

Exercise 3h

Find the common factors, other than 1, of the following.

1. 10 and 14
2. 15 and 24
3. 21 and 28
4. 9 and 15
5. 8 and 18
6. 27 and 30
7. 6 and 12
8. 10 and 20
9. 30 and 36
10. 9 and 18
11. 8 and 100
12. 24 and 40
13. 14, 24, and 38
14. 21, 35 and 56
15. 27, 36 and 39
16. 44, 66 and 88
17. 15, 60 and 75
18. 48, 60 and 72

Highest common factor (HCF)

$\{1, 2, 7, 14\}$ is the set of common factors of 28, 42 and 70. 14 is the greatest element of the set. We say that 14 is the **highest common factor** of 28, 42 and 70. **HCF** is short for highest common factor.

Alternatively: Express the numbers as products of prime factors and multiply the common prime factors together to give the HCF:

$$28 = 2 \times 2 \times 7$$
$$42 = 2 \times 3 \times 7$$
$$70 = 2 \times 5 \times 7$$

The common prime factors are 2 and 7.
$$\text{HCF} = 2 \times 7 = 14$$

Example 6

Find the HCF of 18, 24 and 42.

Compare the sets of factors of the given numbers.

$$F_{18} = \{1, 2, 3, 6, 9, 18\}$$
$$F_{24} = \{1, 2, 3, 4, 6, 8, 12, 24\}$$
$$F_{42} = \{1, 2, 3, 6, 7, 14, 21, 42\}$$

By inspection, HCF = 6.

Alternatively:
$$18 = \mathbf{2} \times \mathbf{3} \times 3$$
$$24 = \mathbf{2} \times 2 \times 2 \times \mathbf{3}$$
$$42 = \mathbf{2} \times \mathbf{3} \times 7$$

The common prime factors are 2 and 3.
$$\text{HCF} = 2 \times 3 = 6.$$

Example 7

Find the HCF of 216, 288 and 360.

The given numbers are large. Express each number as a product of prime factors.

working:

2	216
2	108
2	54
3	27
3	9
3	3
	1

2	288
2	144
2	72
2	36
2	18
3	9
3	3
	1

2	360
2	180
2	90
3	45
3	15
5	5
	1

In index notation,
$$216 = 2^3 \times 3^3$$
$$288 = 2^5 \times 3^2$$
$$360 = 2^3 \times 3^2 \times 5$$

2^3 is the greatest power of 2 contained by all three numbers. Hence the HCF contains 2^3.
3^2 is the greatest power of 3 contained by all three numbers. Hence the HCF also contains 3^2.

$$216 = (2^3 \times 3^2) \times 3$$
$$288 = (2^3 \times 3^2) \times 2^2$$
$$360 = (2^3 \times 3^2) \times 5$$

$$\text{HCF} = 2^3 \times 3^2 = 8 \times 9 = 72$$

Exercise 3i

1. Find the HCF of the following. Leave the answers in prime factors.

   (a) $2 \times 3 \times 3 \times 5$
   $2 \times 2 \times 3 \times 3 \times 3$
   $3 \times 3 \times 5 \times 7$

   (b) $2 \times 2 \times 3 \times 5$
   $2 \times 2 \times 3 \times 3 \times 7$
   $2 \times 3 \times 3 \times 7$

   (c) $2 \times 3 \times 3 \times 5$
   $2 \times 2 \times 2 \times 5$
   $2 \times 2 \times 3 \times 5 \times 7$

   (d) $2 \times 2 \times 2 \times 3 \times 3 \times 7$
   $2 \times 2 \times 3 \times 3 \times 3$
   $2 \times 3 \times 3 \times 3 \times 7$

   (e) $2 \times 3 \times 3 \times 5 \times 7$
   $3 \times 5 \times 5 \times 7 \times 7$
   $2 \times 2 \times 3 \times 5 \times 5 \times 7$

② Find the HCF of the following. Leave the answers in prime factors and use index notation.

(a) $2^3 \times 3^2 \times 7$
$2^2 \times 3 \times 5^2$
$2^2 \times 3^3 \times 5^2$

(b) $2^2 \times 3^3 \times 5$
$2^3 \times 3^4 \times 5$
$2 \times 3^5 \times 7^2$

(c) $2^2 \times 3^3 \times 5^2$
$3^2 \times 5^3 \times 7$
$2 \times 3^2 \times 5^2 \times 7$

(d) $2^3 \times 3^3 \times 5^2$
$2^4 \times 3 \times 5^2 \times 7$
$2^5 \times 3^2 \times 5 \times 7$

(e) $2^3 \times 5^2 \times 7$
$2^2 \times 3^2 \times 5$
$3^3 \times 5^3 \times 7^2$

③ Find the HCF of the following.

(a) 9, 15 and 24
(b) 18, 24 and 32
(c) 12, 30 and 42
(d) 24, 40 and 64
(e) 35, 50 and 65
(f) 30, 45 and 75
(g) 42, 70 and 56
(h) 36, 72 and 63
(i) 63, 42 and 21
(j) 144, 216 and 360
(k) 280, 105 and 175
(l) 126, 234 and 90
(m) 160, 96 and 224
(n) 189, 270 and 108
(o) 288, 180 and 108
(p) 324, 432 and 540

Multiples

Each number in the set {6, 12, 18, 24, 30, ...} has 6 as a factor. We say that the set contains **multiples** of 6.

Multiples of 6 = {6, 12, 18, 24, 30, ...}

In the same way,

multiples of 4 = {4, 8, 12, 16, 20, 24, ...}

Exercise 3j (Oral)

Give five multiples of each of the following.

① 2   ② 3   ③ 5   ④ 7   ⑤ 8
⑥ 9   ⑦ 10   ⑧ 11   ⑨ 12   ⑩ 20

Common multiples

Let M_6 be the set of multiples of 6, then
M_6 = {6, 12, 18, 24, 30, ...}
and M_4 = {4, 8, 12, 16, 20, 24, ...}

Notice that 12 and 24 are in both sets. 12 and 24 are **common multiples** of 6 and 4. If K is the set of common multiples, then

K = {12, 24, ...}

Exercise 3k (Oral)

Give three common multiples of the following.

① 3 and 4   ② 2 and 5   ③ 3 and 7
④ 2, 3 and 5   ⑤ 3, 4 and 5   ⑥ 2, 3 and 7

Lowest common multiple (LCM)

30, 60 and 90 are all common multiples of 6, 10 and 15. 30 is the lowest number that 6, 10 and 15 will divide into with no remainder.

We say that 30 is the **lowest common multiple** of 6, 10 and 15. **LCM** is short for lowest common multiple.

Example 8

Find the LCM of 8, 9 and 12.

$8 = 2 \times 2 \times 2$

Any multiple of 8 must contain $2 \times 2 \times 2$.

$9 = 3 \times 3$

Any multiple of 9 must contain 3×3.

$12 = 2 \times 2 \times 3$

Any multiple of 12 must contain $2 \times 2 \times 3$.

The lowest product containing all three is
$2 \times 2 \times 2 \times 3 \times 3$

The LCM of 8, 9 and 12 is
$2 \times 2 \times 2 \times 3 \times 3 = 72$

Example 9

Find the LCM of 24, 28, 36 and 50

In index notation $24 = 2^3 \times 3$
$28 = 2^2 \times 7$
$36 = 2^2 \times 3^2$
$50 = 2 \times 5^2$

The prime factors are 2, 3, 5 and 7. The LCM contains the highest powers of 2, 3, 5 and 7 that appear in the factorised numbers.

LCM $= 2^3 \times 3^2 \times 5^2 \times 7 = 12\ 600$

To find the LCM of a set of numbers:

(a) express the numbers as a product of prime factors;

(b) for each prime factor, select the term with the highest power;

(c) find the product of these terms; the product is the LCM.

Exercise 3l

1. Make two 1−100 number squares like that of Fig. 3.1.

   (a) On the first number square, shade all the multiples of 2 using shading like this: ▨

   Then shade all the multiples of 3 using shading like this: ▨

   What can you say about those numbers which have been cross-shaded, ▨ ?

   (b) Repeat the method of part (a) for multiples of 3 and 5. Hence find (i) the LCM of 3 and 5, (ii) the 4th common multiple of 3 and 5.

2. Find the LCM of the following. Leave the answers in prime factors.

   (a) $2 \times 2 \times 3$ (b) $2 \times 3 \times 3$
   $2 \times 3 \times 3 \times 5$ $2 \times 2 \times 2 \times 3$
   $2 \times 2 \times 5$ $2 \times 2 \times 3 \times 5$

   (c) $2 \times 2 \times 2 \times 3 \times 3$ (d) $3 \times 3 \times 5$
   $2 \times 3 \times 5 \times 5$ $2 \times 5 \times 5 \times 7$
   $2 \times 2 \times 3 \times 3 \times 5$ $2 \times 3 \times 7$

   (e) $2 \times 3 \times 5$ $3 \times 5 \times 7$
   $3 \times 5 \times 7$
   $2 \times 3 \times 3 \times 3$
   $3 \times 5 \times 5 \times 7$

3. Find the LCM of the following. Leave the answers in prime factors, in index form.

   (a) $2 \times 3 \times 5$; $2^2 \times 3 \times 5$; $2 \times 3 \times 5^2$
   (b) $2 \times 3^2 \times 7$; $2^3 \times 3 \times 5$; $2 \times 5^2 \times 7$
   (c) $3 \times 5 \times 7^2$; $2 \times 3^3 \times 5$; $2^2 \times 3 \times 7$
   (d) $2^2 \times 3 \times 5^3$; $2^3 \times 3 \times 7^2$; $3^2 \times 5^2 \times 7$
   (e) $2^3 \times 3^2 \times 5$; $3 \times 5^3 \times 7^2$;
   $2^4 \times 3 \times 7^2$; $3^2 \times 5^2 \times 7^3$

4. Find the LCM of the following.

   (a) 4 and 6   (b) 6 and 8
   (c) 6 and 9   (d) 7 and 8
   (e) 8 and 12   (f) 9 and 12
   (g) 2, 3 and 4   (h) 3, 4 and 6
   (i) 4, 6 and 9   (j) 6, 8 and 12
   (k) 8, 10 and 15   (l) 6, 8 and 10
   (m) 10, 12 and 15   (n) 5, 6 and 9
   (o) 10, 16 and 18   (p) 4, 5, 6 and 9
   (q) 6, 8, 10 and 12   (r) 9, 10, 12 and 15

Rules for divisibility

Table 3.1 gives some rules for divisors of whole numbers.

Notice the following:

(a) If a number is divisible by another number, it is also divisible by the factors of that number. For example, a number divisible by 8 is also divisible by 2 and 4.

(b) If a number is divisible by two or more numbers, it is also divisible by the LCM of those numbers. For example, a number divisible by both 6 and 9 is also divisible by 18. 18 is the LCM of 6 and 9.

Table 3.1

	Any whole number is exactly divisible by
2	if its last digit is even
3	if the sum of its digits is divisible by 3
4	if its last two digits form a number divisible by 4
5	if its last digit is 5 or 0
6	if it is even and the sum of its digits is divisible by 3
8	if its last three digits form a number divisible by 8
9	if the sum of its digits is divisible by 9
10	if its last digit is 0

Example 10

Test the following numbers to see which are exactly divisible by 9.

(a) 51 066   (b) 9039   (c) 48 681

method: Find the sum of the digits of each number.

(a) $5 + 1 + 0 + 6 + 6 = 18$
   18 is divisible by 9.
   Hence 51 066 is also divisible by 9.

(b) $9 + 0 + 3 + 9 = 21$
   21 is *not* divisible by 9.
   Hence 9039 is *not* divisible by 9.

(c) $4 + 8 + 6 + 8 + 1 = 27$
   27 is divisible by 9.
   Hence 48 681 is divisible by 9.

Example 11

Which of the numbers 5, 6 and 8 will divide into 2328 without remainder?

(a) Test for division by 5:
2328 does not end in 5 or 0.
2328 is *not* divisible by 5.
(b) Test for division by 6.
$2 + 3 + 2 + 8 = 15$
15 is divisible by 3 and 2328 is even.
2328 is divisible by 6.
(c) Test for division by 8.
The last three digits of 2328 form the number 328.
$328 \div 8 = 41$, hence 328 is divisible by 8
2328 is divisible by 8.

2328 is divisible by 6 and 8; it is *not* divisible by 5. (Notice that since 2328 is divisible by 6 and 8, it is also divisible by 24. 24 is the LCM of 6 and 8.)

The rules for divisibility can be useful when finding square roots.

Example 12

Find $\sqrt{5184}$

Use the rules for divisibility. Since 84 is divisible by 4, 5184 is divisible by 4.
Since $5 + 1 + 8 + 4 = 18$, 5184 is divisible by 9.

4	5184
4	1296
4	324
9	81
9	9
	1

$$5184 = 4^2 \times 4 \times 9^2$$
$$= 4^2 \times 2^2 \times 9^2$$

Thus $\sqrt{5184} = 4 \times 2 \times 9 = 72$

Exercise 3m

1. Find out which of the following numbers are exactly divisible by 4.

92	138	547	1926
2756	9428	10 782	56 016

2. Find out which of the following numbers are exactly divisible by 3.

184	267	534	917
1287	3506	10 523	28 104

3. Find out which of the following numbers are exactly divisible by 8.

1756	3042	2872	19 176
48 225	307 138	172 744	563 362

4. Find out which of the following numbers are exactly divisible by 9.

253	579	736	2331
5675	17 326	35 406	52 866

5. Copy and complete Table 3.2. Make a tick, ✓, in a box if the given number is divisible by the divisor. Make a cross, ✗, if it is not. The first row of boxes has been done.

Table 3.2

Number		2	3	4	5	6	8	9	10
(a)	13 545	✗	✓	✗	✓	✗	✗	✓	✗
(b)	32 550								
(c)	83 241								
(d)	2005								
(e)	6328								
(f)	3240								
(g)	51 294								
(h)	4770								
(i)	435								
(j)	360								
(k)	7138								
(l)	1848								

6. Write down those of the following numbers which are exactly divisible by *all three* of the numbers 5, 8 and 9.

3825	3960	5720
91 848	51 840	87 360

7 (a) Find out which of the numbers 5, 8 and 9 will divide into 51 768 exactly.

(b) Hence state which of the numbers 36, 40, 45, 72 will also divide into 51 768 exactly.

8 76 356   36 116   16 869   22 374

(a) Which of the numbers are divisible by 3?

(b) Which of the numbers are divisible by 4?

(c) Which of the numbers are divisible by 6?

(d) Hence state which of the numbers are divisible by 12.

9 (a) A student writes $5044 \times 9 = 45\,295$. Explain how the teacher knows that the student has made a mistake.

(b) Show that $141 \times 9 = 2699$ is false without doing the multiplication.

10 In the number 81 34*, the * is a missing digit.

(a) Find the digit which will make the number divisible by 9.

(b) Find the digit which will make the number divisible by 8.

(c) Find the digit which will make the number divisible by 15.

11 Using the rules for divisibility, find $\sqrt{9216}$.

12 Show that 192 cannot be a perfect square.

Problem solving using HCF and LCM

Example 13

Square paving stones are used to cover an area measuring 16.5 m by 12.75 m. If the stones are all alike, and only whole ones are used, what is the greatest size they can be? How many are there?

method: Find the HCF of 16.5 m and 12.75 m.

working in cm:

2	1650		3	1275
3	825		5	425
5	275		5	85
5	55		17	17
11	11			1
	1			

$1650 = 2 \times 3 \times 5 \times 5 \times 11$
$1275 = 3 \times 5 \times 5 \times 17$

HCF of 1650 and 1275 $= 3 \times 5 \times 5 = 75$

The greatest stones measure 75 cm by 75 cm.

Number of paving stones $= \dfrac{1650 \times 1275}{75 \times 75}$

$= 2 \times 11 \times 17 = 374$

Example 14

A number of oranges can be divided into equal heaps, each containing 10 or 15 or 24 oranges. Find the smallest number of oranges for which this is possible.

method: Find the LCM of 10, 15 and 24.

working: $10 = 2 \times 5$
$15 = 3 \times 5$
$24 = 2^3 \times 3$
$\text{LCM} = 2^3 \times 3 \times 5 = 120$

The smallest number of oranges is 120.

Exercise 3n

1 Find the greatest mass that can be taken an exact number of times from 360 g, 504 g and 672 g.

2 Find the smallest mass that can be measured out in equal amounts of 6 g or 8 g or 9 g.

3 Equal squares, as large as possible, are ruled off on a rectangular board 54 cm by 78 cm. Find the size of each square. How many squares are there?

4 A piece of string can be cut into equal lengths, either 20 cm or 24 cm or 30 cm long. Find the length of the shortest piece of string for which this is possible.

5 Find the smallest sum of money that is an exact multiple of $1.12, 64c and 96c.

6 Find the greatest number which when divided into 179 and 234 will leave a remainder of 3 in each case. (*Hint*: First subtract 3 from both numbers.)

⑦ A certain mass of sugar can be divided into equal heaps. Each heap contains either 27 g, 45 g, 30 g or 20 g. Find the smallest mass of sugar for which this is possible.

⑧ Find the smallest number of sweets that can be put into bags which all contain either 9, 15, 20 or 24 sweets, with none left over.

⑨ Four cars go round a racing track in 1min 12s, 1min 28s, 1min 39s and 1min 48s respectively. They start together. How long will it be until they all pass the starting point together again?

⑩ Find the greatest number which when divided into 1250 and 1000 leaves remainders of 26 and 28 respectively.

Summary

A **factor** F of a number N is such that N ÷ F leaves no remainder, e.g. 12 ÷ 3 = 4, so 3 is a factor of 12; *note*, 4 is also a factor of 12.

A **prime number** has only *two* factors, itself and 1. The **prime factors** of a number are the factors which are prime numbers. 1 is *not* a prime number.

The **index** or **power** of a number N is the number of times N is multiplied by itself.

The **square** of a number is the result of multiplying the number by itself. Since $25 = 5^2$, 25 is the square of 5, and 5 is the **square root** of 25.

A **perfect square** is a whole number whose square root is also a whole number.

The **HCF** or **highest common factor** of a set of numbers is the largest factor common to the set.

A **multiple** of a number has that number as a factor.

The **LCM** or **lowest common multiple** of a set of numbers is the smallest multiple common to the set.

Practice exercise P3.1

① For each of the following sets of numbers
   (a) describe its properties
   (b) write down the first ten numbers of each set
   (i) even numbers
   (ii) odd numbers
   (iii) prime numbers
   (iv) composite numbers
   (v) square numbers

② From the following list,

2	9	11	12	16	17
23	24	30	36	37	48
49	64	67	81	88	100
101	102	144	221	239	256

write down all the numbers that are
   (a) even   (b) odd
   (c) prime   (d) composite
   (e) square

Practice exercise P3.2

① For each of the following, state and use the rules of divisibility to show whether the statements are *true* or *false*.
   (a) 2 is a factor of 72
   (b) 5 is a factor of 81
   (c) 6 is a factor of 25
   (d) 9 is a factor of 63
   (e) 8 is a factor of 64
   (f) 12 is a factor of 142
   (g) 11 is a factor of 232
   (h) 15 is a factor of 255
   (i) 18 is a factor of 396
   (j) 16 is a factor of 82
   (k) 25 is a factor of 275
   (l) 30 is a factor of 410
   (m) 20 is a factor of 160
   (n) 100 is a factor of 450
   (o) 60 is a factor of 1000

2 For each of the following numbers,
  (i) find all its factors
  (ii) list the prime factors.
  (a) 24   (b) 25
  (c) 42   (d) 95
  (e) 100   (f) 132
  (g) 625   (h) 1024
  (i) 1296   (j) 2744

Practice exercise P3.3

1 Write down five
  (a) two-digit numbers that are multiples of
    (i) 5   (ii) 2   (iii) 7   (iv) 3
  (b) three-digit numbers that are multiples of
    (i) 10   (ii) 8   (iii) 19   (iv) 50

2 Which numbers in each list are multiples of the circled number?

  (a)   18   31   199   633   728
    ②   161   220   4147   1950   70
         225   161   399

  (b)   31   199   399   225   728
    ⑤   635   70   399   220   18
         1950   4145

  (c)   441   12   8117   17   21
    ③   8113   3000   125   225   37
         111

  (d)   132   225   111   3000   98
    ⑦   21   8113   126   17   441
         400   35

3 Replace the asterisk * in each number with a digit to make the number
  (a) a multiple of 2
    (i) 3*   (ii) 10*   (iii) 4*   (iv) 35*
  (b) a multiple of 5
    (i) 32*   (ii) *125   (iii) 19 99*   (iv) 14**
  (c) a multiple of 3
    (i) 26*   (ii) 12*5   (iii) 7*89   (iv) *357

Practice exercise P3.4

For each question, calculate the power.

1 2^5 **2** 4^3

3 30^2 **4** 10^6

Practice exercise P3.5

1 Use prime factors to find the (a) HCF and (b) LCM of 210 and 216.

2 Find the LCM of each pair of numbers.
  (a) 15 and 35   (b) 80 and 72

3 Find the HCF of each pair of numbers.
  (a) 66 and 88   (b) 126 and 231

4 Decide whether to use the HCF or LCM to solve these problems. Solve the problems.

  (a) Identical pieces of water piping are laid in two trenches. Each trench uses a whole number of pieces.
    Trench X is 280 m long.
    Trench Y is 360 m long.
    What is the longest a single piece of pipe can be?

  (b) Three drummers play as follows:
    • bass drum every 4 beats
    • snare drum every 6 beats
    • tambourine every 10 beats.
    They start together. After how many beats are they together again?

  (c) Two walls have the same height. One is built using bricks 8 cm high, the other using blocks 14 cm high.
    (i) Find the minimum possible height of the walls.
    (ii) Calculate the number of rows of bricks and of blocks that would have to be in each wall.

  (d) Three flashing lights mark a road accident. One flashes every 4 seconds, another every 10 seconds, another every 15 seconds. They all start flashing at the same time. How often do they flash together?

(e) 40 pupils from Year 8 are divided into equal teams. 36 pupils from Year 9 are divided into equal teams of the same size as Year 8.
   (i) What is the greatest possible size of a team?
   (ii) How many such teams would there be altogether?

(f) 420 m of flex is wound onto a number of drums, with none left over. 600 m of flex is wound onto more drums, with none left over. All the drums have the same length of flex.
   (i) What is the greatest possible length of flex on a drum?
   (ii) How many such drums would there be altogether?

(g) 45 roses are planted in a number of equal rows. 120 gerberas are planted in a number of equal rows. 300 anthuriums are planted in a number of equal rows. Each row contains the same number of plants.
   (i) What is the greatest number of plants that a row could contain?
   (ii) How many such rows of each type of flower would there be?

5 The LCM of two numbers is 504. One of them is 36. The other number is less than 100. Find this number.

6 The HCF of two numbers is 45. The smaller number is 1350. Write down a possible larger number.

Basic units of measurement

The **SI system of units** is an internationally agreed method of measuring quantities such as length, mass and time. SI is short for *Le Système International d'Unités* (International System of Units). Nearly every country in the world uses the SI system.

Except for the measurement of time, the SI system uses decimal multiples to build up tables connecting the units for each quantity. The basic quantities are length, mass and time. Other quantities, such as area, volume, speed and density, can be expressed in terms of the basic quantities.

Length

The **metre** is the basic unit of length. The metre was first taken as one ten-millionth part of the distance between the North Pole and the Equator. In Table 4.1, Greek prefixes are used for multiples of a metre (distances greater than a metre). Latin prefixes are used for sub-multiples of a metre (distances less than a metre).

Table 4.1

Length	Abbreviation	Relationship to basic unit
1 *kilo*metre	1 km	1000 m
1 *hecto*metre	1 hm	100 m
1 *deca*metre	1 dam	10 m
1 metre	1 m	1 m
1 *deci*metre	1 dm	0.1 m
1 *centi*metre	1 cm	0.01 m
1 *milli*metre	1 mm	0.001 m

Notice that the value of each unit is ten times that of the unit just below it. 1 km = 10 hm, 1 cm = 10 mm, etc. Compare this with the decimal place-value system in which the value of a digit is ten times the value of the same digit in the next place to it on the right.

Using a ruler

Example 1

Use a ruler to measure the line l in Fig. 4.1.

Fig. 4.1 ————————————— *l*

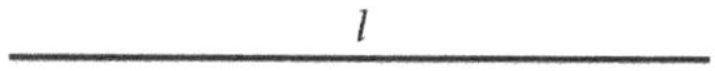

Give the length of *l* in cm and in mm.

Fig. 4.2 shows the line *l* and part of a ruler beside it.

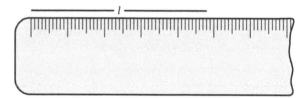

Fig. 4.2

The length of the line starting to read from zero is

length of *l* = 4.8 cm
= 48 mm

Fig. 4.3 shows another method of measuring the same line *l*.

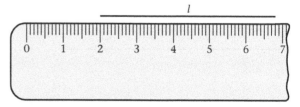

Fig. 4.3

Notice that one end of the line is opposite 6.8 cm on the ruler.

The other end is opposite 2 cm.

Length of $l = (6.8 - 2)$ cm
$= 4.8$ cm
$= 48$ mm

Look at the rulers in Fig. 4.2 and Fig. 4.3. The longer marks are numbered, 0, 1, 2, 3 … . The distance between two consecutive long marks is 1 cm. The distance between two consecutive short marks is 1 mm.

The method of subtraction shown by Fig. 4.3 has two advantages.

(a) It is often more convenient to measure near the centre of the ruler.

(b) It avoids possible confusion with the 0 cm and 1 cm marks.

You may use either method but always remember the importance of measuring accurately.

Exercise 4a (Group work)

❶ Measure the lines in Fig. 4.4. Give each length first in cm and then in mm. After doing Exercise 4a exchange exercise books with a friend and use a ruler to check the accuracy of each other's work.

(a)

(d)

(c)

(d)

(e)

Fig. 4.4

❷ Measure the lines in Fig. 4.5. Give each length first in mm and then in cm.

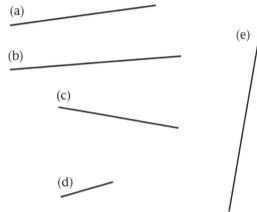

Fig. 4.5

❸ Measure the length and breadth of (a) this textbook, (b) your exercise book. Then copy and complete Table 4.2.

Table 4.2

		Length		Breadth	
		cm	mm	cm	mm
(a)	textbook				
(b)	exercise book				

❹ Measure the length and width of the classroom in metres. Think what type of ruler or measure you need. Convert the result to cm.

❺ Use a ruler to draw straight lines of the following lengths:
(a) 6 cm   (b) 90 mm   (c) 7.5 cm
(d) 45 mm   (e) 5.7 cm   (f) 112 mm

In practice, the hectometre, decametre and decimetre are not used very often. The centimetre is useful for measuring short lengths but industry usually gives such lengths in millimetres. This leaves the kilometre, the metre and the millimetre as the most common units:
1 km $= 1000$ m
1 m $= 1000$ mm

The advantage of a decimal system of measurement is that it is easy to write down compound quantities as decimals without doing any calculation.

Example 2

1 metre 67 centimetres = 1.67 m
= 167 cm
= 1670 mm
= 0.001 67 km

4 km 18 metres = 4.018 km
= 4018 m
= 4 018 000 mm

Exercise 4b

Questions 1–6 are suitable for oral work.

① Express the lengths given in Table 4.3 in
   (a) km,　　　(b) m,　　　(c) mm.

Table 4.3

km	hm	dam	m	dm	cm	mm
	3	6	4			
		2	5	7		
8	0	9	2			
			6	4	0	8
	2	0	5			

② Express the following in metres.
   (a) 3 km　　　(b) 5 km　　　(c) 8 km
   (d) 2 km　　　(e) 6 km　　　(f) 10 km
   (g) 4 km　　　(h) 3.5 km　　(i) 4.2 km
   (j) 6.8 km　　(k) 8.1 km　　(l) 5.9 km

③ Express the following in metres.
   (a) 3.85 km　　　　(b) 8.39 km
   (c) 9.14 km　　　　(d) 9 km 400 m
   (e) 3 km 315 m　　(f) 5 km 50 m

④ Express the following in kilometres.
   (a) 5000 m　　　　(b) 10 000 m
   (c) 2400 m　　　　(d) 6520 m
   (e) 7330 m

⑤ Express the following in metres.
   (a) 173 cm　　　　(b) 458 cm
   (c) 150 cm　　　　(d) 105 cm
   (e) 101 cm　　　　(f) 100 cm
   (g) 53 cm　　　　　(h) 40 cm
   (i) 19 cm　　　　　(j) 5 cm

⑥ Express the following in metres.
   (a) 1000 mm　　　(b) 7000 mm
   (c) 4100 mm　　　(d) 3726 mm
   (e) 8119 mm　　　(f) 300 mm
   (g) 51 mm　　　　(h) 3 mm

⑦ Find the value of the following:
   (a) 93.7 cm + 83 mm + 2.63 m
        (answer in m)
   (b) 1 758 m + 1 347 m − 2.895 km
        (answer in m)
   (c) 213.6 m × 9 (answer in km)
   (d) 69.3 m ÷ 36 (answer in cm)

⑧ A farmer walks round a field with five sides of lengths 376 m, 285 m, 493 m, 329 m and 117 m. How far, in km, does he walk altogether?

⑨ There are 35 bricks in a pile. Each one is 7.6 cm thick. How high is the pile? (Answer in metres.)

⑩ In Fig. 4.6 a pile of six books, each 6.7 cm thick, are stacked underneath a table. If the table is 75 cm high, what is the distance, h cm, between the table and the top of the books?

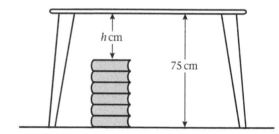

Fig. 4.6

⑪ A piece of ruled paper, 29.7 cm long, contains 33 evenly spaced lines. What is the distance between each line in mm?

⑫ An iron rod is 3.028 m long. It is cut into four pieces of equal length. Find the length of each piece in cm.

Mass

The **gram** is the basic unit of mass. A gram is the mass of 1 cubic centimetre of water at a temperature of 4 °C. This is a very small unit. For this reason, the kilogram (1000 grams) has become the standard unit of mass.

The relationship between the different units of mass is set out in Table 4.4 below. Notice that the prefixes are the same as those for lengths – an advantage of the SI system of units.

Table 4.4

Length	Abbreviation	Relationship to basic unit
1 kilogram	1 kg	1000 g
1 hectogram	1 hg	100 g
1 decagram	1 dag	10 g
1 gram	1 g	1 g
1 decigram	1 dg	0.1 g
1 centigram	1 cg	0.01 g
1 milligram	1 mg	0.001 g

The kilogram, the gram and the milligram are the only units used for practical purposes. A further unit is used for large masses, the **tonne** (t).

$$1 \text{ tonne} = 1000 \text{ kg}$$
$$1 \text{ kg} = 1000 \text{ g}$$
$$1 \text{ g} = 1000 \text{ mg}$$

Exercise 4c

1. Add the following and give the answers in kilograms.
   (a) 2.3 kg; 5.8 kg; 2.1 kg
   (b) 785 g; 97 g; 605 g
   (c) 574 g; 1.706 kg; 605 g
   (d) 2.8 t; 450 kg; 1.37 t

2. Find the value of the following
   (a) 536 g + 65 g − 365.5 g (answer in g)
   (b) 574 g + 8.706 kg − 6.25 kg (answer in kg)
   (c) 7.43 g × 36 (answer in kg)

3. A lorry is filled with 7.1 tonnes of sand. During the journey 210 kg of sand either falls off or blows away. What mass of sand, in tonnes, is delivered at the end of the journey?

4. A sheet of paper has a mass of 4.38 g. What is the mass, in kg, of a packet of 500 sheets of the paper?

5. Six identical textbooks have a total mass of 1.704 kg. What is the mass of one of the books in grams?

6. Four identical tins of biscuits have a mass of 1.14 kg. What is the mass of three of the tins?

7. A wire is bent to form a rectangle 104 mm by 76 mm. 1 cm of the wire has a mass of 1 g. Find the mass of the wire in the rectangle.

Time

The **second** is the basic unit of time. Units of time do not follow the decimal system.

Table 4.5

Time	Abbreviation	Relationship to basic unit
1 second	1 s	1 s
1 minute	1 min	60 s
1 hour	1 h	3600 s (= 60 min)

Example 3

$$2 \text{ min } 15 \text{ s} = (2 \times 60) + 15 \text{ s}$$
$$= 135 \text{ s}$$

Example 4

Find the total time in hours and minutes of the following:
2 h 15 min, 1 h 27 min, 45 min, 1 h 52 min.

working:
2 h	15 min
1 h	27 min
45 min	
1 h	52 min
6 h	19 min

method: (15 + 27 + 45 + 52) min
$$= 139 \text{ min}$$
$$= 2 \text{ h } 19 \text{ min}$$
$$(2 + 1 + 1 + 2) \text{ h} = 6 \text{ h}$$

See Chapter 1 for more examples.

12-hour clock

The clock in Fig. 4.7 shows the time as 8:12. A person reading the time will know whether it is morning or evening, i.e. whether the clock shows 8:12 a.m. or 8:12 p.m. Ordinary 12-hour clocks are satisfactory for most purposes.

Fig. 4.7

24-hour clock

Long train journeys often take more than 12 hours. Journeys by aeroplane often cross into differing time zones. To avoid confusion when reading timetables, railways and airlines use the **24-hour clock**.

Fig. 4.8(a) shows the time 8:12 a.m. on a 24-hour clock. Fig. 4.8(b) shows the time 8:12 p.m.

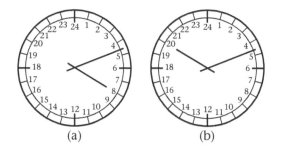

Fig. 4.8   24-hour clock

on a 24-hour clock. This appears as 12 minutes past 20 o'clock. This is written as 20:12 hrs and said as *twenty-twelve hours*. 8:12 a.m. is written as 08:12 hrs and said as *zero-eight-twelve hours*. Table 4.6 gives some ordinary times expressed in terms of the 24-hour clock.

Table 4.6

Ordinary time	24-hour clock	
	digits	words
1:30 p.m.	13:30 hrs	thirteen-thirty
3:54 a.m.	03:54 hrs	zero-three-fifty-four
3:54 p.m.	15:54 hrs	fifteen-fifty-four
midday	12:00 hrs	twelve-hundred
midnight	24:00 hrs	twenty-four-hundred
12:15 a.m.	00:15 hrs	zero-hundred-fifteen

Notice that all times on the 24-hour clock are given using 4 digits. It is sometimes possible to see 24-hour clocks as in Fig. 4.8. However, in airports it is more common to see digital clocks as in Fig. 4.9.

Fig. 4.9

Exercise 4d

1. How many minutes in
   (a) $2\frac{1}{2}$ h   (b) $1\frac{1}{4}$ h   (c) $1\frac{2}{3}$ h
   (d) 180 s   (e) 150 s?

2. How many seconds in
   (a) 5 min   (b) $1\frac{1}{3}$ min   (c) $\frac{1}{4}$ h
   (d) $\frac{1}{2}$ h   (e) 3 h?

3. A motorist took 5 h 11 min for a certain journey. Another motorist took 3 h 54 min for the same journey. What is the difference in their times?

4. What is the difference in hours and minutes between the times 08:30 hrs and 15:12 hrs on the 24-hour clock?

⑤ A car travels 108 km in an hour at a steady speed. How many metres does it go in a minute?

⑥ A man walks at a rate of 96 paces to the minute. Each pace is 0.625 m long. How long does he take to walk 5.4 km?

⑦ Copy and complete Table 4.7.

Table 4.7

Ordinary time	24-hour clock	
	digits	words
2:45 p.m.		
7:30 a.m.		
7:30 p.m.		
6 a.m.		
10 p.m.		
12:40 a.m.		

⑧ Express the following times in terms of the 24-hour clock.
(a) 2 p.m.   (b) 8:30 p.m.
(c) 9 a.m.   (d) 4 p.m.
(e) 5:30 p.m.   (f) 5:45 a.m.
(g) 5:25 a.m.   (h) 11:55 p.m.
(i) 5 to 9 at night
(j) 10 past 6 in the morning   (k) noon
(l) 3 minutes past midnight

⑨ Express the following 24-hour clock times as ordinary times.
(a) 05:00 hrs   (b) 23:00 hrs
(c) 19:30 hrs   (d) 09:30 hrs
(e) 21:45 hrs   (f) 14:15 hrs
(g) 18:19 hrs   (h) 07:40 hrs

⑩ Table 4.8 is the Air Jamaica airline timetable for flights on the given day from Kingston.
(a) What do 'Dep.' and 'Arr.' stand for?
(b) How long does it take to fly from Kingston to Philadelphia?
(c) How long does it take to fly from Kingston to New York?
(d) What does JM057 stand for?
(e) What was the flight number from Kingston to New York?

(f) If flight JM031 was delayed by 1 h 25 min, what was the actual arrival time in Miami?

Table 4.8

Monday					
Flight	to	dep.	arr.	status	
JM057	Atlanta	08:00	12:55	On time	
JM023	Baltimore	08:30	17:15	On time	
JM041	Philadelphia	09:00	18:30	On time	
JM055	Miami	10:20	13:55	On time	
JM031	Miami	14:55	17:30	Delayed	
JM017	New York	15:55	20:40	Delayed	
JM075	Los Angeles	18:15	21:55	On time	
JM263	London	18:40	09:15	On time	

Temperature

The **temperature** of an object tells you how hot or how cold the object is. A **thermometer** is used to measure temperature accurately. There are two main temperature scales. These are the **Celsius** scale and the **Fahrenheit** scale which is widely used in the United States of America. There is a third scale called the **Kelvin** scale which is mainly used in scientific experiments.

The unit of measure is the **degree** in each scale. Each unit in one scale is equal to every other unit on that scale and measures the same change of temperature. Also one Celsius degree is equal to one Kelvin.

There are two points on all three temperature scales that are used to compare the temperature when the different temperature scales are used. These are the temperature of the freezing point and the boiling point of water, as shown in Table 4.9.

Table 4.9

	°C	°F	K
Freezing point of water	0	32	273
Boiling point of water	100	212	373

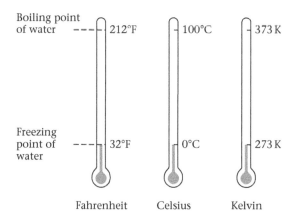

Boiling point of water --- 212°F   100°C   373 K

Freezing point of water --- 32°F   0°C   273 K

Fahrenheit   Celsius   Kelvin

Fig. 4.10   Temperature scales

From Fig 4.10, you can see that the difference in temperature between the freezing point and the boiling point of water is 100 degrees on the Celsius and Kelvin scales, but 180 degrees on the Fahrenheit scale, so that

100 Celsius degrees = 180 Fahrenheit degrees

We can write this equation as

$$100 \, C° = 180 \, F°$$

and $$5 \, C° = 9 \, F°$$

Note that $$1 \, C° = \tfrac{9}{5} F°$$

that is, a change of one degree on the Celsius scale is equal to $\frac{9}{5}$ degrees on the Fahrenheit scale. Therefore, to convert a temperature from Celsius degrees to Fahrenheit degrees, you must multiply by $\frac{9}{5}$ and then add 32 to the result.

Also: $1 \, F° = \frac{5}{9} C°$

that is, a change of one degree on the Fahrenheit scale is equal to only $\frac{5}{9}$ of one degree on the Celsius scale. Therefore, to convert a temperature from Fahrenheit degrees to Celsius degrees, you must first subtract 32 degrees and then multiply the result by $\frac{5}{9}$.

Note that the average temperature of the human body is 98.6 °F.

Example 5

Find the average temperature of the human body in degrees Celsius.

Temperature $= (98.6 - 32) \times \frac{5}{9}$ °C
$= 66.6 \times \frac{5}{9}$ °C
$= 37$ °C

Example 6

One day the temperature changed from 25 °C in the morning to 30 °C at midday.
(a) State the change in temperature in Celsius degrees.
(b) Find the change in Fahrenheit degrees.
(c) Calculate the temperature in °F at midday.

(a) Change in temperature = 5 °C
(b) Change in temperature = 9 °F
(c) 25 °C $= [(25 \times \frac{9}{5}) + 32]$
$= 77$ °F
Temperature at midday $= [77 + 9]$ °F
$= 86$ °F

or

Temperature at midday $= [(25 + 5) \times \frac{9}{5}] + 32$ °F
$= [54 + 32]$ °F
$= 86$ °F

Exercise 4e

1. Convert the following temperature changes
   (a) to degrees Fahrenheit
      (i) 15 °C   (ii) 20 °C   (iii) 45 °C
   (b) to degrees Celsius
      (i) 9 °F   (ii) 27 °F   (iii) 45 °F

2. Convert the following temperatures
   (a) to °C
      (i) 59 °F   (ii) 140 °F
   (b) to °F
      (i) 25 °C   (ii) 40 °C

3. The temperature of a hot oven was 455 °F. The temperature cooled to 410 °F.
   (a) State the change in temperature in °F.
   (b) Calculate the change in °C.
   (c) Convert the given temperatures to °C.

Summary

The basic quantities length, mass and time are now more often measured using the **SI system of units**. For each quantity there is a basic unit of measurement.

quantity	basic unit	relationship to other commonly used units
length	metre	= 1000 mm = 100 cm = 0.001 km
mass	gram	= 1000 mg = 0.001 kg
time	second	= $\frac{1}{60}$ min = $\frac{1}{3600}$ hour

The unit of time does not follow the decimal system. Time is written in different ways. Many countries use the **24-hour clock** but

15:30 hrs, 1530, 15:30, 3:30 p.m., 3.30 p.m.

all refer to the same time of day.

The unit used to measure temperature with the SI system of units is the degree Celsius (°C). Other temperature scales used are the Fahrenheit (°F) and the Kelvin (K).

Practice exercise P4.1

1. The lengths of the sides of a quadrilateral are: 185 cm, 89 cm, 124 cm, 79 cm. Calculate the perimeter of the quadrilateral, giving your answer (a) in cm (b) in metres.

2. Jimmy jogs a distance of 2.5 km three times each day. What is the total distance has he jogged in 5 days?

3. The weight of five babies is recorded as 4.5 kg, 4.2 kg, 5 kg, 3.9 kg, 4.3 kg. What is the total weight of the five babies?

4. Joan walks a total distance of 0.86 km to and from school each day. On Saturdays she walks 2.4 km to and from the store, and on Sundays she walks 1.8 km to and from church.

If Joan walks each distance once each day, how far does she walk in a week from Sunday to Saturday?

5. The weight of a truck carrying a load of sand is 32.7 tonnes. If the weight of the truck with the driver is 24.9 tonnes, what is the weight of the sand?

6. If the weight of the driver in question 5 is 100 kg, what is the weight of the truck?

7. Complete the times in the table using the 12-hour clock and the 24-hour clock.

12-hour clock	24-hour clock
7:40 a.m.	07:40
1:00 p.m.	
	14:30
	09:45
5:45 p.m.	
12:00 midday	

8. Use the flight schedule in Table 4.8 on page 32 to answer these.
   (a) What is the flight time from Kingston to Miami on flight JM031?
   (b) If departure time is delayed by 1 h 45 min, at what time should the flight arrive in Miami?
   (c) If the flight arrives 15 minutes earlier than the time expected at (b), what time did the plane arrive in Miami? (Give the answer on the 24-hour clock and on the 12-hour clock.)

9. Complete the table below to show temperatures on the Fahrenheit and Celsius scales.

Celsius scale	Fahrenheit scale
0 °C	
180 °C	
25 °C	
	86 °F
45 °C	

10 Rachel left for school at 7:42 am and arrived at 8:37 a.m.

(a) How long did it take Rachel to get to school?

Rachel was 22 minutes late for her first class.

(b) At what time did the first class begin?

11 A telephone bill shows date and times of calls and length of calls as listed below:

Date / time	Length of call (min)
14 April 07:07	1
14 Apri 07:07	17
14 Apri 07:54	18
15 Apri 08:01	12
17 April 14:51	5
22 Apri 08:07	10
23 April 08:06	15
25 April 18:05	14
26 April 22:56	5

(a) On which day was most calls made?

(b) At what time were most calls made, a.m or p.m.?

(c) How many calls were made in the afternoon (p.m.)?

(d) What is the total time, in mins, for which the bill is calculated?

(e) If the company gives a reduced rate for calls made after 10:00 p.m. on how many calls would the customer benefit from this reduction?

Chapter 5

Fractions

Pre-requisites

■ base ten number system; factors and multiples

It is not always possible to use whole numbers to describe quantities.

We use **fractions** to describe
(a) equal parts of a whole (Figs 5.1, 5.2, 5.3)

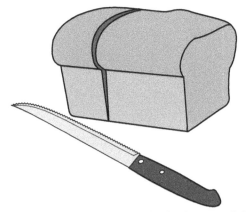

Fig. 5.1 I will eat one-third of the loaf now and the rest later

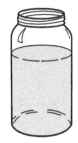

Fig. 5.2 The jar is seven-tenths full

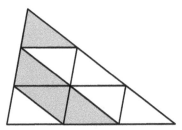

Fig. 5.3 Four-ninths of the triangle has been shaded

(b) a subset of a set of equal elements (Figs 5.4, 5.5).

Fig. 5.4 Two-fifths of the oranges in this basket are peeled

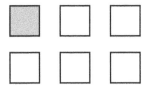

Fig. 5.5 Five-sixths of the squares are unshaded

We write fractions as follows:

one-third $\frac{1}{3}$; seven-tenths $\frac{7}{10}$;
four-ninths $\frac{4}{9}$; two-fifths $\frac{2}{5}$.

The number below the line is called the **denominator**. The denominator shows the number of equal parts. The number above the line is called the **numerator**. The numerator shows the number of parts in the fraction.

Equivalent fractions

Look at the diagrams and the fractions in Table 5.1.

Table 5.1

Diagram	Number of equal parts	Number of parts shaded	Fraction shaded
	3	1	$\frac{1}{3}$
	6	2	$\frac{2}{6}$
	9	3	$\frac{3}{9}$
	12	4	$\frac{4}{12}$
	15	5	$\frac{5}{15}$

In each diagram the whole is the same and the same amount has been shaded:

The fractions $\frac{1}{3}, \frac{2}{6}, \frac{3}{9}, \frac{4}{12}, \frac{5}{15}$ all represent the same amount. Thus $\frac{1}{3} = \frac{2}{6} = \frac{3}{9} = \frac{4}{12} = \frac{5}{15}$. We say that fractions which are equal in this way are **equivalent fractions**. $\frac{2}{6}$ is equivalent to $\frac{5}{15}$ or, simply, $\frac{2}{6} = \frac{5}{15}$.

Exercise 5a (Oral)

Find the missing numbers.

① $\frac{1}{3} = \frac{3}{9} = \frac{6}{\quad} = \frac{\quad}{24} = \frac{50}{\quad} = \frac{\quad}{900} = \frac{100}{\quad}$

② $\frac{1}{2} = \frac{2}{4} = \frac{3}{\quad} = \frac{\quad}{8} = \frac{\quad}{10} = \frac{\quad}{100} = \frac{25}{\quad}$

③ $\frac{2}{3} = \frac{8}{12} = \frac{20}{\quad} = \frac{\quad}{6} = \frac{\quad}{900} = \frac{\quad}{24} = \frac{10}{\quad}$

④ $\frac{1}{4} = \frac{5}{20} = \frac{\quad}{8} = \frac{3}{\quad} = \frac{\quad}{60} = \frac{\quad}{100} = \frac{7}{\quad}$

⑤ $\frac{3}{4} = \frac{6}{8} = \frac{15}{\quad} = \frac{24}{\quad} = \frac{\quad}{28} = \frac{\quad}{100} = \frac{\quad}{24}$

⑥ $\frac{1}{5} = \frac{10}{50} = \frac{\quad}{10} = \frac{4}{\quad} = \frac{\quad}{100} = \frac{24}{\quad} = \frac{5}{\quad}$

Sets of equivalent fractions can be made by multiplying the numerator and denominator by the same number. For example:

$$\frac{1}{4} = \frac{1 \times 5}{4 \times 5} = \frac{5}{20}$$

$$\frac{7}{9} = \frac{7 \times 4}{9 \times 4} = \frac{28}{36}$$

Example 1

Express each of the fractions $\frac{3}{4}, \frac{5}{8}, \frac{7}{12}, \frac{2}{3}$ with a denominator of 24. Hence arrange the fractions in ascending order (i.e. from lowest to highest).

$$\frac{3}{4} = \frac{3 \times 6}{4 \times 6} = \frac{18}{24}$$

$$\frac{5}{8} = \frac{5 \times 3}{8 \times 3} = \frac{15}{24}$$

$$\frac{7}{12} = \frac{7 \times 2}{12 \times 2} = \frac{14}{24}$$

$$\frac{2}{3} = \frac{2 \times 8}{3 \times 8} = \frac{16}{24}$$

The order is $\frac{14}{24}, \frac{15}{24}, \frac{16}{24}, \frac{18}{24}$, i.e. $\frac{7}{12}, \frac{5}{8}, \frac{2}{3}, \frac{3}{4}$.

Note: 24 is the LCM of 4, 8, 12 and 3. We say that 24 is the **lowest common denominator** of the set of fractions.

Exercise 5b

① Find the missing numbers.

(a) $\frac{2}{3} = \frac{\quad}{18}$ (b) $\frac{5}{6} = \frac{20}{\quad}$

(c) $\frac{3}{5} = \frac{\quad}{15}$ (d) $\frac{7}{9} = \frac{14}{\quad}$

(e) $\frac{3}{7} = \frac{9}{\quad}$ (f) $\frac{9}{11} = \frac{18}{\quad}$

(g) $\frac{1}{6} = \frac{\quad}{18}$ (h) $\frac{3}{5} = \frac{\quad}{20}$

(i) $\frac{5}{8} = \frac{25}{\quad}$ (j) $\frac{5}{7} = \frac{35}{\quad}$

(k) $\frac{3}{8} = \frac{\quad}{48}$ (l) $\frac{5}{9} = \frac{\quad}{36}$

② Use the given common denominators to arrange the fractions in ascending order (i.e. from lowest to highest).

(a) $\frac{2}{3}, \frac{5}{6}, \frac{7}{12}, \frac{3}{4}$ common denominator 12

(b) $\frac{3}{4}, \frac{4}{5}, \frac{9}{10}, \frac{17}{20}$ common denominator 20

(c) $\frac{2}{3}, \frac{5}{9}, \frac{7}{12}, \frac{11}{18}$ common denominator 36

③ Find the lowest common denominator of the following sets of fractions. Use this to arrange the fractions in ascending order of size.

(a) $\frac{1}{3}, \frac{2}{9}, \frac{5}{18}$

(b) $\frac{1}{2}, \frac{3}{5}, \frac{8}{15}, \frac{17}{30}$

(c) $\frac{3}{5}, \frac{5}{8}, \frac{7}{10}, \frac{13}{20}$

Lowest terms

The value of a fraction stays the same if both the numerator and denominator are divided by the same number. For example:

$$\frac{16}{24} = \frac{16 \div 8}{24 \div 8} = \frac{2}{3}$$

$$\frac{15}{60} = \frac{15 \div 15}{60 \div 15} = \frac{1}{4}$$

When the numerator and denominator have no common factor, we say that the fraction is in its **lowest terms**, or in its **simplest form**. Thus $\frac{16}{24}$ in its lowest terms is $\frac{2}{3}$; $\frac{1}{4}$ is the simplest form of $\frac{15}{60}$.

To express a fraction in its lowest terms:
(a) look for common factors of the numerator and denominator;
(b) divide the numerator and denominator by their common factors;
(c) repeat until there are no more common factors.

Example 2

Express (a) $\frac{42}{70}$ and (b) $\frac{26}{78}$ in their lowest terms.

(a) by inspection:

$$\frac{42}{70} = \frac{42 \div 7}{70 \div 7} = \frac{6}{10} = \frac{6 \div 2}{10 \div 2} = \frac{3}{5}$$

or by using prime factors:

$$\frac{42}{70} = \frac{2 \times 3 \times 7}{2 \times 5 \times 7}$$

Cancel factors which are both in the numerator and in the denominator, i.e.

$$\frac{42}{70} = \frac{\cancel{2} \times 3 \times \cancel{7}}{\cancel{2} \times 5 \times \cancel{7}}$$

$$= \frac{1 \times 3 \times 1}{1 \times 5 \times 1} = \frac{3}{5}$$

(b) by inspection:

$$\frac{26}{78} = \frac{26 \div 2}{78 \div 2} = \frac{13}{29} = \frac{13 \div 13}{39 \div 13} = \frac{1}{3}$$

or by using prime factors:

$$\frac{26}{78} = \frac{\cancel{2} \times \cancel{13}}{\cancel{2} \times 3 \times \cancel{13}} = \frac{1 \times 1}{1 \times 3 \times 1} = \frac{1}{3}$$

After practice, you will be able to leave out many of the steps shown in the above example.

Exercise 5c

1 Find the missing numbers.

(a) $\dfrac{15}{20} = \dfrac{15 \div 5}{20 \div 5} = \dfrac{}{4}$

(b) $\dfrac{12}{21} = \dfrac{12 \div 3}{21 \div } = \dfrac{4}{}$

(c) $\dfrac{18}{24} = \dfrac{2 \times 3 \times 3}{2 \times 2 \times 2 \times 3} = \dfrac{3}{}$

(d) $\dfrac{80}{100} = \dfrac{4 \times 2 \times 2 \times 5}{2 \times 2 \times 5 \times 5} = \dfrac{4}{}$

(e) $\dfrac{25}{100} = \dfrac{}{20} = \dfrac{}{4}$

(f) $\dfrac{42}{56} = \dfrac{21}{} = \dfrac{3}{}$

(g) $\dfrac{18}{27} = \dfrac{2}{}$ (h) $\dfrac{10}{16} = \dfrac{}{8}$

(i) $\dfrac{28}{36} = \dfrac{7}{}$ (j) $\dfrac{48}{100} = \dfrac{12}{}$

(k) $\dfrac{32}{40} = \dfrac{4}{}$ (l) $\dfrac{16}{48} = \dfrac{}{3}$

2 Reduce the following fractions to their lowest terms,

(a) $\frac{7}{35}$ (b) $\frac{3}{18}$ (c) $\frac{5}{100}$

(d) $\frac{30}{80}$ (e) $\frac{14}{21}$ (f) $\frac{24}{32}$

(g) $\frac{30}{36}$ (h) $\frac{48}{60}$ (i) $\frac{24}{54}$

(j) $\frac{45}{105}$ (k) $\frac{90}{126}$ (l) $\frac{128}{176}$

Mixed numbers

Some quantities need whole numbers and fractions to describe them (Fig. 5.6).

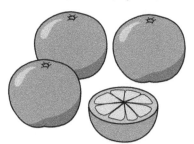

Fig. 5.6 Three and a half oranges

The number three and a half, written $3\frac{1}{2}$, is a **mixed number**. A mixed number has a whole number part and a fractional part: $3\frac{1}{2} = 3 + \frac{1}{2}$. It is possible to express a mixed number as a single fraction:

$$3\frac{1}{2} = 3 + \frac{1}{2} = \frac{3}{1} + \frac{1}{2} = \frac{3 \times 2}{1 \times 2} + \frac{1}{2}$$

$$= \frac{6}{2} + \frac{1}{2} = \frac{6 + 1}{2} = \frac{7}{2}$$

$$3\frac{1}{2} = \frac{7}{2}$$

The numerator of the fraction $\frac{7}{2}$ is greater than the denominator. This is an example of an **improper fraction**. If the numerator is less than the denominator, the fraction is a **proper fraction**. For example, $\frac{2}{7}$ is a proper fraction.

Example 3

Express $4\frac{5}{6}$ as an improper fraction.

$$4\frac{5}{6} = 4 + \frac{5}{6} = \frac{24}{6} + \frac{5}{6} = \frac{24 + 5}{6} = \frac{29}{6}$$

or, more quickly,

$$4\frac{5}{6} = \frac{(4 \times 6) + 5}{6} = \frac{24 + 5}{6} = \frac{29}{6}$$

Example 4

Express $\frac{19}{8}$ as a mixed number.

$$\frac{19}{8} = \frac{16 + 3}{8} = \frac{16}{8} + \frac{3}{8} = 2 + \frac{3}{8} = 2\frac{3}{8}$$

or, more quickly by division,

$$\frac{19}{8} = 2 + \frac{3}{8} = 2\frac{3}{8}$$

Exercise 5d (Oral)

1 Express as mixed numbers.

(a) $\frac{4}{3}$ (b) $\frac{5}{2}$ (c) $\frac{11}{4}$ (d) $\frac{7}{5}$

(e) $\frac{11}{5}$ (f) $\frac{25}{7}$ (g) $\frac{25}{6}$ (h) $\frac{19}{8}$

(i) $\frac{26}{9}$ (j) $\frac{41}{5}$ (k) $\frac{60}{7}$ (l) $\frac{55}{6}$

2 Express as improper fractions.

(a) $1\frac{3}{4}$ (b) $3\frac{1}{3}$ (c) $6\frac{1}{2}$ (d) $4\frac{3}{3}$

(e) $9\frac{1}{4}$ (f) $3\frac{5}{6}$ (g) $4\frac{2}{7}$ (h) $7\frac{3}{5}$

(i) $5\frac{3}{8}$ (j) $6\frac{5}{7}$ (k) $12\frac{2}{3}$ (l) $13\frac{1}{3}$

Addition and subtraction of fractions

We can only add or subtract fractions if they have the same denominators, that is, if the parts are equal. Then we add or subtract the numerators only, that is, the number of equal parts.

Example 5

$$\frac{2}{9} + \frac{5}{9} = \frac{2 + 5}{9} = \frac{7}{9}$$

Example 6

$$\frac{13}{15} - \frac{2}{15} = \frac{13 - 2}{15} = \frac{11}{15}$$

If fractions have different denominators, we need to use equivalent fractions. So we

(a) find a common denominator (preferably the lowest common multiple of the denominators);

(b) express each fraction as an equivalent fraction using that denominator;

(c) add or subtract the new numerators.

Example 7

$$\frac{5}{6} + \frac{3}{8}$$

The LCM of 6 and 8 is 24.

$$\frac{5}{6} + \frac{3}{8} = \frac{5 \times 4}{6 \times 4} + \frac{3 \times 3}{8 \times 3} \qquad \text{[step (a)]}$$

$$= \frac{20}{24} + \frac{9}{24} \qquad \text{[step (b)]}$$

$$= \frac{20 + 9}{24} = \frac{29}{24} = 1\frac{5}{24} \qquad \text{[step (c)]}$$

Example 8

$$\frac{7}{10} - \frac{4}{15}$$

The LCM of 10 and 15 is 30.

$$\frac{7}{10} - \frac{4}{15} = \frac{7 \times 3}{10 \times 3} - \frac{4 \times 2}{15 \times 2} \qquad \text{[Step (a)]}$$

$$= \frac{21}{30} - \frac{8}{30} \qquad \text{[step (b)]}$$

$$= \frac{21 - 8}{30} = \frac{13}{30} \qquad \text{[step (c)]}$$

With practice, steps (a) and (b) may be omitted.

If the numbers are mixed, express them as improper fractions.

Example 9

$$6\tfrac{5}{12} - 3\tfrac{3}{4} = \frac{77}{12} - \frac{15}{4} = \frac{77}{12} - \frac{45}{12} = \frac{32}{12}$$

$$= \frac{32 \div 4}{12 \div 4} = \frac{8}{3} = 2\tfrac{2}{3}$$

Note that $\frac{32}{12}$ was reduced to its lowest terms by dividing the numerator and the denominator by 4.

Example 10

$$3\tfrac{2}{5} + 2\tfrac{2}{3} = \frac{17}{5} + \frac{8}{3} = \frac{51}{15} + \frac{40}{15} = \frac{91}{15} = 6\tfrac{1}{15}$$

Exercise 5e

Simplify the following.

① $\frac{3}{4} + \frac{2}{3}$ ② $\frac{3}{4} - \frac{2}{3}$

③ $\frac{5}{6} + \frac{2}{9}$ ④ $\frac{5}{6} - \frac{2}{9}$

⑤ $\frac{7}{12} - \frac{3}{8}$ ⑥ $\frac{7}{12} + \frac{3}{8}$

⑦ $\frac{11}{15} - \frac{2}{5}$ ⑧ $\frac{11}{15} + \frac{1}{5}$

⑨ $1\tfrac{1}{2} + \frac{3}{4}$ ⑩ $1\tfrac{3}{4} - \frac{1}{2}$

⑪ $1\tfrac{1}{2} + 2\tfrac{1}{3}$ ⑫ $1\tfrac{1}{4} - \frac{2}{3}$

⑬ $3\tfrac{7}{8} + 2\tfrac{3}{4}$ ⑭ $2\tfrac{5}{6} + 5\tfrac{7}{9}$

⑮ $5\tfrac{3}{4} - 2\tfrac{4}{5}$ ⑯ $4\tfrac{1}{6} - 1\tfrac{5}{8}$

⑰ $1\tfrac{1}{2} + 2\tfrac{1}{3} + 3\tfrac{1}{4}$ ⑱ $5\tfrac{3}{4} - 2\tfrac{7}{8} + 1\tfrac{1}{2}$

⑲ $3\tfrac{7}{8} + 7\tfrac{3}{4} - 6\tfrac{1}{2}$ ⑳ $5\tfrac{1}{6} - 3\tfrac{2}{3} + 6\tfrac{7}{12}$

Exercise 5f

① What is the sum of $1\tfrac{7}{12}$ and $3\tfrac{5}{8}$?

② What is the difference between $4\tfrac{3}{8}$ and $3\tfrac{11}{12}$?

③ Find the sum of $2\tfrac{3}{4}$ and $2\tfrac{4}{5}$. Find the difference between this sum and 6.

④ A girl spends $\frac{1}{4}$ of her pocket money on Monday and $\frac{3}{8}$ on Wednesday. What fraction of her money is left?

⑤ A fruit grower uses $\frac{1}{3}$ of his land for bananas, $\frac{3}{8}$ for pineapples, $\frac{1}{6}$ for mangoes and the remainder for oranges. What fraction of his land is used for oranges?

⑥ $\frac{1}{5}$ of a class's timetable is given to English and $\frac{1}{4}$ to Mathematics. What fraction is left for the other subjects?

⑦ During a week, a student spends $\frac{1}{3}$ of his time in bed, $\frac{5}{24}$ of his time in lessons and $\frac{1}{8}$ of his time doing homework.

What fraction of his time is left for doing other things?

⑧ A man goes on a journey. He does $\frac{1}{6}$ of it on a bicycle, $\frac{4}{5}$ on a lorry and walks the rest. What fraction of the journey does he walk?

⑨ A boy plays football for $1\tfrac{3}{4}$ hours, listens to the radio for $\frac{3}{4}$ hour and then spends $1\tfrac{1}{4}$ hours doing his homework. How much time does he spend altogether doing these things?

⑩ By how much is the sum of $2\tfrac{4}{5}$ and $4\tfrac{1}{2}$ less than $8\tfrac{1}{10}$?

Multiplication of fractions

Whole number × fraction

$4 \times \frac{1}{3}$ means 4 lots of $\frac{1}{3}$ of something. We can show this in pictures and numbers (Fig. 5.7):

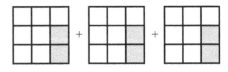

Fig. 5.7

$$4 \times \tfrac{1}{3} = \tfrac{1}{3} + \tfrac{1}{3} + \tfrac{1}{3} + \tfrac{1}{3} = \tfrac{4}{3} = 1\tfrac{1}{3}$$

$$or \; 4 \times \tfrac{1}{3} = \frac{4 \times 1}{3}$$

$$= \tfrac{4}{3} = 1\tfrac{1}{3}$$

Similarly, $3 \times \frac{2}{9}$, (Fig. 5.8):

Fig. 5.8

$$3 \times \frac{2}{9} = \frac{2}{9} + \frac{2}{9} + \frac{2}{9} = \frac{6}{9} = \frac{2}{3}$$

$$or\ 3 \times \frac{2}{9} = \frac{3 \times 2}{9}$$
$$= \frac{6}{9} = \frac{2}{3}$$

⑤ $9 \times \frac{1}{2}$ **⑥** $\frac{2}{3}$ of 15

⑦ $\frac{5}{6} \times 8$ **⑧** $\frac{3}{4}$ of 10

⑨ $\frac{1}{3}$ of 2 **⑩** $\frac{3}{8} \times 20$

Fraction × whole number

$\frac{2}{5} \times 3$ means $\frac{2}{5}$ of 3 objects. Let the objects be 3 squares (Fig. 5.9).

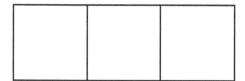

Fig. 5.9

The 3 squares can be divided into fifths. Two of the fifths are shaded in each square (Fig. 5.10).

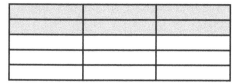

Fig. 5.10

The fifths can be rearranged to make 1 square and $\frac{1}{5}$ of a square altogether (Fig. 5.11).

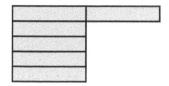

Fig. 5.11

Thus $\frac{2}{5}$ of $3 = \frac{2}{5} \times 3 = 1\frac{1}{5}$.

Also notice that

$$\frac{2}{5} \times 3 = \frac{2 \times 3}{5}$$
$$= \frac{6}{5} = 1\frac{1}{5}$$

as before.

Exercise 5g

Simplify the following.

① $8 \times \frac{2}{3}$ **②** $12 \times \frac{3}{4}$

③ $5 \times \frac{3}{10}$ **④** $2 \times \frac{2}{5}$

Fraction × fraction

$\frac{2}{3} \times \frac{5}{7}$ means $\frac{2}{3}$ of $\frac{5}{7}$ of something. The pictures show how to find $\frac{2}{3}$ of $\frac{5}{7}$ of a rectangle.

First, divide the rectangle into seven equal strips. (Fig. 5.12).

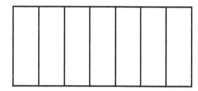

Fig. 5.12

Next, divide each strip into three equal parts (Fig. 5.13).

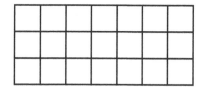

Fig 5.13

Next, shade $\frac{5}{7}$ of the rectangle (Fig. 5.14).

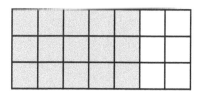

Fig. 5.14.

Then cross-shade $\frac{2}{3}$ of the shaded part (Fig. 5.15).

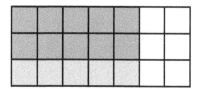

Fig. 5.15

Ten small squares represent $\frac{2}{3}$ of $\frac{5}{7}$ of the rectangle. Since the rectangle contained 21 small squares, then $\frac{2}{3} \times \frac{5}{7} = \frac{10}{21}$. Notice that

$$\frac{2}{3} \times \frac{5}{7} = \frac{2 \times 5}{3 \times 7} = \frac{10}{21}$$

To multiply a fraction by a fraction:
(a) multiply the numerators to make the numerator of the product;
(b) multiply the denominators to make the denominator of the product.

It is important to note that we *must always* change mixed numbers to improper fractions *before* multiplying.

Example 11

Simplify $\frac{3}{8}$ of $2\frac{2}{9} \times 1\frac{3}{5}$

$$\frac{3}{8} \text{ of } 2\frac{2}{9} \times 1\frac{3}{5} = \frac{3}{8} \times \frac{20}{9} \times \frac{8}{5}$$

$$= \frac{3 \times 20 \times 8}{8 \times 9 \times 5}$$

$$= \frac{\cancel{3} \times \cancel{5} \times 4 \times \cancel{8}}{\cancel{8} \times \cancel{3} \times 3 \times \cancel{5}}$$

$$= \frac{4}{3} = 1\frac{1}{3}$$

*Divide the numerator and denominator by the common factors 3, 5 and 8 to simplify.

Exercise 5h

Simplify the following.

1. $\frac{1}{2} \times \frac{3}{4}$
2. $\frac{2}{3}$ of $\frac{3}{4}$
3. $\frac{3}{4}$ of $\frac{1}{3}$
4. $\frac{3}{5} \times \frac{2}{3}$
5. $2\frac{1}{2} \times \frac{2}{5}$
6. $1\frac{1}{2} \times 1\frac{2}{3}$
7. $2\frac{3}{4} \times \frac{4}{5}$
8. $\frac{4}{11}$ of $3\frac{2}{3}$
9. $\frac{6}{7}$ of $5\frac{1}{4}$
10. $5\frac{1}{3} \times 2\frac{1}{4}$
11. $2\frac{1}{3} \times \frac{1}{14}$
12. $\frac{5}{14} \times \frac{21}{15}$
13. $\frac{12}{10} \times \frac{25}{18}$
14. $\frac{16}{21} \times \frac{20}{24} \times \frac{28}{15}$
15. $\frac{4}{7}$ of $8\frac{3}{4}$
16. $1\frac{7}{8} \times 3\frac{1}{5}$
17. $(2\frac{1}{2})^2$
18. $3\frac{3}{4} \times \frac{4}{9} \times 1\frac{1}{5}$
19. $1\frac{7}{9} \times 3\frac{3}{4} \times \frac{3}{8}$
20. $\frac{12}{25}$ of $(1\frac{1}{4})^2$

Division of fractions

Look at the following working very carefully.

$$2 \div \frac{5}{8} = \frac{2}{\frac{5}{8}} = \frac{2 \times \frac{8}{5}}{\frac{5}{8} \times \frac{8}{5}} = \frac{2 \times \frac{8}{5}}{1} = 2 \times \frac{8}{5}$$

Division by 1 does not alter the value of a number. In this example, the divisor, $\frac{5}{8}$, is multiplied by $\frac{8}{5}$ so that $\frac{5}{8} \times \frac{8}{5} = 1$. In order to keep the value of the division problem unchanged, also multiply the dividend, 2, by $\frac{8}{5}$, i.e. $2 \times \frac{8}{5}$.

Hence, $2 \div \frac{5}{8}$ becomes $\dfrac{2 \times \frac{8}{5}}{\frac{5}{8} \times \frac{8}{5}}$

Thus, $2 \div \frac{5}{8} = 2 \times \frac{8}{5}$

$\frac{8}{5}$ is the **reciprocal** of $\frac{5}{8}$; i.e. it is the same fraction turned upside down.

To divide by a fraction, simply multiply by its reciprocal.

Note that

$$2\frac{1}{2} \div 5 = 2\frac{1}{2} \div \frac{5}{1} = 2\frac{1}{2} \times \frac{1}{5}$$

$$= \frac{5}{2} \times \frac{1}{5} = \frac{1}{2}$$

Example 12

Find the value of $2\frac{1}{4} \div \frac{3}{7}$

$$2\frac{1}{4} \div \frac{3}{7} = 2\frac{1}{4} \times \frac{7}{3} = \frac{9}{4} \times \frac{7}{3} = \frac{9 \times 7}{4 \times 3}$$

$$= \frac{^3\cancel{9} \times 7}{4 \times \cancel{3}} = \frac{21}{4} = 5\frac{1}{4}$$

Notice that this example shows that it is possible to get a result which is greater than either the dividend or the divisor.

Another method of dividing by fractions is to multiply both the dividend and the divisor by the same number. Use a number which is a multiple of the denominators of all the fractions in the problem. To make sure that the numbers are as small as possible, use the LCM of the denominators.

$$2\frac{1}{4} \div \frac{3}{7} = \frac{\frac{9}{4}}{\frac{3}{7}} = \frac{\frac{9}{4} \times 28^*}{\frac{3}{7} \times 28} = \frac{9 \times 7}{3 \times 4} = \frac{21}{4} = 5\frac{1}{4}$$

*28 is the LCM of the denominators of the two fractions $\frac{9}{4}$ and $\frac{3}{7}$.

Example 13

Simplify $\dfrac{2\frac{2}{3} \times 1\frac{1}{2}}{4\frac{4}{5}}$

Multiplying by the reciprocal:

$$\frac{2\frac{2}{3} \times 1\frac{1}{2}}{4\frac{4}{5}} = \frac{8}{3} \times \frac{3}{2} \div \frac{24}{5} = \frac{8}{3} \times \frac{3}{2} \times \frac{5}{24}$$

$$= \frac{8 \times 3 \times 5}{3 \times 2 \times 24} = \frac{5}{6}$$

or by equal multiplication

$$\frac{2\frac{2}{3} \times 1\frac{1}{2}}{4\frac{4}{5}} = \frac{\frac{8}{3} \times \frac{3}{2}}{\frac{24}{5}} = \frac{\frac{8}{3} \times \frac{3}{2}}{\frac{24}{5}} = \frac{\frac{8}{3} \times \frac{3}{2} \times 30^*}{\frac{24}{5} \times 30}$$

$$= \frac{8 \times 3 \times 5}{24 \times 6} = \frac{5}{6}$$

*30 is the LCM of the denominators 2, 3 and 5.

Exercise 5i

Simplify the following.

① $1\frac{2}{3} \div 5$

② $2\frac{1}{4} \div 3$

③ $4\frac{1}{5} \div 3$

④ $10 \div \frac{2}{7}$

⑤ $4\frac{1}{5} \div 7$

⑥ $13\frac{3}{4} \div 11$

⑦ $\frac{9}{14} \div \frac{3}{7}$

⑧ $\frac{12}{25} \div \frac{9}{10}$

⑨ $\frac{16}{33} \div \frac{8}{11}$

⑩ $1\frac{1}{7} \div \frac{4}{21}$

⑪ $2\frac{2}{5} \div 1\frac{1}{3}$

⑫ $\frac{5}{6} \div 2\frac{1}{12}$

⑬ $1\frac{4}{5} \div 6\frac{3}{10}$

⑭ $7\frac{1}{8} \div 4\frac{3}{4}$

⑮ $\dfrac{\frac{3}{10} \times \frac{35}{36}}{\frac{14}{15}}$

⑯ $\dfrac{\frac{9}{10} \div 3\frac{1}{5}}{\frac{3}{8}}$

⑰ $\dfrac{7\frac{3}{7} \div \frac{5}{21}}{9\frac{3}{4}}$

⑱ $\dfrac{8\frac{1}{6} \times 3\frac{3}{7}}{11\frac{2}{3}}$

⑲ $\dfrac{5\frac{1}{4} \div 2\frac{4}{5}}{3\frac{3}{4}}$

⑳ $\dfrac{9\frac{3}{5}}{5\frac{1}{7}} \times 1\frac{1}{4}$

Exercise 5j

① Find the product of $3\frac{1}{4}$ and $2\frac{2}{5}$.

② Find the value of $6\frac{2}{3}$ divided by $1\frac{7}{9}$.

③ What is three-quarters of $3\frac{3}{7}$?

④ Find the values of $4\frac{1}{2} \div 1\frac{4}{5}$ and $1\frac{4}{5} \div 4\frac{1}{2}$. Which one is $4\frac{1}{2}$ divided into $1\frac{4}{5}$?

⑤ The mass of each book of an encyclopaedia is $1\frac{3}{4}$kg. There are 20 books in the encyclopaedia. Find the total mass of the encyclopaedia.

⑥ A tank holds 15 litres of water. A cup holds $\frac{3}{10}$ of a litre. How many cups of water does the tank hold?

⑦ LP records turn round $33\frac{1}{3}$ times every minute. If one side of an LP plays for 24 minutes, how many times does it turn round?

⑧ Some shop-soiled goods are sold for $\frac{2}{5}$ of their original price. What will be the selling price of a pen originally costing $2.85?

⑨ In a school, $\frac{9}{10}$ of the students take part in sports. $\frac{2}{3}$ of these play football. What fraction of the students play football?

⑩ It takes $1\frac{3}{4}$ m of cloth to make a skirt. How many skirts can be made from $10\frac{1}{2}$ m of cloth?

Summary

Fractions are used to describe parts of quantities. The total number of **equal parts** into which the quantity is divided is called the **denominator**; the number of parts in the fraction is the **numerator**. Hence the fraction, three-fifths, that is three parts out of a total of five equal parts, is written $\frac{3}{5}$.

Fractions which have the same value are called **equivalent fractions**, e.g.

$$\frac{1}{2} = \frac{2}{4} = \frac{3}{6} = \frac{14}{28} = \frac{50}{100}$$

When the numerator is less than the denominator, the fraction is a **proper fraction**; when the numerator is greater than the denominator, the fraction is an **improper fraction**. When there is a whole number part and a fraction in a number, the number is said to be a **mixed number**.

When there are no factors common to both the numerator and denominator, the fraction is said to be in its **lowest terms**, e.g. $\frac{15}{30}$ in its lowest terms is $\frac{1}{2}$.

Fractions must all have the same denominator in order to be added (or subtracted). If there are different denominators, the LCM of all the denominators is used to write the equivalent fractions in order to add (or subtract) fractions, e.g.

$$\frac{5}{8} + \frac{3}{4} - \frac{1}{3} = \frac{15 + 18 - 8}{24}$$

$$= \frac{25}{24} = 1\frac{1}{24}$$

To multiply fractions, multiply the numerators together to get the numerator of the product; and then multiply the denominators to get the denominator, e.g.

$$\frac{2}{3} \times \frac{5}{7} \times \frac{8}{11} = \frac{80}{231}$$

To divide by a fraction, multiply by its **reciprocal**, e.g.

$$4 \div \frac{2}{3} = 4 \times \frac{3}{2} = 6$$

Practice exercise P5.1

① The shaded part of each diagram represents a fraction. Write the fraction shown in numbers and words.

(a) 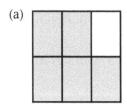 (b)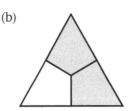

(c) 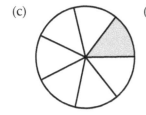 (d)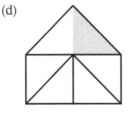

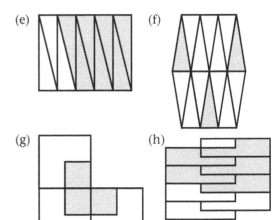

(e)   (f)

(g)   (h)

② Cancel these fractions to their lowest terms.
(a) $\frac{5}{10}$ (b) $\frac{3}{12}$ (c) $\frac{3}{9}$ (d) $\frac{15}{20}$
(e) $\frac{12}{14}$ (f) $\frac{12}{20}$ (g) $\frac{16}{28}$ (h) $\frac{15}{24}$

③ Write in the missing numbers to make these equivalent fraction chains.
(a) $\frac{1}{4} = \frac{3}{} = \frac{5}{} = \frac{}{24} = \frac{}{32} = \frac{10}{}$
(b) $\frac{2}{5} = \frac{}{10} = \frac{}{20} = \frac{10}{} = \frac{16}{} = \frac{}{60}$
(c) $\frac{5}{8} = \frac{10}{} = \frac{15}{} = \frac{}{40} = \frac{75}{}$

④ Two of the fractions are equivalent and one is different. Write down the odd one out.
(a) (i) $\frac{2}{8}$ (ii) $\frac{1}{8}$ (iii) $\frac{2}{16}$
(b) (i) $\frac{6}{8}$ (ii) $\frac{3}{5}$ (iii) $\frac{3}{4}$
(c) (i) $\frac{5}{8}$ (ii) $\frac{10}{14}$ (iii) $\frac{20}{28}$
(d) (i) $\frac{15}{40}$ (ii) $\frac{14}{42}$ (iii) $\frac{12}{32}$
(e) (i) $\frac{8}{12}$ (ii) $\frac{7}{8}$ (iii) $\frac{6}{9}$

Practice exercise P5.2

① Find the value of these.
(a) $\frac{1}{2}$ of $90   (b) $\frac{1}{3}$ of $90
(c) $\frac{1}{4}$ of 80 m   (d) $\frac{1}{5}$ of $90
(e) $\frac{1}{3}$ of 72 h   (f) $\frac{1}{8}$ of 60 kg
(g) $\frac{3}{4}$ of $24   (h) $\frac{4}{9}$ of $240

② Cancel each fraction to its lowest terms and then work out the problem.
(a) $\frac{2}{4}$ of $42   (b) $\frac{2}{8}$ of $42
(c) $\frac{3}{15}$ of 80 cm   (d) $\frac{4}{12}$ of 96 litres

3 Two answers are the same and the other is different. Write down the odd one out.

(a) (i) $\frac{1}{4}$ of 100 ml

(ii) $\frac{1}{5}$ of 20 ml

(iii) $\frac{1}{10}$ of 40 ml

(b) (i) $\frac{2}{3}$ of 30 days

(ii) $\frac{2}{5}$ of 50 days

(iii) $\frac{3}{5}$ of 40 days

(c) (i) $\frac{5}{6}$ of 12 books

(ii) $\frac{5}{8}$ of 24 books

(iii) $\frac{3}{8}$ of 40 books

(d) (i) $\frac{2}{15}$ of 60 cm

(ii) $\frac{1}{10}$ of 120 cm

(iii) $\frac{3}{16}$ of 64 cm

(e) (i) $\frac{5}{6}$ of 54 Mb

(ii) $\frac{5}{8}$ of 64 Mb

(iii) $\frac{5}{7}$ of 56 Mb

Practice exercise P5.3

1 Write each of these as an improper fraction.

(a) $2\frac{1}{2}$ (b) $2\frac{3}{5}$ (c) $1\frac{1}{6}$ (d) $6\frac{3}{7}$

(e) $3\frac{3}{4}$ (f) $1\frac{5}{8}$ (g) $6\frac{1}{10}$ (h) $4\frac{5}{12}$

2 Write each of these as a mixed number.

(a) $\frac{7}{2}$ (b) $\frac{15}{7}$ (c) $\frac{5}{3}$ (d) $\frac{25}{8}$

(e) $\frac{9}{4}$ (f) $\frac{23}{12}$ (g) $\frac{22}{9}$ (h) $\frac{33}{10}$

Practice exercise P5.4

1 Simplify the following.

(a) $\frac{1}{10} + \frac{3}{10}$ (b) $\frac{3}{8} + \frac{3}{8}$ (c) $\frac{2}{7} + \frac{4}{7}$

(d) $\frac{2}{5} + \frac{4}{5}$ (e) $\frac{7}{13} + \frac{7}{13}$ (f) $\frac{7}{10} + \frac{9}{10}$

(g) $\frac{5}{6} - \frac{1}{6}$ (h) $\frac{7}{9} - \frac{4}{9}$ (i) $\frac{13}{16} - \frac{5}{16}$

(j) $1\frac{3}{5} - \frac{4}{5}$ (k) $1\frac{1}{8} - \frac{5}{8}$ (l) $2\frac{1}{4} - \frac{3}{4}$

2 Change the fractions to the type in brackets, then answer the questions. Simplify your answer.

(a) $\frac{1}{2} + \frac{1}{6} \left(\frac{*}{6}\right)$ (b) $\frac{1}{4} + \frac{1}{6} \left(\frac{*}{12}\right)$

(c) $\frac{5}{8} - \frac{1}{4} \left(\frac{*}{8}\right)$ (d) $\frac{5}{6} - \frac{7}{18} \left(\frac{*}{18}\right)$

(e) $\frac{3}{5} + \frac{3}{10} \left(\frac{*}{10}\right)$ (f) $1\frac{1}{2} - \frac{1}{14} \left(\frac{*}{14}\right)$

3 Simplify the following.

(a) $\frac{3}{5} + \frac{3}{10}$ (b) $\frac{1}{4} + \frac{3}{8}$ (c) $\frac{2}{7} - \frac{1}{14}$

(d) $\frac{2}{5} + \frac{1}{3}$ (e) $2\frac{7}{8} - 1\frac{3}{4}$ (f) $\frac{7}{10} + 1\frac{2}{3}$

(g) $3\frac{1}{4} + 1\frac{2}{3} + 2\frac{1}{12}$ (h) $\frac{5}{8} + 3\frac{1}{4} - \frac{5}{6}$

4 By adding and subtracting these fractions, find as many answers as possible.

$\frac{1}{2}$ $\frac{1}{3}$ $\frac{1}{4}$ $\frac{1}{5}$

5 Simplify the following.

(a) $\frac{2}{3} \times \frac{3}{5}$ (b) $\frac{7}{8} \times \frac{4}{21}$ (c) $\frac{7}{10} \times \frac{5}{6}$

(d) $\frac{5}{6} \times \frac{1}{6}$ (e) $1\frac{7}{9} \times \frac{3}{8}$ (f) $2\frac{3}{16} \times 1\frac{5}{7}$

6 Simplify the following.

(a) $\frac{3}{5} \div \frac{3}{10}$ (b) $\frac{1}{4} \div \frac{3}{8}$ (c) $\frac{2}{7} \div \frac{1}{14}$

(d) $\frac{2}{5} \div \frac{1}{3}$ (e) $2\frac{5}{8} \div 1\frac{3}{4}$ (f) $\frac{7}{9} \div 1\frac{2}{3}$

7 Calculate the following.

(a) $2\frac{1}{5} + 1\frac{3}{10} + 3\frac{2}{3}$ (b) $2\frac{5}{8} - 2\frac{1}{6} + 3\frac{5}{12}$

(c) $\frac{3}{4} \times 2\frac{2}{7}$ (d) $2\frac{1}{2} \times 2\frac{3}{5}$

(e) $4\frac{1}{12} \div 7$ (f) $1\frac{5}{16} \div \frac{5}{8}$

(g) $\dfrac{5\frac{2}{3}}{3\frac{1}{3} + 1\frac{5}{8}}$ (h) $\dfrac{2\frac{1}{4} \times 4\frac{1}{2}}{3\frac{3}{4}}$

8 Evaluate the following.

(a) $\frac{1}{9} + \frac{2}{3}$ (b) $1\frac{2}{3} - \frac{3}{8}$

(c) $9\frac{1}{4} - 6\frac{1}{3} + 1\frac{5}{9}$ (d) $3(3\frac{7}{9} + 2\frac{2}{3})$

(e) $3\frac{1}{7}(6\frac{1}{4} - 5\frac{1}{3})$ (f) $3\frac{4}{7} \times \frac{3}{10}$

(g) $6\frac{2}{3} \times 2\frac{5}{8}$ (h) $5\frac{1}{2} \div 1\frac{5}{8}$

(i) $11\frac{1}{4} \div 3\frac{1}{3} \times \frac{2}{5}$ (j) $1\frac{1}{2} + 2\frac{3}{4} \times 1\frac{2}{3}$

(k) $\dfrac{4\frac{1}{3} - 3\frac{5}{6}}{1\frac{5}{6} + 2\frac{3}{4}}$ (l) $5\frac{1}{2} \times \frac{3}{11}$

9 In a box there are footballs, tennis balls and basketballs. $\frac{1}{4}$ of the balls are footballs and $\frac{1}{3}$ are basketballs. There are 30 tennis balls. Find the total number of balls in the box.

Chapter 6

Algebra: basic processes

Numbers in boxes

$14 + \square = 17$. What number in the box will make this true? You may have seen problems like this in Primary School. $14 + \square = 17$ will be true if 3 goes in the box: $14 + \boxed{3} = 17$ is true.

Letters for numbers

In mathematics, instead of boxes we use letters of the alphabet to stand for numbers. We write $14 + x$ instead of $14 + \square$. Any letter can be used. For example, $14 + a$ would be just as good as $14 + x$. Capital letters are not generally used; small letters are more usual.

When using a letter instead of a number, the letter can stand for any number in general. Thus the value of $14 + x$ depends on the value of x. For example,

if x stands for 2; $14 + x$ has the value 16;
if x stands for 12; $14 + x$ has the value 26;
if x stands for 5; $14 + x$ has the value 19.

When letters and numbers are used together in this way, the mathematics is called **generalised arithmetic** or **algebra**. The word 'algebra' comes to us from an important book written around AD 830 by Mohammed Musa al Khowarizmi, a noted Moslem mathematician. The title of the book was *Al-jabr wa'l Muqābalah*.

Algebraic expressions

An expression with both numbers and letters is called an **algebraic expression**. $14 + x$, $3a$, $7 - r$, $2(b + l)$ are all examples of algebraic expressions.

What is the value of $x + 6$? The value of $x + 6$ depends on what x stands for.

If $x = 3$, then $x + 6 = 3 + 6 = 9$
If $x = 8$, then $x + 6 = 8 + 6 = 14$

Exercise 6a (Oral)

① Find the value of the following when $x = 4$.
(a) $x + 5$ (b) $x + 8$ (c) $x - 1$
(d) $x - 4$ (e) $9 + x$ (f) $0 + x$
(g) $10 - x$ (h) $19 - x$

② Find the value of the following when $a = 7$.
(a) $a + a$ (b) $a - a$
(c) $a + a + 6$ (d) $a + a + 3$
(e) $a + 5 + a$ (f) $a - 2 + a$
(g) $9 + a + a$ (h) $16 + a - a$
(i) $11 - a + a$ (j) $a - 8 + a$
(k) $a - 13 + a$ (l) $2 - a + a$

Algebra from words

Here is a story in pictures (Fig. 6.1):

Fig. 6.1

Here is the story in words.
(a) John has a bag with some mangoes in it.
(b) He buys two more mangoes.
(c) John now has a bag with some mangoes and two more mangoes.

Here is the same story with algebraic symbols.

(a) John has *m* mangoes.

(b) John buys 2 mangoes.

(c) John now has $m + 2$ mangoes.

In this story, *m* represents the number of mangoes in the bag. $m + 2$ represents the number of mangoes that John has altogether.

Flow charts

To obtain an algebraic expression for a story in words, we can represent the information, including the arithmetic operations, by a **flow chart**. The flow chart shows the order in which you do the operations.

Information or data is put in this shape of box:

The arithmetic operations (that is, add, subtract, multiply or divide) are put in rectangles, that is:

Arrows from to [ ] show the order of doing the operations. The final result is put in a circle:

◯

The story of John and the mangoes is represented by the simple flow chart below.

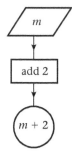

Fig. 6.2

Example 1

A boy has *m* mangoes. He sells two of them. How many mangoes does he now have?

He sells 2 mangoes. He must have 2 less than he had before. He now has $m - 2$ mangoes.

Also:

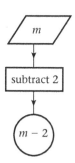

Fig. 6.3

Sometimes a problem may be difficult because letters are used for numbers. If this happens, first change the letter to an easy number. Use that number instead of the letter in solving the problem. Use other easy numbers until the method of solving the problem is clear.

Example 2

A girl is 14 years old. How old will she be in *x* years time?

Use 2 years instead of *x* years.
In 2 years, the girl will be $14 + 2$ years old (16 years old).
Then, use 10 years instead of *x* years.
In 10 years, the girl will be $14 + 10$ years old (24 years old).
So, in *x* years, the girl will be $14 + x$ years old.
$14 + x$ will not simplify.
The girl will be $14 + x$ years old.

Also:

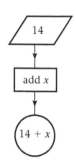

Fig. 6.4

In algebra, letters represent numbers only. If a problem contains units, then units must show in the working. For example, the length of a rope cannot be 7 or x. The units are missing. The rope might be 7 metres long or x centimetres long. Always use units when necessary.

Example 3

How many cents are there in n dollars?

Use \$2 and \$5 instead in n dollars.
Multiply by 100 to change dollars to cents.

$\quad$ \$2 = 2 × 100 cents = 200 cents
$\quad$ \$5 = 5 × 100 cents = 500 cents
So, \$$n$ = n × 100 cents

We usually write 100n cents.

Exercise 6b (Oral)

1. In Fig. 6.5 there are x oranges in the bag. How many oranges are there altogether?

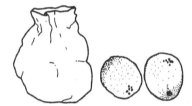

Fig. 6.5

2. The light poles in Fig. 6.6 are d metres apart. How far is the boy from the right-hand pole?

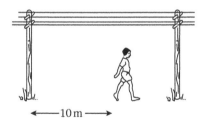

←——10 m——→

Fig. 6.6

3. The mass of one of the dictionaries in Fig. 6.7 is b grams. What is the total mass of the books in the picture (a) in grams, (b) in kilograms?

Fig. 6.7

4. Eggs cost x cents each. Find the total cost of the eggs in Fig. 6.8 (a) in cents, (b) in dollars.

Fig. 6.8

The following exercises will give you more practice at making algebraic expressions from English sentences.

Exercise 6c

Use addition in this exercise.

1. Simba is 13 years old. How old will he be 8 years from now? How old will he be y years from now?

2. What number is 4 more than 3? What number is 4 more than n? What number is 7 more than x?

3. A farmer has 6 goats and z cows. How many animals has she altogether?

4. Last year the body-mass of a boy was 48 kg. The increase in his body-mass this year is m kg. What is the boy's body-mass this year?

5. A man has k coconuts. A friend gives him one more coconut. How many coconuts does the man have now?

6. If today is the 6th of September, what will be the date in n days time? (Assume that the month will still be September.)

7 Each side of a square is *d* centimetres long. What is the total length of the sides of the square?

8 In a test, Richard got 7 marks more than Nancy. Nancy got *x* marks. How many marks did Richard get? David got *y* marks more than Richard. How many marks did David get?

Exercise 6d

Use subtraction in this exercise.

1 What number is 2 less than 9? What number is 2 less than *n*? What number is 5 less than *r*?

2 A woman is 30 years old. What was her age 10 years ago? What was her age *y* years ago?

3 A book has *x* pages. A student reads 30 pages of the book. How many pages has the student not read?

4 A boy has *m* mangoes. If he eats 5 of them, how many does he have left?

5 A girl guesses that a line is 15 cm long. She measures the line. She finds that her guess is *x* cm too big. What is the length of the line?

6 A woman bought a diamond for $*x*. She sold it for $16 000. What was her profit?

7 A man cycles *x* km towards a village which is 20 km away. How far does he still have to cycle?

8 Clive has *s* dollars in his savings account. He uses some of this to buy a suit which costs *t* dollars. How much does he have left in his savings?

Exercise 6e

Use multiplication in this exercise.

1 How many metres are there in 1 kilometre? How many metres are there in 4 kilometres? How many metres are there in *x* kilometres?

2 How many days are there in one week? How many days are there in 5 weeks? How many days are there in *w* weeks?

3 How many calendar months are there in *x* years?

4 What is the number which is 9 times as great as *d*?

5 What is the cost of 8 books at $4 each? What is the cost of *x* books at $4 each? What is the cost of 5 books at $*t* each?

6 A loaf of bread has a mass of *x* kg. What is its mass in grams?

7 A rubber band is *x* mm long when unstretched. It is stretched to 3 times its unstretched length. What is the stretched length?

8 What is the cost of 10 boxes of matches at *m* cents each?

Exercise 6f

Use division in this exercise.

1 How many weeks are there in 14 days? How many weeks are there in *d* days?

2 How many metres are there in 3000 mm? How many metres are there in *k* mm?

3 Six equal blocks have a total mass of *b* kg. What is the mass of each block?

4 Five pencils cost $1.50. How much does one pencil cost? In another shop 5 pencils cost *p* cents. How much does one pencil cost in this shop?

5 A chicken lays *h* eggs in 9 days. What is the average number of eggs that the chicken lays in one day?

6 A stick is 140 cm long. The stick is cut into *x* pieces, each the same length. What is the length of each piece?

7 What is the number which is one-third as big as *n*?

8 Fig. 6.9 shows 5 equal bricks built in a column. The height of the column is *x* cm. How thick is each brick?

Fig. 6.9

Coefficients

In arithmetic 3×4 is a short way of writing $4 + 4 + 4$. Similarly, in algebra, $3 \times a$ is a short way of writing $a + a + a$. $3 \times a$ is usually shortened to $3a$. Say this as 'three a'.

$$3a = 3 \times a = a + a + a$$

Notice the difference between 34 in arithmetic and $3a$ in algebra. 34 means 3 tens and 4 units or $30 + 4$. 34 does *not* mean 3×4. In algebra, $3a$ *always* means $3 \times a$.

Exercise 6g

Write the following in the shortest possible way.

1. $a + a + a$
2. $x + x + x$
3. $p + p$
4. $r + r + r + r + r$
5. $t + t + t + t + t + t + t + t$
6. $m + m + m + m$
7. $z + z + z + z + z + z + z + z + z + z$
8. $y + y$
9. $c + c + c + c + c + c$
10. $k + k + k + k + k$

Exercise 6h (Oral)

1. What do the following mean in arithmetic?
   (a) 3×4 (b) 34 (c) 3×6 (d) 36
   (e) 2×9 (f) 29 (g) 5×1 (h) 51

2. What do the following mean in algebra?
   (a) $3a$ (b) $5x$ (c) $2y$ (d) $4n$
   (e) $3m$ (f) $6d$ (g) $2f$ (h) $10e$

$3a$, $x + 1$, $7 - r$, $7y$ are all examples of algebraic expressions. In the expression $3a$, the 3 is called the **coefficient** of a. The 3 shows that three 'a's have been added together. For example,

(a) in $7y$, 7 is the coefficient of y; seven 'y's have been added together;

(b) in x there is only *one* x; x is the same as $1x$; the coefficient of x is 1.

Note: it is usual to write x not $1x$.

Coefficients are not always whole numbers. Coefficients can also be fractions. In arithmetic $\frac{1}{3} \times 12$ or $\frac{12}{3}$ are short ways of writing $\frac{1}{3}$ of 12 or $12 \div 3$. In algebra $\frac{1}{3}a$ or $\frac{a}{3}$ are short ways of writing $\frac{1}{3}$ of a or $a \div 3$.

In the expression $\frac{1}{3}a$ the coefficient of a is $\frac{1}{3}$. $\frac{1}{3}a = \frac{a}{3}$, so the coefficient of a in $\frac{a}{3}$ is also $\frac{1}{3}$.

For example
(c) in $\frac{3}{4}x$, $\frac{3}{4}$ is the coefficient of x;

(d) in $\frac{2a}{3}$, $\frac{2}{3}$ is the coefficient of a
$\left(\text{since } \frac{2a}{3} = \frac{2}{3}a\right)$;

(e) in $\frac{z}{5}$, $\frac{1}{5}$ is the coefficient of z.

Exercise 6i (Oral)

What is the coefficient of x in each of the following?

1. $3x$
2. $7x$
3. $4x$
4. $8x$
5. $15x$
6. $9x$
7. $18x$
8. $2x$
9. x
10. $10x$
11. $\frac{1}{3}x$
12. $\frac{1}{2}x$
13. $\frac{1}{4}x$
14. $\frac{2}{3}x$
15. $\frac{3}{4}x$
16. $\frac{x}{5}$
17. $\frac{x}{10}$
18. $\frac{2x}{3}$
19. $\frac{3x}{4}$
20. $\frac{3x}{5}$

Exercise 6j

1. A radio programme has x minutes of talking and 20 minutes of music. How long is the programme?

2. A teacher has 6 pens, n of them are red, the rest are blue. How many blue pens does the teacher have?

3. Find the total distance around a rectangle which is a m long and 5 m wide.

4. How many kilograms are there in x grams?

5 A packet of sugar costs 50c. How much would x packets of sugar cost? Give your answer
(a) in cents, (b) in dollars.

6 A school paid $3 800 to buy desks. It bought d desks altogether. What was the cost of one desk?

7 Mary has 20 oranges. Noami has x oranges less than Mary. How many oranges does Naomi have?

8 A man walks $3x$ km on the first day, 12 km on the second day and $2x$ km on the third day. How far does he walk in the three days?

9 Team A scored three times as many points as Team B. Which team scored more points? If Team A scored n points, how many points did Team B score?

10 The area of a square mat is $9\,\text{m}^2$. What is the total area of x mats?

Summary

Algebraic expressions generally include both numbers and letters, where the letter stands for an unknown value. In any problem the same letter can only be used to represent quantities of the same kind. The letter replaces **only the numerical value** of the quantity and not the unit in which the quantity is measured.

When a letter is multiplied by a number, the number is called the **coefficient**. A coefficient may be positive or negative, a whole number or a fraction, e.g.

in $4c$,   the coefficient is 4;
in $\dfrac{y}{3}$,   the coefficient is $\frac{1}{3}$;
in $-5ab$,   the coefficient is -5.

Practice exercise P6.1

If $x = 5$, find the value of

1 $x + x$ **2** $x - x$ **3** $x + 3$

4 $2 + x$ **5** $3 - x$ **6** $4 + x - x$

7 $x + x + 2$ **8** $9 - x + 3$ **9** $x - 2 + x$

Practice exercise P6.2

Write down how to solve these problems.

1 What is Marlon's age if he is 3 years older than Jenny?

2 What is Rachel's age if she 5 years younger than Mona?

3 What is Stan's age if he is twice as old as his brother Toby?

4 What is Karen's age if she is four times as old as Tony?

5 What is Tammy's age if she is half her mother's age?

6 What is Phil's weight if he is 3 kg heavier than Ken?

7 What is Jackie's weight if she is 4 kg lighter than Sarah?

8 What is the total number of legs on all the chairs in a hall?

9 What is the weight of sand delivered to a building site if each lorry delivers 3 tons?

10 If Pam is 23 years younger than her mother, write
(a) Pam's age if you know her mother's age,
(b) her mother's age if you know Pam's age.

Practice exercise P6.3

In each question there is one rule. This rule tells you how to find the bottom number from the top one. Write the rule in words.

1
1	2	3	4
↓	↓	↓	↓
2	4	6	8

②
1	2	3	4
↓	↓	↓	↓
3	4	5	6

③
1	2	3	4
↓	↓	↓	↓
1	3	5	7

④
2	4	6	8
↓	↓	↓	↓
1	2	3	4

⑤
1	2	3	4
↓	↓	↓	↓
1	4	9	16

⑥
3	6	9	12
↓	↓	↓	↓
1	2	3	4

Practice exercise P6.4

(a) Simplify the following algebraic expressions.
(b) Write the coefficient of each result.

① $x + x + x$　　② $x - x + x$

③ $x + x - x - x$　　④ $x - x - x - x$

⑤ $x + x - x + x$　　⑥ $x + x + x - x$

⑦ $x + x + x + x$　　⑧ $x - x - x + x$

⑨ $x - x + x - x$　　⑩ $x - x + x + x$

Practice exercise P6.5

Write the coefficient of each of the following terms

① $2a$　　② $-4x$　　③ $\frac{1}{3}d$

④ $-c$　　⑤ $4b$　　⑥ $\frac{3x}{7}$

⑦ $7x$　　⑧ $-\frac{5}{8}y$　　⑨ $\frac{t}{2}$

⑩ $\frac{p}{3}$　　⑪ $\frac{1}{3}u$　　⑫ $-3x$

Practice exercise P6.6

Write the information in each of the following as an algebraic expression using the letters and numbers shown in **bold**.

① The number of tins of dog food needed each week by the **d**ogs if each dog eats **6** tins a week.

② The weight of each pie when some pie **m**ixture is split into **5** equal parts.

③ The price of a box of biscuits when you know that the price of a box is $**3** more than the price of a **t**in.

④ The weight of a person if they are **8** kg below their old **w**eight.

⑤ The number of children in a hall if there are the same **n**umber in each row and there are **15 rows**.

Properties of solids

Nearly everything that we can see and touch takes up space. These things are either **gases**, **liquids** or **solids**. You will study some of the properties of liquids and gases in science.

Many solids, such as stones and trees have rough and **irregular** shapes. These usually occur in Nature. Other solids, such as tins and houses, have **regular** shapes. These are usually made by humans. This chapter investigates the properties of regular shapes.

How many of the groups of shapes can you identify in Fig. 7.1?

(a)

(b)

(c)

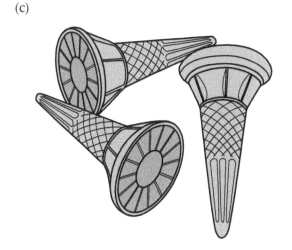

(d)

(e)

Fig. 7.1

Exercise 7a

1. Look at the photographs in Fig. 7.1. Identify as many of the objects as you can.

2. Make a table like the one in Fig. 7.2.

No.	Name of object	Freehand drawing	Name of basic shape
1	tins of beans	 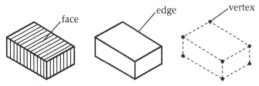	cylinder
2			
3			

Fig. 7.2

Name at least ten objects in the photographs. Make a freehand sketch of each object and name its basic shape. An example is shown in Fig. 7.2.

Exercise 7b (Group work)

1. In small groups, collect as many solids as possible. Many of them you can find at home:

   matchboxes, soap packets, shoe boxes, dice, empty tins, balls, etc.

2. Compare your solids with another group's.

All solids have faces (Fig. 7.3). A **face** may be flat (**plane**) or curved.

Most solids also have edges (Fig. 7.3). An **edge** is a line where two faces meet. Edges may be straight or curved.

Edges meet at a point called a vertex (Fig. 7.3). A **vertex** is a point or corner where three or more edges meet. The plural of vertex is **vertices**.

Fig 7.3

3. Find the faces, edges and vertices of each solid in your collection.

Cuboid

Look at the matchbox in Fig. 7.4. The name of its shape is a **cuboid**.

(a)

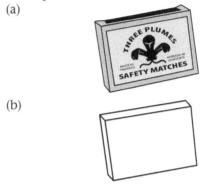

(b)

Fig. 7.4 (a) Matchbox
(b) Outline of matchbox: a cuboid

The cuboid is one of the most common solids.

Identify all the solids in your class collection that are cuboids.

How many faces does the cuboid have? What is the shape of each face?

How many edges does the cuboid have? How many vertices?

Check your results with your classmates.

The cuboid has 6 rectangular faces, 12 straight edges and 8 vertices.

Drawing cuboids

Fig. 7.5 shows two ways of drawing cuboids.

(a)   (b)

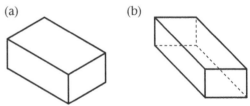

Fig. 7.5 Ways of drawing a cuboid:
(a) projection   (b) skeleton view

In each case it is impossible to show the solid as it really is. However, a **skeleton view** is very useful since it shows all the edges. Fig. 7.6 shows how to draw a skeleton view. Notice that some edges are hidden from view. We usually show these as broken lines.

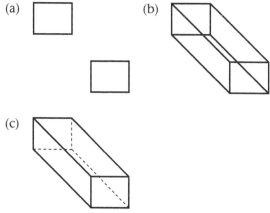

(a) (b) (c)

Fig. 7.6 How to draw a skeleton view of a cuboid
(a) draw two rectangles
(b) join corner to corner
(c) go over the drawing, make the hidden edges broken

Net of a cuboid

Get a tray from a matchbox. Cut the edges as shown in Fig. 7.7. Flatten the shape. The flat shape is called the **net** of the open cuboid.

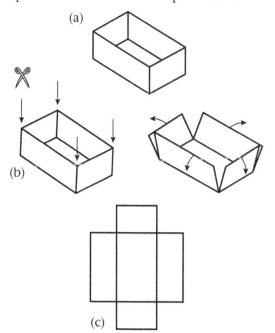

Fig. 7.7 Cutting a matchbox tray to get the net of an open cuboid.
(a) this is an open cuboid
(b) cut along the edges shown
(c) this is the net of the open cuboid

The shape of the net depends on where you make the cuts. You should be able to make the nets shown in Fig. 7.8 from a matchbox tray.

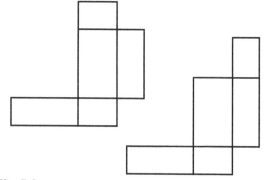

Fig. 7.8
Notice in both cases that the net is made up of five rectangles. The net of a closed cuboid would contain six rectangles.

Cube

A **cube** is a cuboid in which all six faces are squares (Fig. 7.9).

Fig. 7.9 Cube:
(a) projection
(b) skeleton view

Exercise 7c (Group work)

1 Work in small groups. Write down five everyday objects which are cuboids.

2 The solids in Fig. 7.10 are made by placing cuboids together. Decide together how many cuboids there are in each solid.

(a) 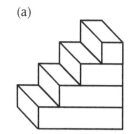 (b)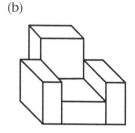

(c)   (d)

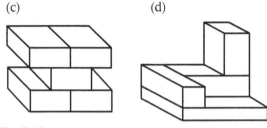

Fig. 7.10

③ Draw a skeleton view of a cuboid and a cube.

④ What shape is the face of a cuboid?

⑤ What shape is the face of a cube?

⑥ The cuboid in Fig. 7.11 is made by building up four layers of cubes.

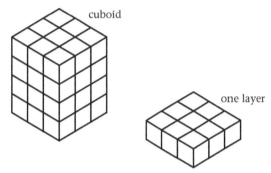

cuboid

one layer

Fig. 7.11

Working in small groups, find out:
(a) How many cubes there are in one layer.
(b) How many cubes there are in the whole cuboid.

If the outside of the cuboid is painted red:
(c) How many cubes have three red faces?
(d) How many cubes have only two red faces?
(e) How many cubes have only one red face?
(f) How many cubes have no red faces?

⑦ Working in pairs copy the following shapes on to paper and try to fold them. Which of the diagrams in Fig. 7.12 can be folded to make a cuboid?

(a) 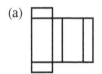   (b)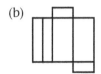

(c)   (d)

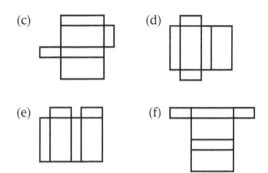

(e)   (f)

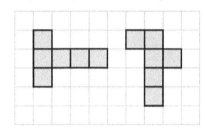

Fig. 7.12

⑧ Work in pairs. Use squared paper to draw as many different nets of cubes as you can. Fig. 7.13 gives two examples.

Fig. 7.13

⑨ Work in pairs. Draw the net of a cube of edge 5 cm on stiff card. Use sellotape to make the cube from its net.

⑩ A cuboid is made from wire so that it is 15 cm long, 12 cm wide and 8 cm high. What length of wire is needed? (Make a sketch.)

⑪ A piece of wire of length 66 cm is used to make a skeleton cube. No wire is wasted and the largest possible cube is made. What will be the length of one of the edges of the cube?

⑫ In Fig. 7.14,
(a) how many faces can you see?
(b) how many edges can you see?
(c) how many vertices can you see?

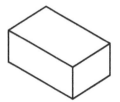

Fig. 7.14

⑬ Copy and complete Table 7.1.

Table 7.1

	Number of vertices	Number of faces	Number of edges
cuboid			
cube			

Cylinder

Look at the tin of milk powder in Fig. 7.15. The name of its shape is a **cylinder**.

Fig. 7.15 (a) Tin of milk powder
(b) Outline of tin: cylinder

A cylinder has two plane faces and one curved face. It has two curved edges and no vertices. The two plane faces are both circles.

Drawing cylinders

Fig. 7.16 shows a skeleton view of a cylinder.

Fig. 7.16 Cylinder:
skeleton view

A skeleton view of a cylinder is drawn in much the same way as that of a cuboid. This is shown in Fig. 7.17.

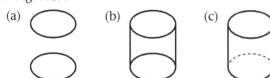

Fig. 7.17 How to draw a skeleton view of a cylinder:
(a) draw two 'flattened' circles (ellipses)
(b) join the ellipses by two straight lines, as shown
(c) draw the hidden edge as a broken line

Net of a cylinder

Fig. 7.18 shows how a cylinder can be cut to give its net. The net is made up of two circles and one rectangle.

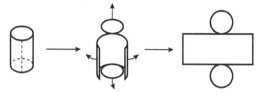

Fig. 7.18 How to cut a cylinder to give its net

Prism

Fig. 7.19 gives some examples of **prisms**.

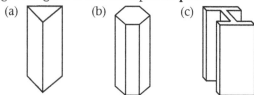

Fig. 7.19 (a) Triangular prism
(b) Hexagonal prism
(c) I-shaped prism

Notice that the base and top face of a prism are always the same shape and size, i.e. the cross-section of a prism is uniform. The names of prisms come from the shape of their cross-sections.

Fig. 7.20 shows prisms with their cross-sections shaded.

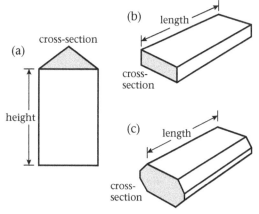

Fig. 7.20

The cuboid is a rectangular prism and the cylinder can be thought of as a special prism.

Exercise 7d (Group work)

1. Work in small groups. Write down five everyday objects which are cylinders.

2. Write down five everyday objects which are prisms.

3. The solids in Fig. 7.21 each include cylinders. Decide together how many cylinders there are in each solid.

(a)

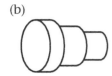

(b) 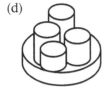 (c)

(d)

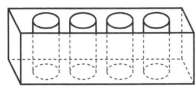

Fig. 7.21

4. Draw a skeleton view of a cylinder and a triangular prism.

5. What is the shape of the plane faces of a cylinder?

6. What is the shape of the end faces of a triangular prism? What is the shape of the other faces of a triangular prism?

7. Working in pairs, name the solids whose nets are shown in Fig. 7.22.

(a) 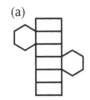 (b) 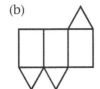 (c)

Fig. 7.22

8. Working in pairs, decide how many faces, edges and vertices a triangular prism has.

9. Sketch three different nets of a triangular prism.

10. A right prism is such that its top and bottom faces are equilateral triangles of side 8 cm. It takes 90 cm of wire to make a skeleton model of the prism. What is the height of the prism?

11. Fig. 7.23 shows two views of a triangular prism. How many vertices, faces and edges can you see in each view? Copy and complete Table 7.2.

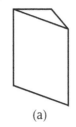 

(a)   (b)

Fig. 7.23

Table 7.2

View	Number of vertices seen	Number of faces seen	Number of edges seen
(a)			
(b)			

12. Copy and complete Table 7.3 for the prisms given in Fig. 7.19. Include faces, edges and vertices which cannot be seen.

Table 7.3

Prism	Number of vertices	Number of faces	Number of edges
triangular			
hexagonal			
I-shaped			

Cone

Cones are quite common shapes. You may be familiar with the ice cream cone. The sharp end of a pencil has the shape of a cone. Fig. 7.24 shows these two cones.

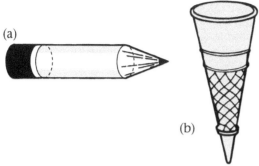

Fig. 7.24
(a) The sharp end of a pencil
(b) An ice cream cone

Fig. 7.25 shows how a cone can be cut to give its net. The figure also shows that the base of a cone is circular.

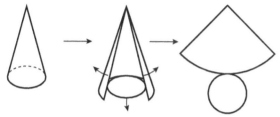

Fig. 7.25 How to cut a cone to give its net

Pyramid

The names of pyramids come from the shapes of their base faces (Fig. 7.26).

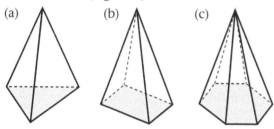

Fig. 7.26 (a) Triangular-based pyramid
            (tetrahedron)
         (b) Square-based pyramid
         (c) Hexagonal-based pyramid

A pyramid with an *n*-sided base will have *n* triangular faces meeting at a common point. This point is usually called **the vertex** of the pyramid. Notice, however, that a pyramid has other vertices around its base.

Fig. 7.27 gives two nets of a square-based pyramid. The nets show that a square-based pyramid has four triangular faces.

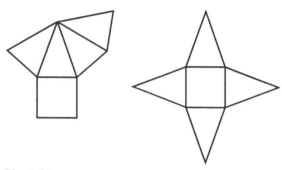

Fig. 7.27

A **right pyramid** is one in which the vertex is above the centre of the base.

Exercise 7e (Group work)

1 Work in small groups. Count the number of sides of the base and the number of faces (not including the base) in each pyramid in Fig. 7.26.
   (a) Discuss what you notice.
   (b) What is the shape of each face (not including the base)?

2 Write down five everyday objects which are wholly or partly cone-shaped.

3 Write down five everyday objects which are wholly or partly pyramid-shaped.

4 The solids in Fig. 7.28 include either cones or pyramids. Decide together how many cones or pyramids you can see in each solid.

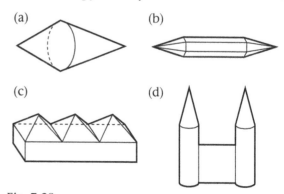

Fig. 7.28

⑤ Draw a skeleton view of a cone and a square-based pyramid.

⑥ Work in small groups. The faces of a pyramid are all triangular. What kind of pyramid is it? Find out the special name for such a pyramid.

⑦ Work in small groups. A pyramid has a total of six triangular faces. Decide together how many faces it has altogether. What is the shape of its other face?

⑧ Work in pairs. Name the solids which have the nets shown in Fig. 7.29.

(a) (b)

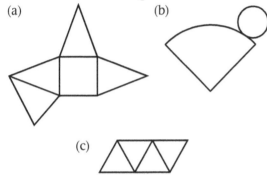

(c)

Fig. 7.29

⑨ Work in small groups. A pyramid is made from the net shown in Fig. 7.30. Discuss and decide which edge will join to
(a) DE, (b) EF, (c) AB, (d) HG.
(e) Which letters will be at the vertex of the pyramid?

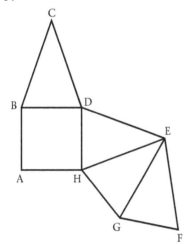

Fig. 7.30

⑩ Fig. 7.31 shows a tower made from a cube and a pyramid. If each edge is 3 cm long, what length of wire will be needed to make a skeleton model of the tower?

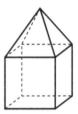

Fig. 7.31

⑪ Fig. 7.32 shows a view of a square-based pyramid.

Fig. 7.32

How many vertices, faces and edges can you see in this view of the pyramid?

⑫ Copy and complete Table 7.4.

Table 7.4

Solid	Number of vertices	Number of faces	Number of edges
triangular pyramid			
square pyramid			
hexagonal pyramid			
triangular prism			
solid in Fig 7.31			

Can you see any pattern in the numbers?

Sphere

Nearly every ball is sphere-shaped (Fig. 7.33).

(a)
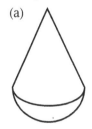

(b)
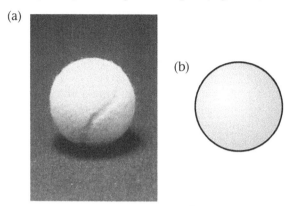

Fig. 7.33 (a) A tennis ball
         (b) Outline of a ball: a sphere

Any plane cut through the centre of a sphere will give two parts which are identical to each other. Half a sphere is called a **hemisphere** (Fig. 7.34).

Fig. 7.34

Exercise 7f (Group work)

1 Work in small groups. Write down five everyday objects each of which is either a sphere or contains part of a sphere in its shape.

2 Working in pairs, discuss which shapes make up the solids in Fig. 7.35 (a)−(d).

(a)  (b)

(c)
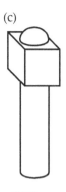

(d)
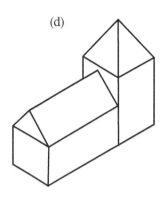

Fig. 7.35

3 Work in pairs. A solid has f plane faces and v vertices.
   Compare and discuss your results in Tables 7.1, 7.3 and 7.4 and write down an expression in f and v for the number of edges that the solid has.

4 Working in small groups, discuss which of the following solids will roll smoothly.
   (a) cube   (b) cone
   (c) sphere   (d) square-based pyramid
   (e) cuboid   (f) cylinder

5 A skeleton square-based pyramid is made from wire. All its edges are the same length. If 40 cm of wire is used, find the length of the edge of each pyramid. (Make a sketch of the pyramid.)

6 Name two solids that have five plane faces.

7 If the net shown in Fig. 7.36 is folded to make a cuboid, which edge will join to
   (a) AB,   (b) CD,   (c) FG,   (d) KL?

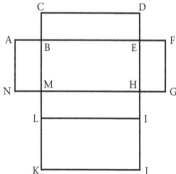

Fig. 7.36

Which points will join to
(e) point A,   (f) point D?

8 Name two solids that have no vertices.

9 Sketch four different nets of a square-based pyramid.

10 Name the solids which have the following shapes for *all* their faces.
(a) triangle   (b) rectangle   (c) square

Summary

Most geometrical solids have **faces**, **edges** and **vertices**. Flat faces meet along straight edges; curved faces meet along curved edges. These and other **properties** were shown for solids such as cubes, cuboids, cylinders, prisms, pyramids and spheres.

A **net** of a solid is the plane shape obtained when a solid is cut along its edges then flattened out.

Practice exercise P7.1

1 For each of the shapes listed below, name one example in everyday life.
(a) cube
(b) cone
(c) sphere
(d) cuboid
(e) cylinder

2 Write true or false for each of the following statements.
(a) The faces of a cuboid are squares
(b) A cube has six square faces
(c) The edges of a cuboid meet at a point called the vertex.
(d) A cylinder has three curved surfaces.
(e) All prisms have six faces.
(f) All sides of a pyramid meet at a point.
(g) Pyramids are named according to the shapes of their bases.
(h) A square-based pyramid has four triangular faces.
(i) A sphere has one curved face.
(j) A sphere does not have a vertex.

3 The net of a cube could be used to make a dice. Look carefully at the number of dots on the faces of at least four die and then answer these questions.
(a) How many dots are on the face opposite to the face with six dots?
(b) How many dots are on the face opposite to the face with four dots?
(c) How are the other two faces arranged?
(d) Are the faces of the die arranged in the same order?

Directed numbers (1)
Addition and multiplication

Pre-requisites
■ natural numbers

Whole numbers on the number line

A **number line** is a picture of the real numbers. In Fig. 8.1 the number line starts at 0 (zero) and increases in equal steps to the right.

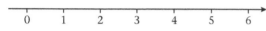

Fig. 8.1

The number line can be used to show addition and subtraction.

Example 1

Use the number line to complete 1 + 4 = .

Fig. 8.2 shows the addition.

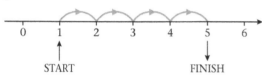

Fig. 8.2

Follow the arrows. Start at 1. Move 4 places to the *right*. Finish at 5.

$$1 + 4 = 5$$

Example 2

Use the number line to complete 6 + (−2) = .

Fig. 8.3 shows the addition.

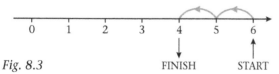

Fig. 8.3

Follow the arrows. Start at 6. Move 2 places to the *left*. Finish at 4.

$$6 + (−2) = 4$$

Exercise 8a

1. Sketch a number line as given in Fig. 8.4.

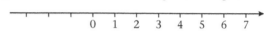

Fig. 8.4

2. Use your number line to complete the following additions.
   (a) 1 + 5 =   (b) 2 + 4 =
   (c) 4 + 2 =   (d) 3 + 1 =
   (e) 4 + 3 =   (f) 2 + 5 =

3. Use your number line to complete the following pattern.
   5 + (−2) =
   5 + (−3) =
   5 + (−4) =
   5 + (−5) =

Positive and negative numbers

Fig. 8.5 shows the addition 5 + (−6) on the number line.

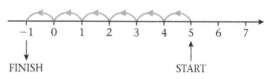

Fig. 8.5

In Fig. 8.5, the finishing point is 1 unit to the *left* of 0.

i.e. 5 + (−6) = 1 unit to the left of 0

Similarly,

5 + (−7) = 2 units to the left of 0
5 + (−8) = 3 units to the left of 0

In practice, we say that numbers to the left of 0 are **negative numbers** and numbers to the right of 0 are **positive numbers**. There are many different symbols for positive and negative numbers. For example, **positive five** can be written +5, (+5), ⁺5 and **negative three** can be written −3, (−3), ⁻3. Fig. 8.6 shows how positive and negative numbers are written on the number line.

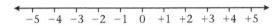

Fig. 8.6

The number line extends in two directions from 0: to the right for positive numbers, to the left for negative numbers.

Positive and negative numbers in daily life

Consider the following examples.

Example 3

Fig. 8.7 shows a man climbing a tree and a man digging a well.

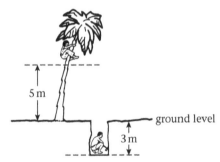

Fig. 8.7

The climber is 5 m *above* ground level.
The digger is 3 m *below* ground level.

We can say that their distances from ground level are +5 m and −3 m. Positive distances are above ground level. Negative distances are below ground level.

Example 4

Fig. 8.8 shows the readings on two thermometers (a) and (b).

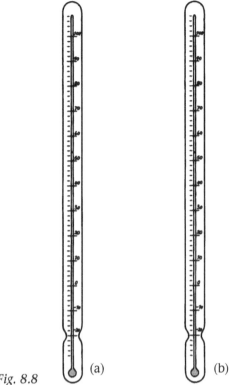

Fig. 8.8

Thermometer (a) shows the shade temperature on an October day. It reads 26 °C.
Thermometer (b) shows the temperature inside a refrigerator on the same day. It reads 3 °C *below* 0, i.e. −3 °C.

Example 5

A path runs in a straight line from east to west. O is a point on the path. This is shown in Fig. 8.9.

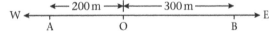

Fig. 8.9

Point A is 200 m west of O. Point B is 300 m east of O. If we take distances east of O to be positive, then the position of A is −200 m and the position of B is +300 m.

Example 6

When an important event is due to take place, the time that it starts is often called 'zero hour'. Times *before* zero hour are *negative*. Times *after* zero hour are *positive*.

For example, a President may decide to broadcast to the nation at 3 p.m. one day. 3 p.m. is zero hour. On that day, the President may have lunch with his Ministers at 1 p.m. We say that he has lunch at −2 hours. He may finish his speech at 4 p.m. He finishes his speech at +1 hour.

Examples 3, 4, 5 and 6 show that we sometimes use terms like *above, below, after, before* to describe quantities such as distances, temperature and time. In these cases we can also describe the quantities by using positive and negative numbers.

Exercise 8b (Oral)

1 Fig. 8.10 is a picture of a football team.

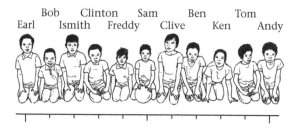

Bob   Clinton   Sam   Ben   Tom
Earl   Ismith   Freddy   Clive   Ken   Andy

Fig. 8.10

The captain, Sam, sits in the middle.
Ben is 2 places to the *right* of Sam. We can say his position is (+2).
Earl is 5 places to the *left* of Sam. We can say his position is (−5).
(a) Use this method to give the position of
    (i) Ken   (ii) Clinton
    (iii) Freddy   (iv) Andy
(b) Name the player in position
    (i) (+4),   (ii) (−4),   (iii) (−3),
    (iv) (+1),   (v) 0.

2 A line is marked off in metres as shown in Fig. 8.11.

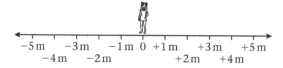

−5 m   −3 m   −1 m 0 +1 m   +3 m   +5 m
    −4 m   −2 m   +2 m   +4 m

Fig. 8.11

A girl moves to the right and left along the line. Find her final position if she always

starts at 0 and moves as follows.
(a) 2 m to the right
(b) 5 m to the left
(c) 3 m to the right, then 1 m to the right
(d) 5 m to the right, then 2 m to the left
(e) 4 m to the right, then 7 m to the left
(f) 2 m to the left, then 2 m to the left
(g) 5 m to the left, then 4 m to the right
(h) 3 m to the left, then 8 m to the right
(i) 3 m to the right, then 3 m to the left
(j) 2 m to the left, then 3 m to the right

Exercise 8c

1 A bird is at the top of a tree 7 m above the surface of a river. A fish swims 2 m below the surface of the river.
(a) Give the positions of the bird and the fish as distances from the surface of the river.
(b) How far is the fish below the bird?

2 The temperature inside a freezer is −3 °C. The temperature falls by a further 12 °C. What is the new temperature?

3 A straight road runs east−west through a point O. Distances east of O are positive. Distances west of O are negative.
(a) Express the positions of the points P, Q, R and T in terms of positive or negative distances.
    (i) P is 8 km west of O
    (ii) Q is 9 km east of O
    (iii) R is 15 km east of O
    (iv) T is 11 km west of O
(b) From the above data, find the distance between these points.
    (i) P and Q   (ii) Q and R
    (iii) R and T   (iv) P and T

4 Zero hour for a meeting is midday (12 noon). Express the following as positive or negative times.
(a) 5 h before zero hour
(b) 7 h after zero hour
(c) $1\frac{1}{2}$ h after zero hour
(d) 45 min before zero hour
(e) 3 p.m.   (f) 10 a.m.   (g) 9 p.m.
(h) 8:30 a.m. (i) 4:45 p.m.

5 A girl arrives at school 10 minutes early. Another girl is 3 minutes late. How long has the first girl been at school before the second girl arrives?

6 A man was born 19 years before Independence. His first son was born 5 years after Independence. How old was the man when his son was born?

7 The top of an oil drill is 10 m above ground level. The drill is 220 m long. What is the distance of the bottom of the drill from ground level?

8 Two pieces of wood should be the same length. One is 5 mm too long. The other is 3 mm too short. What is the difference in the lengths of the two pieces of wood?

9 On Thursday evening Sita has $19. Gino has no money and owes $6 to Sita. We can say that Gino has $(−6). On Friday they each get the same pay, $64. If Gino repays his debt to Sita, how much money does each have?

10 A man has $17. His younger brother has $(−8).
(a) How much does the younger brother owe?
(b) By how much is the elder brother better off?

Directed numbers

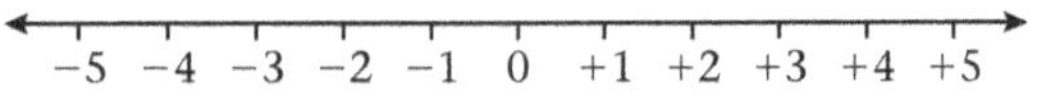

Fig. 8.12

Positive numbers and negative numbers are called **directed numbers**. The sign tells which direction to go from zero to reach the position of the number. Zero is neither positive nor negative. As we move to the right along the number line, the numbers increase. As we move to the left, the numbers decrease (Fig. 8.13).

Thus +2 > −4, i.e. +2 **is greater than** −4 and −5 < +1, i.e. −5 **is less than** +1

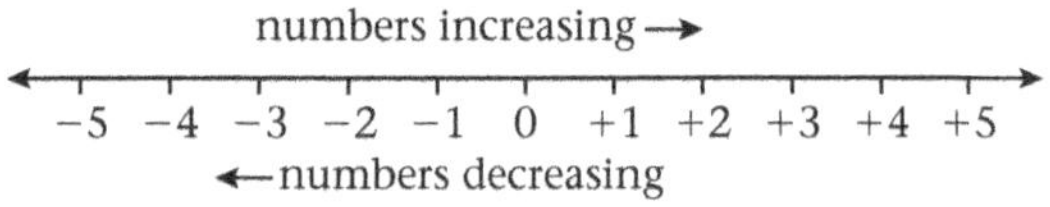

Fig. 8.13

The positive and negative whole numbers together with zero form a set of numbers called **integers**.

If Z = {integers}
then Z = {..., −3, −2, −1, 0, +1, +2, +3, ...}

Any member of Z is an integer.

Exercise 8d (Oral)

Make a number line from −20 to +20 by marking a long strip of paper. Use the number line in this exercise.

1 Which is the greatest integer in each of the following?
(a) +5, −7, +12, −2, −19, +6
(b) −3, −2, +2, +5, −8, −10
(c) −3, −12, −13, −8, −5, −7
(d) +3, −3, +8, −8, +15, −15
(e) +4, 0, −4, −8, −12, −16
(f) −3, −5, 0, −18, −14, −1

2 Arrange the following integers in order from least to greatest.
(a) −3, 0, −2, +1, −8, +5
(b) +5, −2, +9, +7, −5, −4
(c) +6, −8, −20, +5, −14, 0
(d) −6, +6, −18, +18, −11, +11
(e) −1, 0, −5, −3, −14, −10
(f) −9, −11, +1, −10, +17, +16

3 Place either > (is greater than) or < (is less than) between each of the two integers to make each a true statement. The first one has been done.
(a) −9 < +3 (b) −15 +8
(c) +4 −13 (d) +18 +1
(e) −18 −1 (f) 0 −2
(g) −1 +1 (h) −7 −17
(i) −13 +13 (j) 0 −4
(k) +5 −5 (l) −9 0

④ For each of the following write out the complete line. The first line is complete.

	start	move	finish
(a)	+2	5 units to the left	−3
(b)	+5	7 units to the left	
(c)	−1	7 units to the left	
(d)	+2	4 units to the right	
(e)	−3	5 units to the right	
(f)	−10	10 units to the right	
(g)	+12	9 units to the left	
(h)	+14	30 units to the left	
(i)	−20	28 units to the right	
(j)	+1	2 units to the left	
(k)	+2		−4
(l)	+5		+3
(m)	+6		−6
(n)	−2		0
(o)	−15		+15
(p)	+15		−15
(q)	−11		+13
(r)	+12		−2
(s)	0		+6
(t)	0		−17
(u)		2 units to the right	+5
(v)		8 units to the left	+2
(w)		10 units to the left	−4
(x)		35 units to the right	+20
(y)		11 units to the right	+7
(z)		13 units to the left	−5

Adding positive numbers

The addition 2 + 3 = 5 can be written in directed numbers as: (+2) + (+3) = +5. We show this addition on the number line in Fig. 8.14. Follow the arrows from START to FINISH.

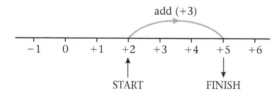

add (+3)

Fig. 8.14

Similarly, we can find the value of −2 + 6. In directed numbers that is: (−2) + (+6). Follow the arrows on the number line in Fig. 8.15 from START to FINISH.

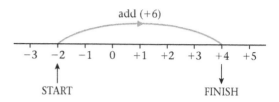

add (+6)

Fig. 8.15

Thus: −2 + 6 = +4

To add a positive number: move that number of places to the right on the number line.

Exercise 8e (Oral)

Complete the following.

① (a) (+6) + (+4) = +10
  (b) (+6) + (+5) =
  (c) (+6) + (+6) =
  (d) (+6) + (+7) =
  (e) (+6) + (+8) =
  (f) (+6) + (+9) =

② (a) (−6) + (+4) = −2
  (b) (−6) + (+5) =
  (c) (−6) + (+6) =
  (d) (−6) + (+7) =
  (e) (−6) + (+8) =
  (f) (−6) + (+9) =

③ (+14) + (+5) =

④ (−4) + (+9) =

⑤ (−8) + (+10) =

⑥ (+8) + (+10) =

⑦ (−5) + (+5) =

⑧ (−3) + (+5) + (+8) =

In many cases, brackets are not given. If a number is not directed, take it to be positive.

Exercise 8f (Group work)

Work in pairs and use the number line to simplify the following.

① (a) 3 + 1   (b) 3 + 2
  (c) 3 + 3   (d) 3 + 4
  (e) 3 + 5

2 (a) $-3 + 1$ (b) $-3 + 2$ (c) $-3 + 3$
    (d) $-3 + 4$ (e) $-3 + 5$

3 $-3 + 7$ **4** $-1 + 8$

5 $-4 + 7$ **6** $-8 + 7$

7 $-4 + 12$ **8** $-11 + 2$

9 $-3 + 4 + 10$ **10** $-4 + 9 + 7$

Adding negative numbers

Consider the addition $(+6) + (-2)$. When adding numbers, it does not matter in which order the numbers are added. For example,

$$3 + 5 = 5 + 3 = 8$$

Thus: $(+6) + (-2) \;=\; (-2) + (+6)$
But $(-2) + (+6) \;=\; +4$ (see Fig. 8.15)
Thus: $(+6) + (-2) \;=\; +4$

On the number line in Fig. 8.16 we know that we START at $+6$ and FINISH at $+4$. The only way this can happen is if we move 2 places to the *left*.

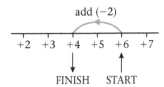

Fig. 8.16

To add a negative number: move that number of places to the left on the number line.

Notice that adding a negative number is equivalent to subtracting a positive number of the same value, for example see Fig. 8.17.

Fig. 8.17

Example 7

Simplify (a) $2 + (-5)$,   (b) $-3 + (-3)$.

(a) $2 + (-5) = (+2) + (-5)$
              $= -3$

(b) $-3 + (-3) = (-3) + (-3)$
                $= -6$

Example 8

Simplify $(-3) + (-9) + 14$.

$$(-3) + (-9) + 14 = (-3) + (-9) + (+14)$$
$$= (-12) + (+14)$$
$$= (+2)$$
$$= 2$$

The working of Example 8 is shown on the number line in Fig. 8.18.

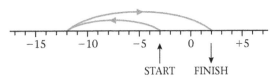

Fig. 8.18

In practice there is no need to start at 0. The sign in front of the first number shows you where to start.

Exercise 8g (Oral)

1 Complete the following number patterns.
   (a) $(+3) + (+2) = +5$
        $(+3) + (+1) =$
        $(+3) + \;\;\;\; 0 \;=$
        $(+3) + (-1) =$
        $(+3) + (-2) =$
        $(+3) + (-3) =$
   (b) $(-1) + (+1) = 0$
        $(-1) + \;\;\;\; 0 \;=$
        $(-1) + (-1) =$
        $(-1) + (-2) =$
        $(-1) + (-3) =$
        $(-1) + (-4) =$
   (c) $4 + (+4) = +8$
        $4 + (+2) =$
        $4 + \;\;\;\; 0 \;=$
        $4 + (-2) =$
        $4 + (-4) =$
        $4 + (-6) =$

2 Simplify the following.
   (a) $(+4) + (-7)$
   (b) $5 + (-3)$
   (c) $(-2) + (-6)$
   (d) $-8 + (-6)$
   (e) $20 + (-20)$

(f) $(+13) + (-10)$
(g) $(-9) + (-9)$
(h) $-18 + (-3)$
(i) $0 + (-5)$
(j) $60 + (-40)$
(k) $(-12) + (-8)$
(l) $43 + (-23)$
(m) $-25 + (-14)$
(n) $100 + (-144)$
(o) $-51 + (-38)$
(p) $16 + (-9) - (+10)$
(q) $-3 + 14 + (-5)$

Exercise 8h

Simplify the following.

1 (a) $3 + (-1)$
   (b) $3 + (-2)$
   (c) $3 + (-3)$
   (d) $3 + (-4)$
   (e) $3 + (-5)$

2 (a) $-3 + (-1)$
   (b) $-3 + (-2)$
   (c) $-3 + (-3)$
   (b) $-3 + (-4)$
   (e) $-3 + (-5)$

3 $13 + (-9)$

4 $9 + (-13)$

5 $7 + (-15)$

6 $3 + (-10)$

7 $11 + (-4)$

8 $12 + (-17)$

9 $19 + (-12)$

10 $12 + 6 + (-4)$

Notice that:

(a) $+7 + (-4) = +3$
(b) $+4 + (-7) = -3$
(c) $+7 + 4 = +11$
(d) $-7 + (-4) = -11$

In (a) and (b), the two signs are different. Disregard signs and take the smaller number from the larger. Then:

(a) if the larger number has a + sign before it, the result is positive;

(b) if the larger number has a − sign before it, the result is negative.

In (c) and (d), the two signs are the same. Disregard the signs and add the two numbers. Then:

(c) if both signs are +, the result is positive;

(d) if both signs are −, the result is negative.

This helps to simplify directed numbers without using the number line.

Exercise 8i

Simplify the following. Try not to use the number line.

1 $19 + (-11)$ **2** $12 + (-17)$

3 $-5 + (-8)$ **4** $18 + (-27)$

5 $31 + (-32)$ **6** $-11 + (-19)$

7 $-13 + (-21)$ **8** $33 + (-18)$

9 $43 + (-51)$ **10** $29 + (-53)$

It is important to understand that the symbols '+' and '−' have two meanings.

1. They indicate **directed numbers**, that is, '+' indicates a positive integer, and '−' indicates a negative integer.

2. They indicate **mathematical operations**, that is, '+' represents the operation of addition, and '−' represents the operation of subtraction.

Note that as directed numbers,

$+4$ is read as positive 4
and -4 is read as negative 4.

There is no space between the symbol and the number.
However, as a mathematical operation,

$5 + 4$ is read as five plus 4 or five add 4;
$5 + (-4)$ is read as five plus negative 4
       or five add negative 4, and
$5 - 4$ or $5 - (+4)$ is read as five minus 4
       or five subtract positive 4.

There are spaces both before and after the symbol.

Multiplication with directed numbers

Positive multipliers

Multiplication is a short way of writing repeated additions. For example,

$3 \times 4 = 3$ lots of 4
$= 4 + 4 + 4$
$= 12$

With directed numbers,

$$(+4) + (+4) + (+4) = 3 \text{ lots of } (+4)$$
$$= 3 \times (+4)$$

The multiplier is 3. It is positive. Thus,

$$(+3) \times (+4) = (+4) + (+4) + (+4)$$
$$= +12$$
$$(+3) \times (+4) = +12$$

Fig. 8.19 shows $1 \times (+4)$ and $(+3) \times (+4)$ as movements on the number line. The movements are in the same direction from 0.

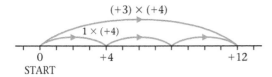

Fig. 8.19

Similarly,

$$(-2) + (-2) + (-2) + (-2) + (-2)$$
$$= 5 \text{ lots of } (-2)$$
$$= 5 \times (-2)$$

The multiplier is 5. It is positive. Thus,

$$(+5) \times (-2) = (-2) + (-2) + (-2) + (-2) + (-2)$$
$$= -10$$
$$(+5) \times (-2) = -10$$

Fig. 8.20 shows $1 \times (-2)$ and $(+5) \times (-2)$ as movements on the number line. The movements are in the same direction from 0.

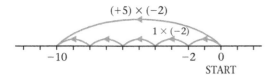

Fig. 8.20

In general: $(+a) \times (+b) = +(a \times b)$
$\qquad\qquad (+a) \times (-b) = -(a \times b)$

Example 9

Simplify the following.
(a) $(+9) \times (+4)$
(b) $(+17) \times (-3)$
(c) $(+\frac{1}{2}) \times (+\frac{1}{4})$
(d) $3 \times (-1.2)$

(a) $(+9) \times (+4) = +(9 \times 4) = +36$
(b) $(+17) \times (-3) = -(17 \times 3) = -51$
(c) $(+\frac{1}{2}) \times (+\frac{1}{4}) = +(\frac{1}{2} \times \frac{1}{4}) = +\frac{1}{8}$
(d) $3 \times (-1.2) = (+3) \times (-1.2)$
$\qquad\qquad\qquad = -(3 \times 1.2) = -3.6$

Exercise 8j

1. Continue the following number pattern as far as the 7th term.
   (a) $+15, +10, +5, 0, -5, \dots$
   (b) $+6, +4, +2, 0 \dots$
   (c) $+30, +20, +10, \dots$

2. Complete the following.
   (a) $(+5) \times (+3) = \dots$
   $\quad\ (+5) \times (+2) = \dots$
   $\quad\ (+5) \times (+1) = \dots$
   $\quad\ (+5) \times \quad 0 \ = \dots$
   $\quad\ (+5) \times (-1) = \dots$
   $\quad\ (+5) \times (-2) = \dots$
   $\quad\ (+5) \times (-3) = \dots$
   (b) $(+2) \times (+3) = \dots$
   $\quad\ (+2) \times (+2) = \dots$
   $\quad\ (+2) \times (+1) = \dots$
   $\quad\ (+2) \times \quad 0 \ = \dots$
   $\quad\ (+2) \times (-1) = \dots$
   $\quad\ (+2) \times (-2) = \dots$
   $\quad\ (+2) \times (-3) = \dots$
   (c) $(+10) \times (+3) = \dots$
   $\quad\ (+10) \times (+2) = \dots$
   $\quad\ (+10) \times (+1) = \dots$
   $\quad\ (+10) \times \quad 0 \ = \dots$
   $\quad\ (+10) \times (-1) = \dots$
   $\quad\ (+10) \times (-2) = \dots$

3. Simplify the following.
   (a) $(+6) \times (+3)$ (b) $(+5) \times (+10)$
   (c) $(+5) \times (-8)$ (d) $(+4) \times (-7)$
   (e) $(+17) \times (+2)$ (f) $(+20) \times (+6)$
   (g) $(+10) \times (-10)$ (h) $(+15) \times (-4)$
   (i) $(+3) \times (+2\frac{1}{2})$ (j) $(+1\frac{1}{4}) \times (+\frac{1}{5})$
   (k) $(+2) \times (-3\frac{1}{2})$ (l) $(+1) \times (-\frac{2}{3})$
   (m) $(+1.5) \times (+2)$ (n) $(+5.1) \times (+3)$
   (o) $(+4) \times (-1.2)$ (p) $(+3.2) \times (-5)$
   (q) $8 \times (+7)$ (r) $8 \times (-7)$
   (s) $7 \times (+8)$ (t) $7 \times (-8)$

Negative multipliers

Exercise 8k

① Find the next four terms in each of the following patterns.

(a) $+15, +12, +9, +6, ..., ..., ..., ...$

(b) $(+5) \times (+3), (+4) \times (+3),$
$(+3) \times (+3), (+2) \times (+3),$
$... \times ..., \quad ... \times ...,$
$... \times ..., \quad ... \times ...$

(c) $(+5) \times (+3) = +15$
$(+4) \times (+3) = +12$
$(+3) \times (+3) = +9$
$(+2) \times (+3) = +6$
$... \times ... = ...$
$... \times ... = ...$
$... \times ... = ...$
$... \times ... = ...$

② Find the next four terms in each of the following patterns.

(a) $-15, -12, -9, -6, ..., ..., ..., ...$

(b) $(+5) \times (-3), (+4) \times (-3),$
$(+3) \times (-3), (+2) \times (-3),$
$... \times ..., \quad ... \times ...,$
$... \times ..., \quad ... \times ...$

(c) $(+5) \times (-3) = -15$
$(+4) \times (-3) = -12$
$(+3) \times (-3) = -9$
$(+2) \times (-3) = -6$
$... \times ... = ...$
$... \times ... = ...$
$... \times ... = ...$
$... \times ... = ...$

Look at your results in question 1 of Exercise 8k. The last row of 1(c) is: $(-2) \times (+3) = -6$. (-2) is the multiplier. It is negative. The number being multiplied, $(+3)$, is positive. The result is negative. This result may not be surprising. We already know that $(+3) \times (-2) = -6$. Thus we would expect that $(-2) \times (+3) = -6$.

Fig. 8.21 shows $1 \times (+3)$ and $(-2) \times (+3)$ as movements on the number line. The movements are in opposite directions from 0.

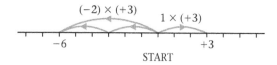

Fig. 8.21

Look at your results in question 2 of Exercise 8k. The last row of 2(c) is: $(-2) \times (-3) = +6$. (-2) is the multiplier. It is negative. The number being multiplied, (-3), is also negative. The result is *positive*. This result may be quite surprising. However, on the number line the movements are again in opposite directions from 0. Fig. 8.22 shows $1 \times (-3)$ and $(-2) \times (-3)$ as movements on the number line.

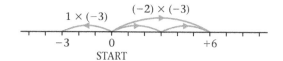

Fig. 8.22

In general, $(-a) \times (+b) = -(a \times b)$
$(-a) \times (-b) = +(a \times b)$

Example 10
Simplify the following.

(a) $(-7) \times (+4)$ (b) $(-5) \times (-18)$
(c) $(-\frac{1}{3}) \times (+\frac{2}{5})$ (d) $(-4) \times (-2.2)$

(a) $(-7) \times (+4) = -(7 \times 4) = -28$
(b) $(-5) \times (-18) = +(5 \times 18) = +90$
(c) $(-\frac{1}{3}) \times (+\frac{2}{5}) = -(\frac{1}{3} \times \frac{2}{5}) = -\frac{2}{15}$
(d) $(-4) \times (-2.2) = +(4 \times 2.2) = +8.8$

Exercise 8l

① Continue the following number patterns as far as the 7th term.

(a) $+15, +10, +5, 0, ...$
(b) $-6, -4, -2, ...$
(c) $-30, -20, -10, ...$

② Complete the following.

(a) $(+3) \times (+5) = ...$
$(+2) \times (+5) = ...$
$(+1) \times (+5) = ...$
$0 \times (+5) = ...$
$(-1) \times (+5) = ...$
$(-2) \times (+5) = ...$
$(-3) \times (+5) = ...$

(b) $(+3) \times (-2) = ...$
$(+2) \times (-2) = ...$
$(+1) \times (-2) = ...$
$0 \times (-2) = ...$
$(-1) \times (-2) = ...$
$(-2) \times (-2) = ...$
$(-3) \times (-2) = ...$

(c) $(+3) \times (-10) = ...$
$(+2) \times (-10) = ...$
$(+1) \times (-10) = ...$
$0 \times (-10) = ...$
$(-1) \times (-10) = ...$
$(-2) \times (-10) = ...$

③ Simplify the following.

(a) $(-6) \times (+3)$ (b) $(+5) \times (+10)$
(c) $(-5) \times (-8)$ (d) $(-4) \times (-7)$
(e) $(-16) \times (+2)$ (f) $(-20) \times (+4)$
(g) $(-8) \times (-8)$ (h) $(-13) \times (-5)$
(i) $(-2) \times (+1\frac{1}{2})$ (j) $(-1\frac{1}{3}) \times (+1)$
(k) $(-\frac{1}{3}) \times (-3)$ (l) $(-4) \times (-\frac{1}{2})$
(m) $(-5) \times (+1.2)$ (n) $(-2.4) \times (+3)$
(o) $(-0.3) \times (-9)$ (p) $(-6) \times (-3.1)$
(q) -8×6 (r) $8 \times (-6)$
(s) $-8 \times (-6)$ (t) $-6 \times (-8)$

Summary

(a) Numbers that differ by one unit may be represented at equal distances apart on a number line.
(b) The number line can be extended for numbers less than zero.
(c) Numbers greater than zero are **positive**. Numbers less than zero are **negative**.
(d) Positive and negative numbers are called **directed numbers**.

(e) As we move right from zero on the number line, numbers are positive and increasing; as we move left from zero on the number line, numbers are negative and decreasing.
(f) Positive and negative whole numbers together with zero form the set of **integers**.
(g) To add a positive number, move to the right on the number line.
(h) To add a negative number, move to the left on the number line.
(i) When two numbers of the same sign are multiplied together, the result is positive.
(j) When two numbers of different sign are multiplied together, the result is negative.

Practice exercise P8.1

① In each of the following sets of numbers, select
(a) the greatest integer,
(b) the smallest integer.
(i) $-2, +4, +1, -6$
(ii) $0, -3, +5, +1, -7$
(iii) $-5, +5, -1, +1, -2, +2$

② Arrange the numbers from the least to the greatest in each of the following sets.
(a) $-5, +3, +1, -1$
(b) $-2, +4, 0, +1, -6$
(c) $-3, +3, -4, +4, -5, +5$

③ In each of the following, replace * with $<$ or $>$ to make the statement true.
(a) $7 * -5$ (b) $2 * 3$
(c) $-5 * -9$ (d) $-23 * 13$
(e) $21 * -19$ (f) $11 * -13$
(g) $-7 * -5$ (h) $-7 * 3$
(i) $-5 * -9$ (j) $-23 * 13$
(k) $21 * -12$ (l) $-1 * -13$

Practice exercise P8.2

① Add these numbers.
(a) $3, 2$ (b) $4, -1$
(c) $5, -3$ (d) $3, -5$
(e) $7, -5$ (f) $-7, 5$
(g) $6, -6$ (h) $-5, -9$
(i) $15, -9, 6$ (j) $-23, 13, -11$
(k) $21, -16, -21$ (l) $17, -13, -9$

2 Multiply these pairs of numbers.

(a) 7, −5   (b) −7, 3   (c) −5, −9
(d) −6, −2   (e) −3, −4   (f) −5, −13
(g) −2, 3   (h) −3, 2   (i) −4, 3
(j) −6, 8   (k) −8, 9   (l) −12, 3
(m) −18, 4   (n) −23, 13   (o) 17, −13

Practice exercise P8.3

Simplify and evaluate these calculations.

1   $5 + (-4)$ 2 $-2 + (-3)$

3   $-9 + (-5)$ 4 $-2 + 6$

5   $9 + (-2) + (-7)$ 6 $-4 + (-3) - 11$

7   $5 \times (-7)$ 8 $-3 \times (-6)$

9   -9×5 10 $-4 \times (-13)$

11   $-5 \times (-3) \times 11$ 12 $5 \times (-4) \times (-6)$

Practice exercise P8.4

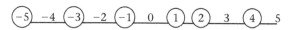

1 Write down the numbers circled on this line which are

(a) positive,   (b) negative,
(c) less than 2,   (d) more than −4.

2 Here are some temperatures.

   2 °C   4 °C   −3 °C   −5 °C
−1 °C   7 °C   −6 °C

(a) Write the temperatures in order, starting with the smallest.
(b) List the temperatures which are below 3 °C.

Revision exercises and tests

Chapters 1–8

Revision exercise 1 (Chapters 1, 3)

1. What is the value of
   (a) the 5 in 253
   (b) the 7 in 367
   (c) the 2 in 2 403
   (d) the 9 in 3 937?

2. What is the value of
   (a) the 6 in 4.62
   (b) the 8 in 0.238
   (c) the 1 in 3.815
   (d) the 5 in 1.05?

3. Expand 2065_{eight} in powers of its base.

4. Convert 63_{ten} to (a) base five, (b) base three.

5. Convert $11\,011_{two}$ to base ten. Convert the result to base eight.

6. Express the following as products of their prime factors.
   (a) 18
   (b) 26
   (c) 45
   (d) 75

7. Calculate the value of the following.
   (a) 7^2
   (b) 5^3
   (c) $2^3 \times 3^2$
   (d) $2^2 \times 3^3$

8. Find the HCF of 63 and 90.

9. Find the LCM of 12 and 15.

10. Find the greatest number which when divided into 593 and 621 will leave a remainder of 5 in each case.

Revision test 1 (Chapters 1, 3)

1. The value of the 8 in 18 214 is
   A 8 tens
   B 8 hundreds
   C 8 thousands
   D 8 ten thousands

2. If $412_{five} = 1*7_{ten}$, what digit does the * stand for?
   A 0
   B 1
   C 2
   D 4

3. Express 22_{ten} in base two.
   A 1010
   B 1110
   C 10 110
   D 11 010

4. The value of 6^2 is
   A 12
   B 24
   C 26
   D 36

5. If M_8 = {multiples of 8} and M_{12} = {multiples of 12}, what is the LCM of 8 and 12?
   A 2
   B 4
   C 24
   D 96

6. Convert 113_{ten} to (a) base two, (b) base eight.

7. Express 60 as a product of prime factors.

8. Find the HCF of 32, 40 and 56.

9. A four-digit square number is divisible by 3. If the last two digits are 8 and 9, what is the four-digit number?

10. (a) Find out which of the numbers 2, 5, 8, 9 will divide into 10 170 exactly.
    (b) Hence state which of the numbers 20, 40, 45, 90 will also divide into 10 170 exactly.

Revision exercise 2 (Chapters 2, 6, 8)

1. Which of the following statements are true?
   (a) $12 \in \{2, 4, 6, 8, \ldots\}$
   (b) $\{a, b\} \not\subset \{f, a, c, e\}$
   (c) If $Z = \{0\}$, $n(Z) = 0$

2. If $A = \{x, y, z\}$, list all the subsets of A.

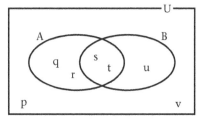

Fig. R1

3. Fig. R1 is a Venn diagram showing the elements of sets U, A, B. Answer the following questions.
   (a) Find $n(U)$.
   (b) Find $n(A)$.
   (c) List the elements of A.
   (d) List the subsets of B which have exactly two elements.

④ Find the values of the following when $x = 6$.
 (a) $3x + 1$ (b) $11 - x$
 (c) $x + 5 - x$ (d) $\frac{2}{3}x$

⑤ Write down the number which is
 (a) 5 more than m (b) 5 less than m
 (c) 5 times m (d) one-fifth of m

⑥ Simplify the following.
 (a) $(-3) + (+8)$
 (b) $(+5) + (-14)$
 (c) $(-7) + (+7)$

⑦ Simplify the following.
 (a) $(-5) \times 3$ (b) $(-8) \times (-2)$
 (c) $6 \times (-5)$ (d) $(-1) \times (-9)$

⑧ Simplify the following.
 (a) $-4 + (-7)$ (b) $6 + (-9)$
 (c) $(-7) \times (-3)$ (d) $29 \times (-2)$

⑨ A boy buys 2 packets each containing x biscuits. He eats 6 biscuits. How many biscuits does he have left?

⑩ A trader sells x pencils at 10 cents each, y pens at 20 cents each and 5 rubbers at x cents each. How much money does the trader get altogether?

Revision test 2 (Chapters 2, 6, 8)

① If N = {odd numbers greater than 11}, which one of the following is an element of N?
 A 10 B 11 C 12 D 13

② If Y = {5, 1, 4, 1, 2, 6, 1}, then $n(Y) =$
 A 1 B 5 C 6 D 7

③ Which one of the following numbers is the greatest?
 A -2 B -30 C -100 D -50

③ The number which is 6 less than m is
 A $m + 6$ B $6 - m$
 C $m - 6$ D $\frac{m}{6}$

④ $-20 + (-70) =$
 A -90 B -50 C $+50$ D $+90$

⑤ If A ⊂ B and B ⊂ C, draw a Venn diagram to show the relationship between A, B and C.

⑦ A man has n cows. He sells half of them. He then gives 8 cows to his daughter's husband. How many cows does he have left?

⑧ The temperature inside a refrigerator is $-3.9\,°C$. What will be the temperature if it rises by $2.4\,°C$?

⑨ Simplify the following.
 (a) $-3 + 11$ (b) $-9 + 4$
 (c) $8 + (-15)$ (d) $-6 + (-6)$

⑩ A trader buys 30 shirts for $\$x$ altogether. She sells them all for $\$y$ each. Write an algebraic expression for the total amount of money she received. If she received more money than she spent, write an expression for this amount.

Revision exercise 3 (Chapters 4, 5, 7)

① A shelf 2.66 m long is used to store copies of a textbook. Each textbook is 2.8 cm thick. How many books can be stored on the shelf?

② A train is supposed to start from P at 11:58 a.m. and to reach Q at 1:49 p.m. If it starts 4 min late and arrives 18 min late, how long does the journey take?

Table R1 is a timetable of flights between two islands.

Table R1

Thu			Thu
FJ251	↓	↑	FJ252
14:15	Lv Maloney Ar		21:15
15:45	Ar St Mark Lv		19:45

Use Table R1 to answer questions 3 and 4.

③ (a) What is the flight number from Maloney to St Mark?
 (b) What time does the aeroplane leave St Mark?
 (c) How long is the aeroplane on the ground at St Mark?

④ (a) How long does it take to fly from St Mark to Maloney?
 (b) If the distance between St Mark and Maloney is 900 km, calculate the average flying speed of the aeroplane in km/h.

⑤ Reduce the following to their lowest terms.
  (a) $\frac{12}{20}$ (b) $\frac{10}{45}$ (c) $\frac{48}{60}$

⑥ Simplify the following.
  (a) $\frac{1}{4} + \frac{1}{3}$ (b) $\frac{3}{5} - \frac{1}{4}$
  (c) $\frac{5}{9} \times \frac{3}{10}$ (d) $\frac{5}{6} \div \frac{2}{3}$
  (e) $2\frac{3}{4} + 5\frac{1}{8} - 3\frac{1}{2}$ (f) $4\frac{3}{8} \times \frac{4}{15} \div 11\frac{2}{3}$

⑦ Calculate the length of wire needed to make the skeleton model of a cuboid which measures 12 cm by 10 cm by 8 cm. Allow 5 cm of wire for joins.

⑧ In a class $\frac{4}{5}$ of the students have mathematical instruments. If $\frac{1}{4}$ of these students have lost their protractors, what fraction of students in the class has protractors?

⑨ How many faces has a triangular prism? How many of these are rectangular?

⑩ When a girl was ill her temperature rose from 98.6 °F to 102 °F.
  (a) By how many degrees did her temperature rise
    (i) in °F   (ii) in °C?
  (b) What was her temperature in °C when she was (a) well, (b) ill?

Revision test 3 (Chapters 4, 5, 7)

① 5300 mm expressed in metres is
  A   0.053   B   0.53
  C   5.3   D   53

② Which one of the following is *not* equivalent to $\frac{1}{2}$?
  A $\frac{9}{18}$ B $\frac{11}{22}$ C $\frac{15}{30}$ D $\frac{24}{42}$

③ If $5\frac{1}{7}$ is expressed as an improper fraction, its numerator will be
  A   8   B   13   C   35   D   36

④ Which net in Fig. R2 is the net of a triangular prism?

Fig. R2

⑤ It takes 72 cm of wire to make a skeleton model of a cube. The length in cm of one edge of the cube is
  A   6   B   9   C   12   D   18

⑥ A motorist travels 108 km in 2 h 15 min. If she travels at a steady speed, find how far she goes in 1 hour.

⑦ Simplify the following.
  (a) $5\frac{1}{4} + 1\frac{1}{6} - 3\frac{2}{3}$
  (b) $6\frac{1}{4} \times 1\frac{3}{5}$
  (c) $6\frac{3}{4} \div 5\frac{5}{8}$

⑧ During a 1-hour radio programme there were 18 min of talking; the rest was music. What fraction of the radio programme was music?

⑨ Fig. R3 shows a cuboid ABCDEFGH.

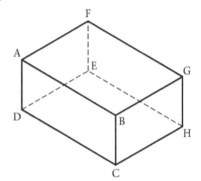

Fig. R3

  (a) Faces ABGF and BCHG meet along which edge?
  (b) Which edges meet at the vertex H?
  (c) Edges BG and AB meet at which vertex?

⑩ Fig. R4 shows the net of a triangular based pyramid.

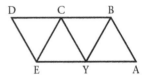

Fig. R4

  If the net is folded to make the pyramid,
  (a) which edge will join to edge BC?
  (b) which point will joint to point A?

General revision test A (Chapters 1–8)

1. The value of the 3 in 24.635 is
   A 3 thousandths B 3 hundredths
   C 3 tenth D 3 tens

2. The HCF of 24 and 60 is
   A 3 B 4 C 6 D 12

3. If $m = 3$, $n = 2$, then $5 \times 3m - (3n + 2n) =$
   A 14 B 22 C 35 D 43

4. The lowest common denominator of $\frac{2}{3}, \frac{4}{5}, \frac{5}{6}$ and $\frac{3}{10}$ is
   A 15 B 30 C 60 D 900

5. 5 added to a certain number gives a result of -6. The number is
   A -11 B -1 C $+1$ D $+11$

6. A temperature of $122\,°F$ is the same as
   A $162\,°C$ B $90\,°C$
   C $68\,°C$ D $50\,°C$

7. If $x = 3$, the value of $7x - 2x$ is
   A 8 B 15 C 21 D 53

8. Which one of the following statements about a cylinder is false?
   A A cylinder has two vertices.
   B A cylinder has two curved edges.
   C A cylinder has one curved face.
   D The net of a cylinder contains a rectangle.

9. I buy 7 metres of ribbon at $\$x$ per metre. My change from $30 will be
   A $\$23x$ B $\$(30 - x)$
   C $\$(30 - 7x)$ D $\$7(30 - x)$

10. P = {letters of the word student}
    R = {letters of the word test}
    Which one of the following statements is true?
    A $e \notin P$ B $P \subset R$
    C $R = \varnothing$ D $P \supset R$

11. If U = {1, 2, 3, ..., 9}, A = {2, 4, 6, 8} and B = {3, 4, 5, 6}
    draw a Venn diagram showing the elements of U, A and B.

12. Simplify the following.
    (a) $1\frac{1}{20} + \frac{3}{4}$ (b) $5\frac{3}{8} - 4\frac{3}{4}$
    (c) $3\frac{3}{4} \times 1\frac{1}{2}$ (d) $3\frac{2}{3} \div 2\frac{2}{9}$

13. A woman gives $\frac{1}{4}$ of a cake to her son, $\frac{1}{4}$ to her daughter and $\frac{1}{3}$ to her husband. What fraction is left for herself?

14. One-sixth of a stick is cut off and then three-tenths of the remaining piece is thrown away. What fraction of the original stick remains?

15. A boy walks for $3\frac{1}{2}$ min and runs for $8\frac{1}{2}$ min. What fraction of his journey time is spent running?

16. John sets out from home at 10:30 a.m. and rides to a village 18 km away at 8 km per hour.
    (a) How long does the journey take him?
    (b) At what time does John arrive at the village?

17. Find the value of the following when $x = 2$, $y = 5$, $z = 1$.
    (a) $x^2 + y^2$ (b) $(x + y)^2$
    (c) $\frac{x + y}{y + z}$ (d) $\frac{x}{y} + \frac{y}{z}$

18. Express the following times in terms of the 24-hour clock.
    (a) 6:25 a.m.
    (b) 6:45 p.m.
    (c) 5 minutes to 8 p.m.
    (d) 10 minutes to midnight

19. Mary leaves home when her digital watch reads 09:55 and arrives at church when it reads 10:36. How long did it take Mary to get to church? What would her digital watch read if she arrives back at home $1\frac{1}{2}$ hours after she reached church?

20. Three clocks are set to ring a bell every 20 minutes, 30 minutes and 35 minutes respectively. If the three bells ring together at 8:00 a.m., at what time would they next ring together?

Chapter 9
Decimals, fractions and percentages

We have seen in Chapter 1 that the place-value system can be extended to include decimal fractions. In Chapter 4 we saw that the SI system makes use of this. For example, 3.54 kg means $3\,kg + \frac{5}{10}\,kg + \frac{4}{100}\,kg$ or $3\,kg + \frac{54}{100}\,kg$. When using measuring instruments, such as a ruler, decimal fractions are used to give intermediate values.

Exercise 9a (Oral)

Each part of Fig. 9.1 shows a scale and an arrow. The reading opposite each arrow is a decimal number. Find the reading in each case, estimating where necessary.

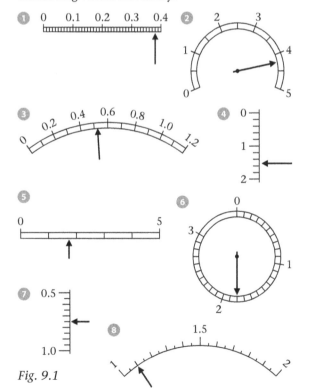

Fig. 9.1

Just as we add, subtract, multiply and divide whole numbers and fractions, we can do the same with decimal numbers and decimal fractions.

Addition and subtraction of decimal numbers

Be very careful to set your work out correctly. Units must be under units, decimal points under decimal points, and so on. For example, 24.8 + 6.5 is set out as

$$\begin{array}{ccc} 24.8 & & 24.8 \\ +\ 6.5 & \text{not} & \underline{6.5} \\ \hline \end{array}$$

After you have set out your work correctly, add and subtract in the same way as you do for whole numbers, but remember to write down the decimal point when you come to it.

Exercise 9b

Questions 1–5 each contain ten short items. Give yourself 10 minutes to do questions 1 and 2. Correct your work. If you make fewer than 5 mistakes, go on to question 6. If you make 5 mistakes or more, do questions 3, 4, 5.

① (*5 min*)

(a) 0.2 + 0.6	(b) 0.6 − 0.5
(c) 1.3 + 0.8	(d) 2 − 0.5
(e) 2.7 + 3.8	(f) 8.8 − 5.4
(g) 15.86 + 5.15	(h) 0.56 − 0.18

(i) 4.23 + 0.78 + 7.52
(j) 59.25 − 49.17

② (*5 min*)

(a) 0.3 + 0.5	(b) 0.9 − 0.4
(c) 1.9 + 0.8	(d) 3 − 1.6
(e) 5.4 + 2.7	(f) 0.47 − 0.25
(g) 6.33 + 4.07	(h) 7.1 − 3.7

(i) 8.4 + 7.5 + 31.9 (j) 25.08 − 14.13

3 (*5 min*)

(a) $0.1 + 0.8$ (b) $0.4 - 0.1$

(c) $0.6 + 1.7$ (d) $2 - 1.7$

(e) $6.1 + 1.9$ (f) $0.59 - 0.55$

(g) $\begin{array}{r} 12.62 \\ + 13.59 \end{array}$ (h) $\begin{array}{r} 8.2 \\ - 5.7 \end{array}$

(i) $3.67 + 2.74 + 5.4 + 0.23$

(j) $18.33 - 16.55$

4 (*5 min*)

(a) $0.3 + 0.2$ (b) $0.8 - 0.6$

(c) $0.5 + 0.9$ (d) $3 - 0.3$

(e) $3.5 + 4.6$ (f) $7.5 - 2.5$

(g) $\begin{array}{r} 13.45 \\ + 7.55 \end{array}$ (h) $\begin{array}{r} 0.85 \\ - 0.26 \end{array}$

(i) $4.5 + 0.76 + 6.4 + 1.06$

(j) $24.11 - 14.07$

5 (*5 min*)

(a) $0.4 + 0.4$ (b) $0.3 - 0.2$

(c) $1.6 + 0.4$ (d) $2 - 1.8$

(e) $7.5 + 1.7$ (f) $9.3 - 6.2$

(g) $\begin{array}{r} 8.92 \\ + 4.09 \end{array}$ (h) $\begin{array}{r} 0.53 \\ - 0.26 \end{array}$

(i) $69.3 + 6.93 + 0.693$

(j) $40.5 - 4.05$

6 Find the sum of (a) $1.62, $2.52 and $3.42, (b) $1.12, 64c and 96c.

7 Find the difference between (a) 28.95 m and 17.58 m, (b) 3.062 kg and 855 g.

8 A ball of string is 13.5 m long when new. Lengths of 2.3 m, 2.37 m and 95 cm are cut off. What length of string is cut off altogether? How much string is left? (Give both answers in metres.)

9 A recipe for bread says, 'Mix 30 g of yeast, 15 g of sugar and 15 g of salt. Add the mixture to 1 kg of flour. Add water and stir.' A cook followed the recipe and the total mass of mixture was 1.65 kg. What mass of water was added?

10 Find the difference between 9.28 and the sum of 2.31 and 3.92.

Multiplication by powers of 10

Table 9.1 shows what happens when 3.07 is multiplied by increasing powers of 10.

Table 9.1

$3.07 \times$	$1 =$	3.07
$3.07 \times$	$10 = 3.07 \times 10^1 =$	30.7
$3.07 \times$	$100 = 3.07 \times 10^2 =$	$307.$
$3.07 \times$	$1\,000 = 3.07 \times 10^3 =$	$3\,070.$
$3.07 \times$	$10\,000 = 3.07 \times 10^4 =$	$30\,700.$

Note that when **multiplying** a number by powers of 10:

(a) as the power of 10 increases, it is as if the decimal point stays where it is and the digits in the number move to the left;

(b) the digits move as many places to the left as the power of 10 (or as the number of zeros in the multiplier);

(c) as each place to the right of the digits and to the left of the decimal point becomes empty, we fill it with a zero to act as a place holder, e.g. $3.07 \times 1000 = 3070$;

(d) if the fraction, that is, the digits to the right of the decimal point, becomes zero there is no need to write anything after the point.

Division by powers of 10

Similarly, Table 9.2 shows what happens when 3.07 is divided by increasing powers of 10.

Table 9.2

$3.07 \div$	$1 =$	3.07
$3.07 \div$	$10 = 3.07 \div 10^1 =$	0.307
$3.07 \div$	$100 = 3.07 \div 10^2 =$	$0.030\,7$
$3.07 \div$	$1\,000 = 3.07 \div 10^3 =$	$0.003\,07$
$3.07 \div$	$10\,000 = 3.07 \div 10^4 =$	$0.000\,307$

Notice that when **dividing** a number by powers of 10:

(a) as the power of 10 increases, it is as if the decimal point stays where it is and the digits including zeros, in the number move to the right;

(b) the digits move as many places to the right as the power of 10 (or as the number of zeros in the divisor);

(c) as the place to the left of the decimal point becomes empty, we fill it with a zero to act as a place holder.

Example 1

Write the following as decimal numbers.

(a) $0.036 \times 10\,000$ (b) $\frac{23}{1000}$

(c) $120 \div 100\,000$ (d) $0.000\,45 \times 100$

(a) $0.036 \times 10\,000 = 0.036 \times 10^4 = 360$

(b) $\frac{23}{1000} = 23 \div 1\,000 = 23 \div 10^3 = 0.023$

(c) $120 \div 100\,000 = 120 \div 10^5 = 0.001\,20$
$$= 0.001\,2$$

It is not necessary to write zeros to the right of a decimal fraction. For example, $0.200\,000$ is just the same as 0.2.

(d) $0.000\,45 \times 100 = 0.000\,45 \times 10^2 = 0.045$

Exercise 9c (Oral)

Give the following as decimal numbers.

① 2.67×100 ② $4.55 \times 10\,000$

③ 8.03×1000 ④ $5.6 \div 1000$

⑤ $8.11 \div 10$ ⑥ $7.04 \div 100$

⑦ $0.027 \times 10\,000$ ⑧ 0.36×100

⑨ $\frac{52}{1000}$ ⑩ $\frac{63}{10\,000}$

⑪ $\frac{423}{100}$ ⑫ 0.05×1000

⑬ 650×100 ⑭ $90 \times 100\,000$

⑮ 4.37×10^3 ⑯ 0.207×10^2

⑰ $4502 \div 10^2$ ⑱ $0.003\,6 \div 10^4$

⑲ $5140 \div 100\,000$ ⑳ $60 \div 100$

㉑ $300 \div 10\,000$ ㉒ $0.000\,26 \times 100$

㉓ 0.07×10 ㉔ $\frac{2.42}{10\,000}$

㉕ $\frac{4000.7}{100}$ ㉖ $\frac{70.63}{1000}$

㉗ Express \$6.20 in cents.

㉘ Express 344c in dollars.

㉙ Express 592 cm in metres.

㉚ Express 6.24 tonnes in (a) kg, (b) g.

㉛ Express 920 kg in (a) tonnes, (b) g.

㉜ Express 420 g in (a) kg, (b) tonnes.

㉝ Express 504 mg in (a) g, (b) kg.

㉞ Express 20.9 m in (a) mm, (b) km.

㉟ Express 7050 metres in (a) km, (b) mm.

Multiplication of decimal numbers

Example 2

$0.7 \times 2 = \frac{7}{10} \times 2 = \frac{14}{10} = 1.4$

Example 3

$2.3 \times 1.2 = \frac{23}{10} \times \frac{12}{10} = \frac{276}{100} = 2.76$

Notice that the total number of decimal places in the two numbers being multiplied equals the number of decimal places in the answer.

Example 4

Find the product of 38.6 and 1.64.

$$38.6 \times 1.64 = \frac{386}{10} \times \frac{164}{100} = \frac{386 \times 164}{1000}$$

Use long multiplication to find the numerator:

```
     386
  ×  164
   1 544
  23 160
  38 600
  63 304
```

$$38.6 \times 1.64 = \frac{63\,304}{1000} = 63.304$$

Again, notice that the number of decimal places in the product equals the total number of places in the two numbers being multiplied. Hence, to multiply decimal numbers using this method:

(a) multiply the given numbers as if there were no decimal points, that is, as if the numbers were whole numbers;

(b) count the total number of digits after the decimal points in the numbers being multiplied;

(c) place the decimal point so that the product has the same number of digits after the point.

Example 5

0.25×0.009

$25 \times 9 = 225$

There are five digits after the decimal points in the given numbers.

So, $0.25 \times 0.009 = 0.00225$

Exercise 9d (Oral)

Find the following products.

1. 0.7×8
2. 0.2×4
3. 0.2×6
4. 0.3×9
5. 0.8×5
6. 9×0.4
7. 7×0.7
8. 7×0.6
9. 5×0.6
10. 8×0.5
11. 0.6×0.3
12. 0.4×0.7
13. 0.1×0.6
14. 0.3×0.4
15. 0.3×0.3
16. 0.09×0.2
17. 0.7×0.06
18. 0.04×0.08
19. 0.5×0.006
20. 0.003×0.05

Exercise 9e

Questions 1–5 each contain five items. Give yourself 10 minutes to do question 1 and 2. Correct your work. If you make fewer than 4 mistakes, go on to question 6. If you make 4 or more mistakes, do questions 3, 4, 5.

1. (5 min)
   (a) 0.6×1.8
   (b) 1.93×0.3
   (c) $(0.05)^2$
   (d) 0.64×2.5
   (e) 6.83×0.62

2. (5 min)
   (a) 8×0.15
   (b) 27.9×0.5
   (c) $(0.2)^2$
   (d) 6.8×3.1
   (e) 7.3×0.144

3. (5 min)
   (a) 0.4×1.4
   (b) 510×0.002
   (c) $(0.8)^2$
   (d) 5.4×0.052
   (e) 2.71×0.25

4. (5 min)
   (a) 5×0.6
   (b) 0.17×3
   (c) $(0.004)^2$
   (d) 410×0.0023
   (e) 0.87×0.306

5. (5 min)
   (a) 3×0.9
   (b) 1.3×0.7
   (c) $(0.3)^2$
   (d) 0.77×0.15
   (e) 3.42×9.9

6. What is 0.25 of 6.36?

7. Find the product of 2.03 and 0.055.

8. A ream of paper contains 480 sheets. Each sheet is 0.014 cm thick. Find the thickness of the ream of paper.

9. A dress needs 3.5 m of cloth. If cloth costs $19.00 per metre, how much will it cost to buy enough cloth for a dress?

10. 1 metre of string has a mass of 2.3 g. What is the mass of 112.8 metres of this string? Give your answer, (a) in g, (b) in kg.

Division of decimal numbers

If the divisor is a whole number, divide in the usual way. Be careful to include the decimal point in the correct place.

Example 6

$32.64 \div 4 = 8.16$

Example 7

$27.17 \div 13$

$$
\begin{array}{r}
2.09 \\
13\overline{)27.17} \\
26 \\
\hline
1.17 \\
1.17 \\
\hline
\end{array}
$$

$27.17 \div 13 = 2.09$

Example 8

$10.71 \div 63$

$$
\begin{array}{r}
0.17 \\
63\overline{)10.71} \\
6.3 \\
\hline
4.41 \\
4.41 \\
\hline
\end{array}
$$

$10.71 \div 63 = 0.17$

If the divisor contains a decimal fraction, make an equivalent division so that the divisor is a whole number.

Example 9

$16 \div 0.2$

To make 0.2 a whole number, multiply by 10. To keep the value of the division the same, multiply 16 by 10, that is, by the same number:

$$16 \div 0.2 = \frac{16}{0.2}$$

$$= \frac{16 \times 10}{0.2 \times 10} = \frac{160}{2} = 80$$

Compare division by a fraction:

$$16 \div \frac{2}{10} = 16 \times \frac{10}{2}$$

Notice also that, as in division by fractions (in Chapter 5), the result may be greater than the dividend or the divisor.

Example 10

$13.05 \div 2.9$

$$13.05 \div 2.9 = \frac{13.05}{2.9} = \frac{13.05 \times 10}{2.9 \times 10} = \frac{130.5}{29}$$

$13.05 \div 2.9 = 130.5 \div 29$

$$
\begin{array}{r}
4.5 \\
29\overline{)130.5} \\
116 \\
\hline
14.5 \\
14.5 \\
\hline
\end{array}
$$

$13.05 \div 2.9 = 130.5 \div 29 = 4.5$

The method in Examples 9 and 10 is as follows:

(a) count how many places the digits in the divisor must move to the left in order to make the divisor a whole number;
(b) move the digits of *both* numbers to the left by this number of places;
(c) divide as before.

Note that it may be necessary to fill empty places in the dividend with zeros as place holders.

Example 11

$4.2 \div 0.006$

$4.2 \div 0.006 = 4\,200 \div 6 = 700$

Example 12

$1.938 \div 0.34$

$1.938 \div 0.34 = 193.8 \div 34$

$$
\begin{array}{r}
5.7 \\
34\overline{)193.8} \\
170 \\
\hline
23.8 \\
23.8 \\
\hline
\end{array}
$$

$1.938 \div 0.34 = 193.8 \div 34 = 5.7$

Exercise 9f (Oral)

Find the result of the following divisions.

① $18 \div 6$ ② $1.8 \div 6$

③ $0.18 \div 6$ ④ $18 \div 0.6$

⑤ $1.8 \div 0.6$ ⑥ $180 \div 0.6$

⑦ $0.18 \div 0.06$ ⑧ $1\,800 \div 0.06$

⑨ $\frac{30}{5}$ ⑩ $\frac{3}{5}$ ⑪ $\frac{30}{0.5}$ ⑫ $\frac{0.3}{5}$

⑬ $\frac{3}{0.005}$ ⑭ $\frac{30}{0.05}$ ⑮ $\frac{300}{0.5}$ ⑯ $\frac{3}{0.05}$

Exercise 9g

Questions 1–5 each contain five items. Give yourself 10 minutes to do questions 1 and 2. Correct your work. If you make fewer than 4 mistakes go on to question 6. If you make 4 mistakes or more, do questions 3, 4, 5.

① (5 *min*)

(a) $\frac{0.09}{3}$ (b) $\frac{3.6}{0.4}$ (c) $\frac{7.3}{0.002}$

(d) $0.042 \div 0.06$ (e) $12.24 \div 3.6$

② (5 *min*)

(a) $\frac{1.4}{7}$ (b) $\frac{0.27}{0.9}$ (c) $\frac{3.68}{0.08}$

(d) $0.012 \div 0.4$ (e) $3.172 \div 0.52$

③ (5 *min*)

(a) $\frac{0.32}{8}$ (b) $\frac{4}{0.5}$ (c) $\frac{0.615}{0.3}$

(d) $0.28 \div 0.007$ (e) $80.5 \div 2.3$

④ (5 *min*)

(a) $\frac{0.6}{3}$ (b) $\frac{1.8}{0.2}$ (c) $\frac{4.77}{0.09}$

(d) $4.8 \div 0.08$ (e) $6.44 \div 0.028$

⑤ (5 *min*)

(a) $\frac{0.7}{7}$ (b) $\frac{0.4}{0.8}$ (c) $\frac{0.432}{0.4}$

(d) $36 \div 0.09$ (e) $351 \div 0.45$

⑥ If $35 \times 67 = 2345$, what is the value of $2.345 \div 6.7$?

⑦ The mass of 6 equal books is 2.82 kg. Calculate the mass of 1 book (a) in kg, (b) in g.

⑧ A pile of boards is 36 cm high. Each board is 0.8 cm thick. How many boards are in the pile?

⑨ A farmer pays $11.70 for 6.5 metres of wire. What is the cost of 1 metre of wire?

⑩ A test car travels 98.6 km on 7.25 litres of petrol. How many km does it travel on 1 litre of petrol?

Changing fractions to decimals

To express a fraction as a decimal, make an equivalent fraction with a denominator which is a power of 10. The decimal fraction is then obtained from the numerator. The position of the decimal point is fixed by having the same number of digits after the point as the number of zeros in the denominator.

Example 13

Express as decimal fractions

(a) $\frac{1}{5}$ (b) $\frac{3}{4}$ (c) $\frac{7}{80}$

(a) $\frac{1}{5} = \frac{1 \times 2}{5 \times 2} = \frac{2}{10} = 0.2$

(b) $\frac{3}{4} = \frac{3 \times 25}{4 \times 25} = \frac{75}{100} = 0.75$

(c) $\frac{7}{80} = \frac{7 \times 125}{80 \times 125} = \frac{875}{10\,000} = 0.087\,5$

It is usually quicker to divide the numerator of the fraction by its denominator, taking care to write down the decimal point as it arises:

(a) $\frac{1}{5} = \frac{1.0}{5} = 0.2$

(b) $\frac{3}{4} = \frac{3.00}{4} = 0.75$

(c) $\frac{7}{80} = \frac{0.7}{8} = \frac{0.7000}{8} = 0.087\,5$

These fractions give exact decimal fractions. We say they are **terminating decimals**.

Exercise 9h

Express the following as terminating decimals.

① $\frac{1}{2}$ **②** $\frac{1}{4}$ **③** $\frac{3}{8}$ **④** $\frac{2}{5}$

⑤ $\frac{1}{20}$ **⑥** $\frac{7}{20}$ **⑦** $\frac{1}{25}$ **⑧** $\frac{3}{25}$

⑨ $\frac{5}{8}$ **⑩** $\frac{13}{50}$ **⑪** $\frac{23}{25}$ **⑫** $\frac{17}{20}$

Example 14

Express $\frac{7}{9}$ as a decimal.

$$\frac{7}{9} = 7.000 \div 9$$
$$= 0.777 \ldots$$

$$\begin{array}{r} 0.777 \ldots \\ 9\overline{)0.777 \ldots} \end{array}$$

$\frac{7}{9}$ is a never ending decimal fraction: 0.777... The digit 7 is repeated as often as we like. We say it is a **recurring decimal**. We say this as 'zero point seven recurring' and write it as $0.\dot{7}$.

Many fractions give rise to recurring decimals. For example:

$\frac{4}{11} = 0.363\,636 \ldots$ $= 0.\dot{3}\dot{6}$

$\frac{5}{7} = 0.714\,285\,714\,285 \ldots$ $= 0.\dot{7}1\dot{4}\,\dot{2}8\dot{5}$

$\frac{1}{6} = 0.166\,666 \ldots$ $= 0.1\dot{6}$

Exercise 9i

Express the following as recurring decimals.

① $\frac{1}{3}$ **②** $\frac{2}{3}$ **③** $\frac{2}{9}$ **④** $\frac{5}{9}$

⑤ $\frac{8}{9}$ **⑥** $\frac{3}{11}$ **⑦** $\frac{5}{11}$ **⑧** $\frac{5}{6}$

⑨ $\frac{7}{12}$ **⑩** $\frac{1}{7}$

Changing decimals to fractions

Any exact decimal fraction can be expressed as a fraction with a denominator which is a power of 10. For example:

$$0.3 = \frac{3}{10}$$

$$0.54 = \frac{5}{10} + \frac{4}{100} = \frac{50}{100} + \frac{4}{100} = \frac{54}{100}$$

$$0.207 = \frac{2}{10} + \frac{0}{100} + \frac{7}{1000}$$

$$= \frac{200}{1000} + \frac{7}{1000}$$

$$= \frac{207}{1000}$$

We may write in its lowest terms by dividing out common factors in the numerator and denominator, that is,

$$\frac{54}{100} = \frac{2 \times 27}{2 \times 50} = \frac{27}{50}$$

Example 15

Express 0.376 as a fraction in its lowest terms.

$$0.376 = \frac{376}{1000} = \frac{376 \div 8}{1000 \div 8} = \frac{47}{125}$$

Exercise 9j

Express the following as fractions in their lowest terms.

1. 0.8
2. 0.75
3. 0.2
4. 0.45
5. 0.44
6. 0.54
7. 0.56
8. 0.66
9. 0.84

To express a recurring decimal as a fraction, put the repeated digits in the numerator with the same number of '9's as the denominator. For example,

$$0.666 = 0.\dot{6}$$
$$= \frac{6}{9} = \frac{2}{3}$$
$$0.363\,636 = 0.\dot{3}\dot{6}$$
$$= \frac{36}{99} = \frac{4}{11}$$

Exercise 9k

Express the following as fractions in their lowest terms.

1. $0.\dot{4}$
2. $0.\dot{1}\dot{8}$
3. $0.8\dot{3}$
4. $0.\dot{1}42\,85\dot{7}$
5. $0.\dot{3}07\,69\dot{2}$

Percentages

A **percentage** is a **fraction with 100 in the denominator**. The symbol % means 'hundredths'; hence, $15\% = \frac{15}{100}$.

$$\frac{1}{5} = \frac{2}{10} = \frac{20}{100}$$

In the same way that the fraction $\frac{2}{10}$ can be given as a decimal fraction 0.2, the fraction $\frac{20}{100}$ can be given as a percentage, 20%. The symbol % means hundredths. Thus 20% is a fraction where 20 is the number of parts out of a total of 100 parts.

Also, $100\% = \frac{100}{100} = 1$. Thus, 100% of something means all of it, that is, the total.

Writing fractions as percentages

To express a fraction as a percentage make an equivalent fraction with a denominator of 100. For example:

$$\frac{2}{5} = \frac{2 \times 20}{5 \times 20} = \frac{40}{100} = 40\%$$

$$\frac{176}{300} = \frac{176 \div 3}{300 \div 3} = \frac{58\frac{2}{3}}{100} = 58\frac{2}{3}\%$$

Note that

$$\frac{2}{5} = \frac{\frac{2}{5}}{1}$$
$$= \frac{\frac{2}{5} \times 100}{1 \times 100}$$
$$\therefore \frac{2}{5} = (\frac{2}{5} \times 100)\%$$
$$= 40\%$$

Hence, since a percentage is a fraction with a denominator of 100, in order to write a fraction as a percentage, multiply the fraction by 100 and write with the % symbol.

Exercise 9l

Express the following fractions as percentages.

1. $\frac{1}{2}$
2. $\frac{1}{4}$
3. $\frac{1}{3}$
4. $\frac{3}{4}$
5. $\frac{1}{5}$
6. $\frac{1}{10}$
7. $\frac{1}{20}$
8. $\frac{1}{25}$
9. $\frac{3}{10}$
10. $\frac{2}{5}$
11. $\frac{2}{15}$
12. $\frac{17}{20}$
13. $\frac{5}{8}$
14. $\frac{1}{6}$
15. $\frac{9}{25}$

Writing percentages as fractions

To express a percentage as a fraction, write it as a fraction with 100 as the denominator. Reduce this fraction to its lowest terms.

For example:

$$15\% = \frac{15}{100} = \frac{3}{20}$$

Exercise 9m

Express the following percentages as fractions in their lowest terms.

1. 5% 2. 50% 3. 25% 4. 40%

5. 70% 6. 4% 7. 35% 8. 12%

9. 80% 10. 75% 11. 24% 12. 65%

13. 64% 14. 60% 15. 45%

Changing percentages to decimals

Remember that 17% means $\frac{17}{100}$. Thus 17% is the same as 0.17.

$$17\% = \frac{17}{100} = 0.17$$

To change a percentage to a decimal fraction, divide the percentage by 100.

For example:

$$16\% = 16 \div 100 = 0.16$$

Changing decimals to percentages

To change a decimal fraction to a percentage, multiply the decimal fraction by 100 and write it with the percentage symbol, %.

For example:

$$0.732 = (0.732 \times 100)\% = 73.2\%$$

Notice also:

$$0.\dot{3} = 0.33333... = (0.33333... \times 100)\%$$
$$= 33.333...\% = 33.\dot{3}\% = 33\tfrac{3}{9}\% = 33\tfrac{1}{3}\%$$

$$0.2\dot{7} = 0.2777...$$
$$= 27.77...\%$$
$$= 27.\dot{7}\%$$
$$= 27\tfrac{7}{9}\%$$

Exercise 9n (Oral)

1. Express the following percentages as decimals.

   (a) 5%   (b) 50%   (c) 25%
   (d) 40%   (e) 70%   (f) 4%
   (g) 15%   (h) 12%   (i) 80%
   (j) 75%   (k) 24%   (l) 35%
   (m) 64%   (n) 60%   (o) 45%

2. Express the following decimals as percentages.

   (a) 0.25   (b) 0.85   (c) 0.36
   (d) 0.81   (e) 0.08   (f) 0.125
   (g) 0.308   (h) 0.7   (i) 0.07
   (j) 0.\dot{6}   (k) 0.5   (l) 0.022
   (m) 0.005   (n) 0.8   (o) 0.1\dot{3}

To express one quantity as a fraction of another

Example 16

Express 5 g as a fraction of 20 g.

$$\frac{5\,g}{20\,g} = \frac{5}{20} = \frac{1}{4}$$

Example 17

Express 7 min 30 s as a fraction of 1 hour.

$$7 \text{ min } 30 \text{ s} = 7\tfrac{1}{2} \text{ min}$$
$$= 450 \text{ s}$$
$$1 \text{ h} = 60 \text{ min}$$
$$= 3600 \text{ s}$$

$$\frac{450\,s}{3600\,s} = \frac{9 \times 50}{72 \times 50} = \frac{9}{72} = \frac{1}{8} \quad *$$

or

$$\frac{7\tfrac{1}{2} \text{ min}}{60 \text{ min}} = \frac{15}{120} = \frac{1 \times 15}{8 \times 15} = \frac{1}{8} \quad *$$

*Notice that both quantities must be in the same units before fractions can be reduced.

Exercise 9o

1. What fraction of 1 minute is 15 seconds?

2. Express 25 cm as a fraction of 3 m.

③ What fraction is 500 g of 2 kg?

④ What fraction of 6 weeks is 6 days?

⑤ What fraction of $1 is 40c?

⑥ Express 40 min as a fraction of 1 hour.

⑦ Express 650 m as a fraction of 1 km.

⑧ What fraction is 4 min 30 s of 12 min?

⑨ Express 13 weeks as a fraction of 1 year.

⑩ What fraction is 4 mm of 10 cm?

To express one quantity as a percentage of another

Make sure that both quantities are in the same units. Express the first quantity as a fraction of the second. Find the equivalent fraction with 100 as the denominator. Simplify the numerator and write as a percentage, that is, with the % symbol.

For example, in Examples 16 and 17, to get the answers as percentages, we may work as follows:

$$\frac{5}{20} = \frac{5 \times 5}{20 \times 5} = \frac{25}{100} = 25\%$$

$$\frac{450}{3600} = \frac{450 \div 36}{3600 \div 36} = \frac{25 \div 2}{100} = 12\frac{1}{2}\%$$

$$\frac{15}{120} = 15 \times \frac{\frac{100}{120}}{120 \times \frac{100}{120}} = \frac{15 \times \frac{5}{6}}{100} = \frac{\frac{25}{2}}{100} = 12\frac{1}{2}\%$$

Example 18

A man hammers a post 2.2 m long into the ground. 66 cm of the post is below ground. What percentage of the post is above the ground?

2.2 m = 220 cm

Fraction of post below the ground

$$= \frac{66\,\text{cm}}{220\,\text{cm}} = \frac{66}{220} = \frac{66 \div 22}{220 \div 22} = \frac{3}{10}$$

$$\frac{3}{10} = \frac{3 \times 10}{10 \times 10} = 30\%$$

Percentage of post above ground
$$= 100\% - 30\% = 70\%$$

① Express the first quantity as a percentage of the second.
(a) 10c, $1   (b) 25c, $1
(c) 400 g, 1 kg   (d) 800 mm, 1 m
(e) 3 mm, 1 cm   (f) 500 mm, 4 m
(g) 45 g, 75 g   (h) 25 m, 1 km
(i) 30c, $1.50   (j) $1, $2.50

② In an exam a student scored 60 marks out of a possible 80. What percentage is this?

③ Three girls are missing from a class of 25 girls. What percentage of the class is missing?

④ 26 cm of wood is cut off a board 130 cm long. What percentage has been cut off?

⑤ In a box of 200 oranges, 18 are bad. What percentage is bad? What percentage is good?

⑥ A piece of elastic is 48 cm long. It is stretched to a length of 60 cm. What is the increase in length? What percentage of the original length is the increase?

⑦ A woman buys a 5 kg bag of flour. She uses 800 g to make some bread. What percentage of flour has she used and what percentage remains?

⑧ The distance between two villages is 8 km. A boy walks from one village to the other. He walks most of the way but runs the last 480 metres. What percentage of the journey did he run?

Summary

Percentages, fractions and decimals are all different ways of representing parts of quantities.

When adding or subtracting decimal numbers, it is essential to ensure that digits with the same place-value are collected together; this can be achieved by writing the numbers so that decimal points are written under one another.

When multiplying a number by powers of 10, the decimal point remains fixed and digits in the number move as many places to the **left** as the power of 10 in the multiplier.

When dividing a number by powers of 10, the decimal point remains fixed and digits in the number move as many places to the **right** as the power of 10 in the divisor.

When two numbers are multiplied together, the product has the same number of digits after the decimal point as the total number of digits after the point in the numbers being multiplied.

When dividing by a number in which there is a decimal fraction, multiply both the divisor and the dividend by the power of 10 which makes the divisor a whole number; then divide by the usual method.

Percentages, that is, numbers written with a % symbol, represent quantities that are hundredths of a whole; thus

$$16\% = \frac{16}{100} = 0.16; \quad 150\% = \frac{150}{100} = 1.5$$

Hence, it is possible to change from one form of representing a part of a quantity to another form.

Practice exercise P9.1

1. Arrange the following numbers in each set from smallest to largest:
   (a) 5.24   4.85   5.18   6.69   5.26
   (b) 6.8   6.08   8.06   8.6   0.68   0.86   0.08
   (c) 9.99   9.09   0.999 9   0.09   0.99
   (d) 0.13   0.31   0.83   0.38   0.1   0.18   0.81
   (e) 9.49   9.41   9.48   9.43   9.45   9.47   9.44
   (f) 0.086 32   0.069 41   0.088 03   0.600 9
       0.069 14   0.080 63   0.086 03

Practice exercise P9.2

In Monimanya they have three types of banknote. A **moni** is worth $1.69. A **manya** is worth $3.38. A **banyam** is worth $8.45.

1. In your pocket, you have 2 banyams, 2 manyas and 1 moni. How much is this in dollars?

2. (a) (i) How many monis would you get for $10?
       (ii) What would be left over?

(b) (i) How many manyas would you get for $10?
    (ii) What would be left over?
(c) (i) How many banyams would you get for $10?
    (ii) What would be left over?

Practice exercise P9.3

In each of these questions, work out the answers. Write the answers in order of size, smallest to largest.

1. (a) 6.2×5 (b) 8.3×4
   (c) 12.7×2 (d) 15.4×3
   (e) 2.6×8

2. (a) $3.7 \div 2$ (b) $11.7 \div 6$
   (c) $9.3 \div 5$ (d) $19.95 \div 7$
   (e) $7.48 \div 4$

3. (a) 5.62×23 (b) 7.42×17
   (c) 2.4×52 (d) $2824 \div 25$
   (e) $1755 \div 18$

Practice exercise P9.4

In each of these questions, two answers are the same and one is different. Underline the odd one out.

1. (a) $15.48 \div 18$ (b) $13.77 \div 17$
   (c) $15.39 \div 19$

2. (a) 0.28×35 (b) $245 \div 25$
   (c) $349.2 \div 36$

3. (a) 2.13×42 (b) $4032 \div 45$
   (c) 1.42×63

Practice exercise P9.5

Work out the following.

1. (a) $259 \div 1000$ (b) 0.063×100
   (c) $2.9 \div 1000$ (d) $700 \times 10\,000\,000$
   (e) $9200 \div 100\,000$ (f) $0.000\,057 \times 1000$

2. (a) 9×0.8 (b) 0.06×4
   (c) 0.3×0.09 (d) 0.008×7
   (e) 0.1×0.6 (f) 0.007×900
   (g) 0.007×0.04 (h) 5000×0.3

3 (a) $90 \div 0.3$ (b) $0.6 \div 2$
    (c) $4 \div 0.08$ (d) $0.8 \div 0.002$
    (e) $0.4 \div 80$ (f) $0.009 \div 300$
    (g) 0.04^2 (h) $0.000\,6^2$

Practice exercise P9.6

1 Fill in the missing numbers.

    (a) $0.2 \times 100 = 20$ (b) $0.3 \times 200 = 60$
        $0.2 \times ? = 2$ $0.3 \times 20 = ?$
        $0.2 \times ? = 0.2$ $0.3 \times ? = 0.6$
        $0.2 \times 0.1 = ?$ $0.3 \times ? = 0.06$
        $0.2 \times ? = 0.002$ $0.3 \times 0.02 = ?$
        $0.2 \times 0.001 = ?$ $0.3 \times ? = 0.000\,6$

2 Use the results of question 1 to help you to answer the following questions.

    (a) 0.2×0.4 (b) 0.5×0.3
    (c) 0.6×0.7 (d) 0.03×0.08
    (e) 0.02×0.05 (f) 0.001×0.3

3 Calculate 257×33, and use the result to find the value of:

    (a) 2.57×3.3 (b) 25.7×3.3
    (c) 25.7×0.33 (d) 0.257×33
    (e) 0.257×3.3 (f) 0.257×0.33

4 Calculate the following without using a calculator:

    (a) $43.028 + 2.3007$ (b) $34.503 - 3.4503$
    (c) 3.012×0.007 (d) $7.5012 \div 0.012$

Practice exercise P9.7

1 What is the cost of 0.7 kg of potatoes at 80¢ per kg?

2 14 motorcycle batteries weigh 47 600 g. How much does one battery weigh?

3 A book contains 200 pages and is 0.9 cm thick. What is the thickness of one of its pages?

4 A drum contains 1500 m of cable. What length of cable do 400 drums contain?

5 1 byte of data takes 0.000 02 seconds to be transmitted from the Internet to a computer. How many kilobytes (kb) would be transferred in 400 seconds?

6

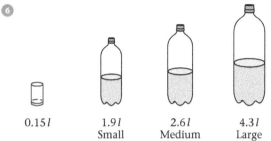

0.15 *l*   1.9 *l*   2.6 *l*   4.3 *l*
              Small   Medium   Large

A glass has a capacity of 0.15 litres. Juice is bottled in 3 different sizes.

(a) How many glasses can be filled from each size of bottle?

(b) How much juice is left over for each size of bottle?

(c) Darren uses the size of bottle which has least left over. How many bottles does he need to buy to fill 277 glasses? *Assume that he uses the leftovers to fill glasses too.*

Practice exercise P9.8

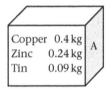 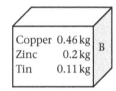

Copper	0.4 kg	A
Zinc	0.24 kg	
Tin	0.09 kg	

Copper	0.46 kg	B
Zinc	0.2 kg	
Tin	0.11 kg	

1 The diagram shows two blocks of metal.

(a) for block A, calculate:
    (i) the total weight of metal.
    (ii) how much more copper than zinc there is in 1 block
    (iii) how much copper is contained in 36 blocks

(b) for block B, calculate:
    (i) the total weight of metal
    (ii) how much metal is not zinc
    (iii) the total weight of 32 blocks.

2 Seven of block A are moulded into a model of the Statue of Liberty. Six of block B are moulded into a model of the Eiffel Tower. Compare the amounts of each metal (copper, zinc and tin) contained in the models.

         Mathematics for Caribbean Schools

③ A statue made from A blocks contains 3.06 kg of tin. Its base is made from B blocks containing 4.29 kg of tin. Which part used more blocks, the statue or the base? What was the difference?

Practice exercise P9.9

① 12 500 people voted in an election. Calculate how many votes went to each party.

Party	Percentages	Number of votes
Packers	48%	
Warriors	32%	
Lifers	17%	
Readers	3%	

② Complete the table.

	Percentages					
Amount (100%)	50%	25%	10%	30%	63%	8%
100 litres						
$20						
1300 dogs						
435 km						

③ Each year a family uses 270 000 litres of water. This table shows how they use it. How many litres are used for each activity?

Type of use	Percentage	Number of litres used
cooking	12%	
toilet	18%	
laundry	15%	
bathing	22%	
gardening	33%	

④ 20% of the cost of a certain product goes to charity. The product costs £35 to buy. How much does the charity receive for each product sold?

⑤ Find the new price when:
(a) a TV set costing $2500 is sold at a discount of 15%
(b) VAT at 15% is added to an item costing $120.

⑥ In a special offer, each bottle of washing-up liquid with an extra 10% contains 550 ml of liquid. What did the original bottle hold?

Practice exercise P9.10

Complete the table of decimal and percentage equivalents for the given fractions:

Fractions	$\frac{3}{5}$	$\frac{5}{8}$	$\frac{2}{3}$	$\frac{7}{10}$	$\frac{3}{4}$	$\frac{4}{5}$	$\frac{7}{8}$	$\frac{9}{10}$	$\frac{1}{1}$
Decimals									
Percentages									

Practice exercise P9.11

① Change these fractions into:
(a) percentages   (b) decimals.
(i) $\frac{9}{100}$ (ii) $\frac{63}{100}$ (iii) $\frac{7}{10}$
(iv) $\frac{2}{5}$ (v) $\frac{3}{20}$ (vi) $\frac{17}{25}$
(vii) $\frac{12}{30}$ (viii) $1\frac{11}{100}$ (ix) $1\frac{3}{10}$

② Change these percentages into:
(a) fractions   (b) decimals.
(i) 31%   (ii) 93%   (iii) 30%
(iv) 35%   (v) 64%   (vi) 140%

③ Express each of the following as:
(a) a fraction   (b) a percentage
(i) 0.5   (ii) 2.6   (iii) 19.6
(iv) 0.045   (v) 15.05

④ Arrange the following numbers from smallest to largest:
$\frac{3}{4}$ 60% $\frac{5}{8}$ $\frac{13}{10}$ 0.875 $\frac{7}{9}$ $\frac{2}{3}$ 2.5

Chapter 10

Measurement of angles

Pre-requisites
■ whole numbers; decimal numbers

Angles

We use the word **angle** for amount of turn, or **rotation**. For example, Fig. 10.1 shows how the hands of a clock move between 9 o'clock and 10 o'clock.

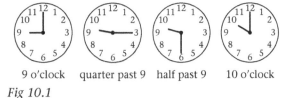

9 o'clock   quarter past 9   half past 9   10 o'clock

Fig 10.1

Both hands turn. In one hour the amount that each hand turns is different.

The minute hand makes one complete **turn**, or one **revolution** (Fig. 10.2). The angle turned = 1 revolution.

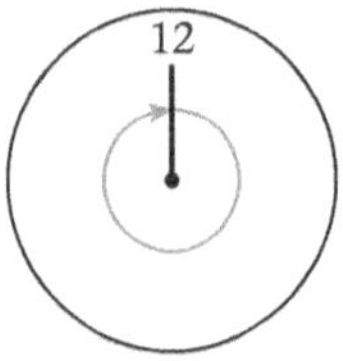

Fig 10.2

The hour hand turns through $\frac{1}{12}$ of a revolution (Fig. 10.3). The angle turned $= \frac{1}{12}$ revolution.

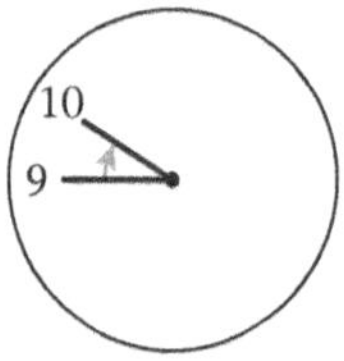

Fig 10.3

Just as we can measure length, that is, the distance along a straight line, so we can measure angle, that is, the amount of turn about a point. To avoid fractions, one revolution is divided into 360 equal parts. Each part is called a **degree**. We use the symbol ° for degree.

1 revolution = 360 degrees or 360°

$1° = \frac{1}{360}$ revolution

Angles between lines

Figs. 10.4 and 10.5 show two lines OA and OB meeting at the point O. As OB turns about the fixed point O, the amount of turn, the angle, increases in the direction of the arrow.

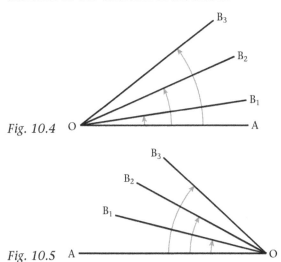

Fig. 10.4

Fig. 10.5

Notice that the line OA remains fixed while OB turns.

Example 1

What is the angle between the hour hand and the minute hand of a clock at 1 o'clock?

Fig. 10.6

The angle between the hands is the amount that one hand must turn to reach the position of the other.

Angle between hands $= \frac{1}{12}$ of a revolution
$$= \frac{1}{12} \text{ of } 360° = 30°$$
Notice, in the previous example, that the size of the angle does not depend on the size of the hands. At 1 o'clock the angle between the hour hand and the minute hand is 30° whether on a watch or on a clock (Fig. 10.7).

watch   clock

Fig. 10.7

Exercise 10a

❶ Copy and complete Table 10.1 ('revs' is short for revolutions).

Table 10.1

Revs	Degrees	Revs	Degrees
1	360°	$\frac{1}{3}$	
2			90°
	1080°	$\frac{1}{10}$	
10		$\frac{1}{8}$	
	180°	$1\frac{1}{2}$	

❷ Find the angles between the hour hand and the minute hand of a clock at the following times. Give your answers both in revolutions and in degrees.
(a) 2 o'clock   (b) 3 o'clock
(c) 4 o'clock   (d) 5 o'clock
(e) 6 o'clock   (f) 7 o'clock
(g) 9 o'clock   (h) 10 o'clock
(i) 11 o'clock

❸ Where does the hour hand of a clock point at half past two?
What is the angle between the hour hand and the minute hand of a clock at half past two?

❹ Find the angles between the hour hand and the minute hand of a clock at the following times. Give your answers in degrees.

(a) half past three   (b) half past ten
(c) half past eight   (d) half past twelve

❺ Where does the hour hand of a clock point at quarter past two?
What is the angle between the hour hand and the minute hand of a clock at quarter past two?

❻ Find the angles between the hour hand and the minute hand of a clock at the following times. Give your answers in degrees.
(a) quarter past eleven
(b) quarter past nine
(c) quarter to nine
(d) quarter past three

Naming angles

When lines OA and OB meet at the point O (Fig. 10.8), we say that angle AOB or angle BOA is the angle between them. AÔB is short for angle AOB. Notice that O, the middle letter, is the **vertex** of the angle. The lines OA and OB are the **arms** of the angle.

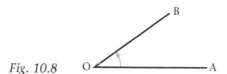

Fig. 10.8   O

Kinds of angles

Angles are often classified according to their size. In Fig. 10.9, AÔB = 90°. AOB is a **right angle** or **quarter turn**.

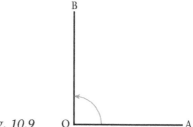

Fig. 10.9   O

The right angle is especially important. There are many examples where lines meet at right angles. A right angle is often shown on a diagram by drawing a small square at the vertex of the angle (Fig. 10.10).

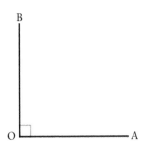

Fig 10.10

In Fig. 10.11, AÔB is less than 90°. AÔB is an **acute angle**.

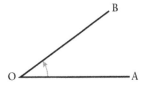

Fig 10.11

In Fig. 10.12, AÔB is greater than 90° but less than 180°. AÔB is an **obtuse angle**.

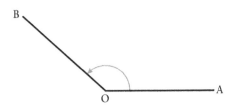

Fig 10.12

In Fig. 10.13, AOB = 180° = 2 right angles. AÔB is a **straight angle** or **half turn**.

Fig 10.13

In both Figs. 10.14 and 10.15, AÔB is less than 360° but greater than 180°. AÔB is a **reflex angle**.

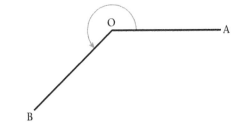

Fig 10.14

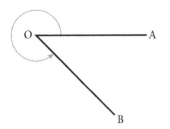

Fig 10.15

Measuring angles

Angles are usually measured to the nearest degree. We use a **protractor** to measure the number of degrees in an angle. There are many kinds of protractor; two are shown in Fig. 10.16.

(a)

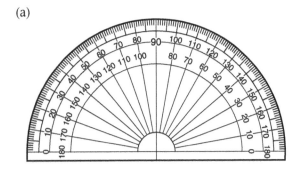

(b)

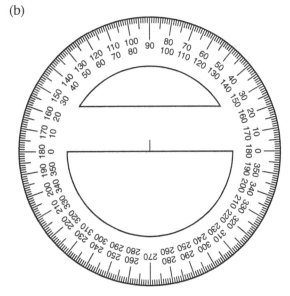

Fig 10.16

Look carefully at your semi-circular protractor. Notice that there are two rows of numbers: the inner row reading from 0 on the right and the outer row reading from 0 on the left. When measuring an angle you must decide whether the size of the angle is increasing from the right or from the left and use the appropriate row of numbers.

To measure an angle:

(a) place the protractor over the angle so that its centre is exactly over the vertex of the angle, O, and the base line is exactly along one arm of the angle;

(b) count the degrees *from* the base line *to* the other arm of the angle (Figs. 10.17(a) and 10.17(b)).

(a)

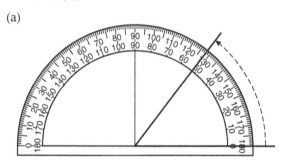

(b)

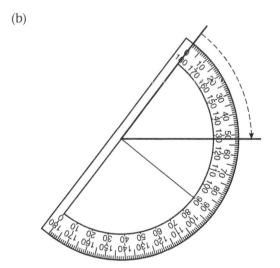

Notice that you read from 0, on the right in Fig. 10.17(a) and on the left in Fig. 10.17(b), in the direction of the arrow. The size of this angle is 52°.

Example 2

Measure the size of the obtuse angle $P\widehat{O}Q$ in Fig. 10.18. Calculate the size of reflex angle $P\widehat{O}Q$.

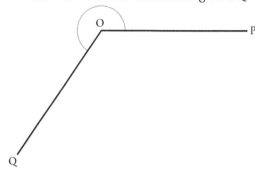

Fig. 10.18

Place the protractor on the angle with its centre on O and its base line on OP as shown in Fig. 10.19.

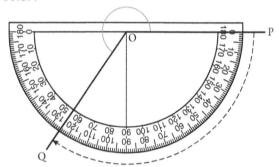

Fig. 10.19

obtuse $P\widehat{O}Q$ = 125° (by measurement)
reflex $P\widehat{O}Q$ = 360° − 125° = 235°

Exercise 10b (Oral)

1 In Fig. 10.20,
   (a) name all the acute angles (5 of them),
   (b) name all the obtuse angles (3 of them),
   (c) name all the reflex angles (8 of them).

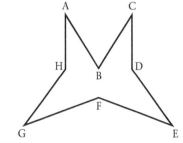

Fig. 10.20

2 In Fig. 10.21, 12 angles have been marked using small letters of the alphabet. State whether each is an acute, right, obtuse, straight or reflex angle.

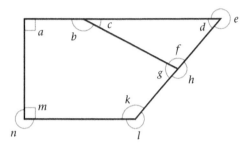

Fig. 10.21

3 Read the sizes of the angles in Fig. 10.22 (a)−(h).

(a)
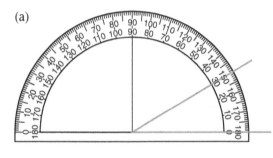

(b)
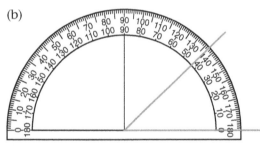

(c)
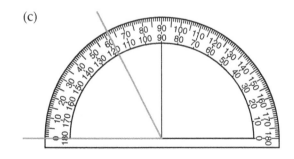

(d)

(e)

(f)

(g)

(h)
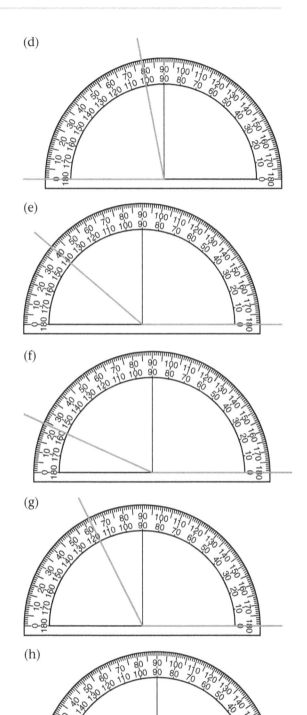

Fig. 10.22

Exercise 10c

You will need a protractor, a ruler and a sheet of thin white paper. Place the sheet of paper over angles 1–10 in Fig. 10.23. Trace the angles on to the paper. Measure each angle (make the arms of the angle longer if necessary).

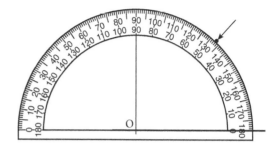

Fig. 10.25

Remove the protractor. Join the mark to O with a ruler (Fig. 10.26).

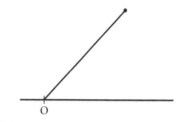

Fig. 10.26

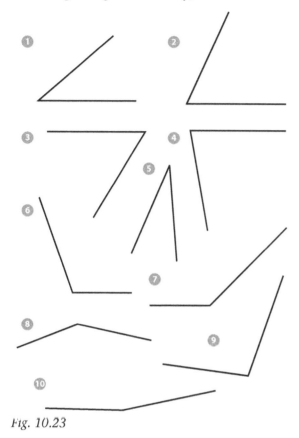

Fig. 10.23

Drawing angles

Example 3

Draw an angle of 47°.

Draw a line and mark a point O on it (Fig. 10.24). Place a protractor so that its centre is over O and its base line is exactly over the line already drawn. Count round until 47° is reached. Make a mark on the paper opposite 47° (Fig. 10.25).

Fig. 10.24

Exercise 10d

1 Use a protractor to draw angles of
   (a) 40°   (b) 65°
   (c) 74°   (d) 130°
   (e) 105°   (f) 143°

2 Use a protractor to draw reflex angles of
   (a) 300°   (b) 285°
   (c) 215°   (d) 238°

   (*Hint*: Subtract the given angles from 360° to get acute or obtuse angles.)

3 Copy Fig. 10.27. Make $A\widehat{C}D = 68°$. Measure $B\widehat{C}D$. What do you notice?

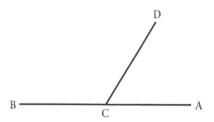

Fig. 10.27

Degrees and minutes

While angles are usually measured to the nearest degree, it is possible to calculate with angles which contain fractions of a degree. There are 60 **minutes** in 1 degree. (These are different minutes from those used for measuring time.)

In symbols: $1° = 60'$

Example 4

Find the value of (a) $15° 56' + 8° 23'$,
    (b) $26° 48' - 15° 59'$.

(a)
$$\begin{array}{r} 15° \quad 56' \\ + \; 8° \quad 23' \\ \hline 24° \quad 19' \end{array}$$

method: $56' + 23' = 79' = 1° 19'$
    Write down 19' and carry 1°

(b)
$$\begin{array}{r} 26° \quad 48' \\ + \; 15° \quad 59' \\ \hline \end{array} \quad \rightarrow \quad \begin{array}{r} 25° \quad 108' \\ + \; 15° \quad 59' \\ \hline 10° \quad 49' \end{array}$$

method: $59'$ cannot be subtracted from $48'$; write $26° 48'$ as $25°$ and add $60'$ to $48'$ giving $25° 108'$ in the top line.

With practice it is not necessary to write down every step when adding or subtracting degrees and minutes.

Example 5

(a) Change $26.8°$ to degrees and minutes.
(b) Express $53° 27'$ as a decimal number of degrees.

(a) $26.8° = 26° + 0.8°$
        $= 26° + 0.8 \times 60'$
        $= 26° + 48'$
        $= 26° 48'$

(b) $53° 27' = 53° + \frac{27°}{60}$
        $= 53° + 0.45°$
        $= 53.45°$

Thus:
(a) to change degrees to minutes, multiply by 60,
(b) to change minutes to degrees, divide by 60.

❶ Change the following to minutes.
    (a) $2°$ (b) $3\frac{1}{2}°$ (c) $5°$
    (d) $8\frac{1}{4}°$ (e) $22\frac{1}{2}°$ (f) $90°$

❷ Change the following to degrees.
    (a) $180'$ (b) $90'$ (c) $240'$
    (d) $20'$ (e) $600'$ (f) $450'$

❸ Find the value of the following.
    (a) $28° 22' + 42° 31'$
    (b) $36° 42' + 18° 53'$
    (c) $44° 43' - 21° 18'$
    (d) $65° 11' - 58° 32'$
    (e) $18° 44' \times 3$
    (f) $25° 52' \div 4$

Exercise 10f

❶ Say whether each of the following angles is acute, obtuse or reflex.
    (a) $93°$ (b) $175°$ (c) $86°$
    (d) $191°$ (e) $347°$ (f) $28°$
    (g) $79°$ (h) $112°$ (i) $63°$
    (j) $156°$ (k) $211°$ (l) $167°$
    (m) $183°$ (n) $72°$ (o) $98°$

❷ Estimate the sizes of the angles in Fig. 10.28 to the nearest $10°$.

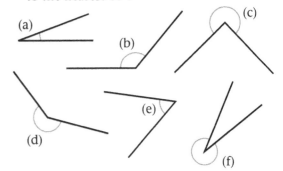

Fig. 10.28

❸ Use the method of Exercise 10c to trace the angles in question 2 on to a piece of paper. Measure each angle. How good were your estimates?

❹ *Without using a protractor*, try to draw angles of the following sizes.
    (a) $90°$ (b) $45°$ (c) $80°$
    (d) $100°$ (e) $120°$

⑤ Measure the angles you drew in question 4. How accurate were your drawings?

⑥ Convert the following into minutes.
0.1°, 0.2°, 0.3°, 0.4°, 0.5°, 0.6°, 0.7°, 0.8°, 0.9°

⑦ Convert the following into decimal parts of a degree.
(a) 6′   (b) 48′   (c) 15′
(d) 21′   (e) 33′   (f) 57′

⑧ Change the following into degrees and minutes.
(a) 18.2°   (b) 77.7°   (c) 45.75°
(d) $67\frac{1}{2}°$ (e) $28\frac{2}{3}°$ (f) $32\frac{1}{6}°$

⑧ Express the following as a decimal number of degrees.
(a) 51° 36′   (b) 13° 54′
(c) 32° 15′   (d) 80° 39′

⑩ Draw any large triangle. Measure its three angles. Find the sum of the three angles of the triangle. What do you notice?

Summary

Angle is a measure of **rotation**. There are four right angles, or 360°, in one full **revolution**. Angles may be **acute** (less than 90°), **right** (90°), **obtuse** (between 90° and 180°), **straight** (180°) or **reflex** (more than 180°). Angles are measured using a **protractor**.

Practice exercise P10.1

Remember when measuring or drawing angles to decide whether the angle is acute or obtuse.

① Estimate the size of each angle in Fig.10.29.
(a)   (b)

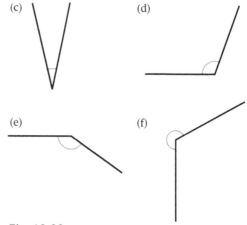

(c)   (d)

(e)   (f)

Fig. 10.29

② Measure all the following angles in Fig. 10.30 and say whether the angle is acute or obtuse.

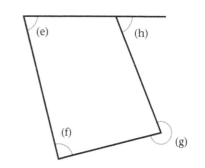

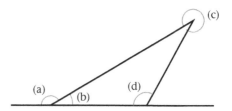

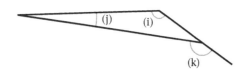

Fig. 10.30

Chapter 11

Simplifying algebraic expressions (1)
Like terms and unlike terms, substitution

Pre-requisites
- basic arithmetic operations; basic algebraic processes; directed numbers

Grouping positive and negative terms

Expressions such as $3x$, x, $8x$, $12x$, $\frac{1}{2}x$ are called **terms** in x. We can add terms in x together. $3x + 2x$ means 3 'x's plus 2 'x's. This gives 5 'x's altogether.

Thus: $3x + 2x = 5x$; $5x$ uses less space, or is simpler, than $3x + 2x$. Thus two terms in x have been **simplified** to one term in x.

We can also subtract terms. $7y - 4y$ means 7 'y's minus 4 'y's. This leaves 3 'y's. Thus $7y - 4y = 3y$; again, two terms have been simplified to one term.

Exercise 11a (Oral)

Simplify the following.

①	$2a + 3a$	②	$5x - 3x$
③	$4b + 4b$	④	$6y - 2y$
⑤	$4c + c$	⑥	$7z - z$
⑦	$p + 2p$	⑧	$8k - 7k$
⑨	$4q + 6q$	⑩	$r + r$
⑪	$3n - 3n$	⑫	$8x + x + 3x$
⑬	$5y + 13y + 2y$	⑭	$z + 9z + z$
⑮	$4a + 5a + a$		

It is possible to simplify expressions which contain many terms. For example, the expression, $3a - 8a + 5a + 9a - 2a$ means: add $3a$, $5a$ and $9a$ together, take away $8a$ and $2a$. This gives $17a$ take away $10a$. The result is $7a$. We can write this as follows.

$$3a - 8a + 5a + 9a - 2a$$
$$= 3a + 5a + 9a - 8a - 2a$$
$$= 17a - 10a$$
$$= 7a$$

The first line in this example is important. We have grouped all the terms with a + sign before them together, and all the terms with a − sign before them together. This method is called **grouping positive and negative terms**. It is usually the best way of simplifying big expressions.

Example 1

Simplify $3m - 8m - 2m + 16m - 4m$.

$$3m - 8m - 2m + 16m - 4m$$
$$= 3m + 16m - 8m - 2m - 4m$$
$$= 19m - 14m$$
$$= 5m$$

Exercise 11b

Simplify the following. Set out your work as in Example 1 above.

① $10b + b - 3b$

② $3c + 5c - 8c$

③ $7d + 4d - 8d$

④ $16e + 12e - 13e$

⑤ $6f - 2f + 3f$

⑥ $4g - 7g + 10g$

⑦ $20h - 15h + 6h$

⑧ $2j - 9j + 9j$

⑨ $11k - 16k + 9k$

⑩ $8m - 3m - 2m$

⑪ $3t + 6t + t + t$

⑫ $4u + u + 6u - 5u$

⑬ $9v + 2v + 3v - 8v$

⑭ $6w + 4w - 3w - 2w$

⑮ $12x + 3x - 5x - 2x$

⑯ $3y - 5y + 10y - 4y$

⑰ $7z - 2z + 7z - 2z$

⑱ $14n + 2n - 6n - 12n + 7n$

⑲ $p - 15p - 5p - 2p + 25p$

⑳ $2q - 20q + 11q + 8q + 5q$

Grouping like and unlike terms

What is the sum of 5 potatoes and 8 potatoes?
13 potatoes. Similarly, in algebra, $5p + 8p = 13p$.
What is the sum of 3 potatoes and 7 torches?
All that can be said is that there is a mixture of
potatoes and torches.

Similarly, in algebra, $3p + 7t$ cannot be simplified.
$5p$ and $8p$ are **like terms**. Their sum is $13p$. $3p$ and
$7t$ are **unlike terms**. Their sum is $3p + 7t$.
Notice that:

3 potatoes + 8 torches + 6 potatoes + 5 torches
    = 9 potatoes + 13 torches

Just as real things can be grouped, so, in algebra,
like terms can be grouped:

$3p + 8t + 6p + 5t = 3p + 6p + 8t + 5t$
$$= 9p + 13t$$

Example 2

Simplify $7h - 4 - 3h + 11$.

Notice that terms in h and number terms are
unlike.
$7h - 4 - 3h + 11 = 7h - 3h + 11 - 4$
$$= 4h + 7$$

The number term 7 is called a **constant**, since its
value is the same for all values of h.

Example 3

Simplify $8x - 6y - 9y - 2x$.

$8x - 6y - 9y \quad 2x - 8x - 2x - 6y - 9y$
$$= 6x - 15y$$

Example 4

Simplify $10r - 3r - 8 - 4r$.

Notice that there are three terms in r.
$10r - 3r - 8 - 4r = 10r - 3r - 4r - 8$
$$= 10r - 7r - 8$$
$$= 3r - 8$$

Exercise 11c

Simplify the following. Set your work out as in
the examples above.

1. $2x + 3x + 7$
2. $6x - 3x + 2$
3. $5x + 6x + 8y$
4. $10x - 4x + 1$
5. $4x + x - 2$
6. $7x - 2x - 2y$
7. $x + 2 + 9x$
8. $5x - 8 - 4x$
9. $6y + 7y + 12x$
10. $8y - 7y - x$
11. $14x + 14 + 3$
12. $3x + 11 - 2$
13. $7y - 3x + 4y$
14. $14y - 9x - 6y$
15. $6a + 2a + 3 + 10$
16. $8a - 5a + 14 - 6$
17. $3x + 8x + 9y - 4y$
18. $11x - 2x + 3y + 5y$
19. $4a - a + 5b - 4b$
20. $7m + 3 + 2m + 1$

Exercise 11d (Further practice)

Simplify the following *where possible*.

1. $2x + 7x$
2. $2x + 6y$
3. $8n - 5n$
4. $14k - 6k$
5. $8e + 5e - 7e$
6. $4g - g + 3g$
7. $15a - 2$
8. $31p - 9q$
9. $28f - 11$
10. $4c - 9c + 15c - 7c$
11. $3m + 5n + 4m + 6n$
12. $9a + 10b - 5a - 4b$
13. $8h - 3 - 5h + 9$
14. $7x + 5y - 4$
15. $11x + 9y - 7x$
16. $13a + 2b - 9a + 7b$
17. $5f - 3g - 7f + 9g + 3f - 4g$
18. $7m - 2n + 6 - 5m + 7n + 3$
19. $a - 2b - 4a + 3c + 4a + 5b$
20. $6p - 2q + 4r - 2p + 3s + 5q$

Example 5

A full matchbox contains x matches. There are three full matchboxes and one matchbox containing y matches. How many matches are there altogether?

No. of matches in three full boxes = $3x$
No. of matches in other box = y
Total no. of matches = $3x + y$

Example 6

How many packets of pins at 15 cents per packet can I buy for x dollars?

Cost of 1 packet = 15 cents
x dollars = $100x$ cents
No. of packets for $100x$ cents = $\dfrac{100x}{15} = \dfrac{20x}{3}$

Exercise 11e

1. A carpenter has two planks of wood. One plank is x cm long, the other plank is $3x$ cm long. What is the total length of the two planks?

2. A shop sells x books at \$2 each, y books at \$3 each and z books at \$5 each. How many books are sold? How much money is paid for the books?

3. 180 people attend a meeting. There are n men and n women and the rest are children. How many children are at the meeting?

4. A man earns \$85 per week. Each week he spends \$3x on food and \$2x on rent. How much money is he left with each week?

5. A man weighs k kg. He goes on a diet and loses one-tenth of his weight. What is his weight after going on the diet?

6. A man has $6x$ sheep and $5y$ goats. He sells $3x$ goats and $2y$ sheep. How many animals is he left with?

7. A man has x matches in a matchbox. On Monday he uses one-quarter of them. On Tuesday he uses 5 more. How many matches has he used? How many matches has he left?

8. A girl gets n dollars pocket money each week. She saves her money for 5 weeks and buys a present for her mother which costs 79 dollars. How much money has she left?

9. A woman has x eggs in a basket. She sells one-third of them. Two of the eggs get broken. How many eggs are left in the basket?

10. Table 11.1 shows the numbers of 'goals for' and 'goals against' a team during four weeks of a month.
In the month calculate the number of
(a) 'goals for'
(b) 'goals against'.

Table 11.1

	Goals for	Goals against
week 1	x	$2y$
week 2	$2x$	y
week 3	$3y$	y
week 4	y	x

Multiplication of algebraic terms

(a) Just as $5a$ is short for $5 \times a$, so ab is short for $a \times b$.

(b) Just as $3 \times 5 = 5 \times 3 = 15$
so $a \times b = b \times a = ab$.
It is usual to write the letters in alphabetical order, but ba would be just as correct as ab.

(c) Just as 5^2 is short for 5×5, so a^2 is short for $a \times a$ and x^3 is short for $x \times x \times x$.

(d) $4x + 4x + 4x = 12x$
$\qquad 3 \times 4x = 12x$
and
$\qquad 3x + 3x + 3x + 3x = 12x$
$\qquad\qquad 4 \times 3x = 12x$
Thus: $3 \times 4x = 4 \times 3x = 12x$
The terms 3, 4 and x can be multiplied in any order:
$3 \times 4x = 4 \times 3x = 3x \times 4 = 4x \times 3$
$\qquad = 4 \times x \times 3 = x \times 3 \times 4 = 12x$
It is usual to write the numbers before the letters. Remember that these numbers are called the coefficients of the terms.

Example 7

Simplify	Working	Result
$2x \times 3$	$= 2 \times x \times 3$ $= 2 \times 3 \times x$ $= 6 \times x$	$= 6x$
$5 \times 2y$	$= 5 \times 2 \times y$ $= 10 \times y$	$= 10y$
$7a \times 3b$	$= 7 \times a \times 3 \times b$ $= 7 \times 3 \times a \times b$ $= 21 \times ab$	$= 21ab$
$6x \times 4x$	$= 6 \times x \times 4 \times x$ $= 6 \times 4 \times x \times x$ $= 24 \times x^2$	$= 24x^2$
$5 \times 6ab$	$= 5 \times 6 \times ab$ $= 30 \times ab$	$= 30ab$
$8ab \times 7a$	$= 8 \times a \times b \times 7 \times a$ $= 8 \times 7 \times a \times a \times b$ $= 56 \times a^2 \times b$	$= 56a^2b$
$y \times 11xy$	$= y \times 11 \times x \times y$ $= 11 \times x \times y \times y$ $= 11 \times xy^2$	$= 11xy^2$

Exercise 11f

Simplify the following.

1. $5 \times a$
2. $x \times 4$
3. $x \times y$
4. $y \times x$
5. $a \times a$
6. $1 \times x^2$
7. $2a \times 3$
8. $3 \times 2a$
9. $3a \times 2$
10. $2 \times 3a$
11. $4x \times 7$
12. $5 \times 8n$
13. $2 \times x^2$
14. $3y^2 \times 4$
15. $16 \times 2x^2$
16. $4ab \times 5$
17. $7 \times 5pq$
18. $9ab \times 3$
19. $6x \times x$
20. $y \times 8y$
21. $3x \times x$
22. $p \times 2q$
23. $6a \times b$
24. $m \times 7n$
25. $4a \times 3a$
26. $5n \times 7n$
27. $3x \times 10x$
28. $9n \times 4n$
29. $10q \times 5p$
30. $4a \times 5b$
31. $4ab \times 7a$
32. $3b \times 11ab$
33. $2xy \times 9y$
34. $6y \times 5xy$
35. $14pq \times p$
36. $3a \times 8ab$

Division of algebraic terms

In algebra, letters stand for numbers. Just as fractions can be reduced to their lowest terms by equal division of the numerator and denominator, so a letter can be divided by the same letter. For example $x \div x = 1$, just as $3 \div 3 = 1$.

Example 8

Simplify	Working	Result
$\dfrac{14a}{7}$	$= \dfrac{7 \times 2a}{7} = \dfrac{1 \times 2a}{1}$	$= 2a$
$\frac{1}{3}$ of $36x$	$= \dfrac{36 \times x}{3} = \dfrac{3 \times 12x}{3}$ $= \dfrac{1 \times 12x}{1}$	$= 12x$
$\frac{1}{5}$ of y	does not simplify	$\frac{1}{5}y$ or $\dfrac{y}{5}$
$5ab \div a$	$= \dfrac{5 \times a \times b}{a}$ $= \dfrac{5 \times 1 \times b}{1}$ $= 5 \times b$	$= 5b$
$\dfrac{6xy}{2y}$	$= \dfrac{6 \times x \times y}{2 \times y} = \dfrac{3 \times x \times 1}{1 \times 1}$ $= 3 \times x$	$= 3x$
$x^2 \div x$	$= \dfrac{x \times x}{x} = \dfrac{1 \times x}{1}$	$= x$
$24x^2y \div 3xy$	$= \dfrac{24 \times x^2 \times y}{3 \times x \times y}$ $= \dfrac{3 \times 8 \times x \times x \times 1}{3 \times 1 \times 1} = 8 \times x$	$= 8x$

Exercise 11g

Simplify the following.

1. $\dfrac{6a}{3}$
2. $12a \div 2$
3. $\frac{1}{4}$ of $24x$
4. $18y \div 6$
5. $\frac{1}{5}$ of $15x$
6. $\dfrac{32c}{8}$
7. $\frac{1}{8}$ of $32x$
8. $\dfrac{21y}{3}$
9. $\frac{1}{7}$ of $35y$
10. $\frac{1}{9} \times x$
11. $\frac{1}{4}$ of d
12. $x \times \frac{1}{2}$

⑬ $28ab \div 4$ ⑭ $16xy \div x$

⑮ $17mn \div n$ ⑯ $22kl \div 11$

⑰ $d^2 \div d$ ⑱ $a \div a^2$

⑲ $\dfrac{z^2}{z}$ ⑳ $\dfrac{3x^2}{x}$

㉑ $6x^2 \div x$ ㉒ $\dfrac{7x^3}{x}$

㉓ $\dfrac{5x^3}{x^2}$ ㉔ $\dfrac{12x^2}{3x}$

㉕ $\dfrac{18a^3}{3a}$ ㉖ $\dfrac{33mn}{3m}$

㉗ $\dfrac{42xy}{7y}$ ㉘ $\dfrac{72a^2b}{8a}$

㉙ $\dfrac{48x^2y}{12xy}$ ㉚ $\dfrac{40pq^2}{8pq}$

Substitution

Example 9

Find the value of (a) $4x$, (b) $5y - xy$
when $x = 2$ and $y = 3$.

(a) Substitute the value 2 for x, i.e. use the value
2 instead of x.
When $x = 2$, $4x = 4 \times x$
$\qquad = 4 \times 2 = 8$
(b) $5y - xy = 5 \times y - x \times y$
When $x = 2$ and $y = 3$,
$5y - xy = 5 \times 3 - 2 \times 3$
$\qquad = 15 - 6 = 9$

Exercise 11h

① Find the value of the following when $a = 1$,
$b = 2$ and $c = 3$.
(a) $5a$ (b) $9c$
(c) $a + c$ (d) $b - a$
(e) ab (f) $ac - b$
(g) $3b - 2c$ (h) $4c - 3b$

② Find the value of the following when $x = 4$,
$y = 5$ and $z = 3$.
(a) $2x$ (b) $10z$
(c) xy (d) xz
(e) z^2 (f) xyz
(g) $x + z$ (h) $xz + y$
(i) $yz - 2x$ (j) $\dfrac{y + z}{x}$
(k) $\dfrac{x + z}{5y - x}$ (l) $\dfrac{2x}{2y + 2z}$

Order of operations

The value of $17 - 5 \times 2$ is 7
but
the value of $(17 - 5) \times 2$ is 24.

Remember the following rules:
(a) if there are no brackets, do multiplication
and division before addition and
subtraction;
(b) if there are brackets, simplify the terms
inside the brackets first.

Example 10

Find the value of $16 \times 2 - 3 + 14 \div 7$.

$16 \times 2 - 3 + 14 \div 7 = 32 - 3 + 2$
$\qquad\qquad\qquad = 34 - 3 = 31$

but note that
$16 \times (2 - 3) + 14 \div 7 = 16 \times (-1) + 14 \div 7$
$\qquad\qquad\qquad = -16 + 2 = -14.$

Example 11

Simplify $7 \times 3a - (3a + 5a) \times 2$.

$7 \times 3a - (3a + 5a) \times 2 = 7 \times 3a - 8a \times 2$
$\qquad\qquad\qquad = 21a - 16a = 5a$

Exercise 11i

Find the value of the following.

① $18 - 6 \times 2$ ② $12 \div 4 + 2$

③ $6 - 18 \div 3$ ④ $4 \times 5 + 8 \div 4$

⑤ $16 \div 2 - 3 \times 2$ ⑥ $7 \times 3 + 27 \div 9$

⑦ $(5 + 3) + 3 \times 5$ ⑧ $5 + (3 + 3) \times 5$

⑨ $4 \times 6 - (7 - 3)$ ⑩ $8 \times 3 - 17 + 15 \div 5$

Exercise 11j

Simplify the following as far as possible.

① $3x \times 2 + 5x$ ② $4 \times 5p - 3p$

③ $(4n + 3n) \times 10$ ④ $2 \times 7b - 3$

⑤ $6 + 2 \times 5m$ ⑥ $4a \times (3 + 5)$

⑦ $(17y - 5y) \times 2$ ⑧ $21a - 2 \times 7a$

⑨ $3x + 8x \div 2$ ⑩ $n + 12n \div 4$

⑪ $15 \div 5 + 6y$

⑫ $7x \div x + 5$

⑬ $5a + 21a \div 7$

⑭ $4x - 6 \div 3$

⑮ $1 - x \times 0$

⑯ $6a \times (4 - 2) \times 7a$

⑰ $3 \times 5x + 4x \div 2$

⑱ $4 \times 8x + 7x \times 3$

⑲ $24x \div 6 + x + 1 \times 5x$

⑳ $5 \times (6x - 4x) \times 0 - 7x \times 4$

Summary

In a simple algebraic expression, the coefficient and/or letter(s) constitute a **term**. Terms are connected by plus or minus signs. Hence, the algebraic expression

$$4c + \frac{y}{3} - 5ab$$

has three terms. There are four terms in

$$4c + \frac{y}{3} - 5ab - 12.$$

When a term is a number only (e.g. 12) that term is called a **constant**.

A very useful strategy in simplifying an algebraic expression, that is, in making the expression easier to work with, is the method of **grouping**. Terms are grouped when they are all of the same kind; these are called **like terms**.

Terms may be grouped so that all positive terms are collected, and then all negative terms; the resulting positive and negative terms are combined according to the rules of arithmetic. The order of operations relating to multiplication and division, and brackets, is the same as in arithmetic.

Practice exercise P11.1

(a) Simplify the following algebraic expressions.

(b) Write the coefficient of each result.

① $3x + 4x + x$

② $5x - x + 2x$

③ $x + 2x - 3x - 4x$

④ $2x - x - 7x - x$

⑤ $5x + x - 4x + x$

⑥ $3x + \frac{2x}{3} + x - \frac{x}{3}$

⑦ $\frac{7}{8}x + 2x + \frac{5}{8}x$

⑧ $5 - \frac{1}{4}x + 3\frac{1}{2}x$

Practice exercise P11.2

Simplify the following expressions.

① $2a + \frac{1}{2}a$

② $-4x - 2\frac{3}{4}x$

③ $\frac{1}{3}d + 3d$

④ $-3c + \frac{1}{8}c$

⑤ $4\frac{3}{8}b - 2b$

⑥ $\frac{13x}{5} - \frac{2x}{5}$

⑦ $7x + 2x - \frac{2}{3}x$

⑧ $-\frac{5}{8}y + 4y$

⑨ $\frac{3t}{2} + \frac{4t}{5}$

⑩ $\frac{12p}{7} - \frac{5p}{7}$

⑪ $\frac{1}{3}u - \frac{17}{3}u$

⑫ $-3x + \frac{14x}{3}$

Practice exercise P11.3

Simplify the following expressions.

① $6a + 3 - 2a$

② $9a - 2b - 4a + b$

③ $c + 7 - 2c - 4b - 3c$

④ $3k + 3m - 2 - 6k - 3m - 2n$

⑤ $\frac{1}{2}f - \frac{2}{3}g + \frac{1}{4}f + \frac{1}{3}g + 2$

⑥ $3x - y + 2y + x$

⑦ $h - 2k + 3 - k + 2h - 1$

Practice exercise P11.4

Write each of the following statements using the letters and numbers shown in bold.

① The wages Bill earns if he is paid $7 for each hour he works.

② The number of stamps Ann has if 45 are added to her collection.

③ The length left on a carpet when 5 m is cut off.

④ The money raised in a sponsored swim at $10 for each length.

⑤ The number of 5 m pieces that can be cut from a length of carpet.

⑥ The total time on some cassettes of 90 minutes each.

⑦ The length remaining on a 20 m hose when a piece has been cut off.

⑧ The number of cans if there is the same number on each of 5 shelves.

⑨ The sale price of a radio with $35 off the usual price.

⑩ The weight of an adult and a child is 120 kg.

⑪ The total length of a car and a truck.

⑫ The distance around the sides of a square is 4 times the length of a side.

Practice exercise P11.5

Write each of the following as an algebraic expression. Select your own letters and make it clear what the letters mean.

① The total cost of a number of books if they are $8 each.

② The total weight of books weighing 250 g each.

③ The amount each person gets when a prize is shared equally between 10 people.

④ The total length of a train with 8 carriages.

⑤ The sale price of a pair of shoes with $20 off the usual price.

Practice exercise P11.6

If $a = 5$, $b = 3$ and $c = 2$, find the value of:

① $a + b$

② $a - b$

③ $2a + 3c$

④ $2b + 4a$

⑤ $\dfrac{3a - 2b}{3c}$

⑥ $\dfrac{4a - 5c}{12}$

⑦ $\dfrac{3c}{a + 3b}$

⑧ $\dfrac{9a - 4b + 3c}{a + 2c}$

Practice exercise P11.7

If $m = -2$, $n = -1$ and $p = 4$, find the value of:

① $m + n$

② $n \times m + p$

③ $4n \div 3p$

④ $2p - 4mn \times p$

⑤ $\dfrac{3n - 2m}{4p}$

⑥ $\dfrac{4m + 5p}{2m}$

⑦ $\dfrac{2p}{p + 2m}$

⑧ $\dfrac{9n - 4m + 3p}{m + n}$

Practice exercise P11.8

Simplify the following expressions:

① $4uv - vu$

② $3u^2 + 2u^2$

③ $6y^2 - 3y^2$

④ $3xy - 2xy$

⑤ $4uv + vu$

⑥ $2uw - 3v + 2wu$

⑦ $\frac{2}{3}ab + \frac{1}{3}ab$

⑧ $2bc - 4bc$

⑨ $a^2 + 2a^2 + 3a$

⑩ $4ac - 8ac + 3a$

⑪ $yx^2 - 5x^2y$

⑫ $4y^2x + 3xy^2 - xy$

Practice exercise P11.9

For each of the following problems:
(a) insert brackets in the given expressions to show the order of operations that gives the correct solution, clearly explaining your reasons;
(b) solve the problem.

① There are 22 people in a school hall when 5 classes each of 31 pupils join them. How many people are now in the hall?
[22 + 31 × 5]

② In a multi-storey car park, one level has 44 cars parked and four levels are full. Each level when full holds 95 cars. If 15 cars leave, how many cars remain in the car park?
[44 + 95 × 4 − 15]

③ Peter has 4 full boxes of CDs with 18 in each box and 2 more boxes with 12 CDs. He puts the CDs in three equal sets. How many CDs are there in each set?
[4 × 18 + 12 × 2 ÷ 3]

④ Sugar is delivered to a shop in boxes of 12 packs and boxes of 24 packs. There are 27 packs of sugar in the shop. If 15 boxes of 12 packs and 8 boxes of 24 packs are delivered, and 20 packs are sold, how many packs does the shop have in stock altogether?
[27 + 15 × 12 + 8 × 24 − 20]

Properties of plane shapes

Pre-requisites
- angles; triangles

Triangles

Tri-angle means three angles. A **triangle** has three sides and three angles. It is the simplest geometrical plane shape. There are many different kinds of triangle (Fig. 12.1).

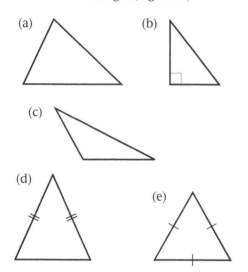

Fig. 12.1

(a) Scalene triangle   *(b) Right-angled triangle*
(c) Obtuse-angled triangle
(d) Isosceles triangle   *(e) Equilateral triangle*

A **scalene** triangle is a triangle in which all three sides are of different length.

In a **right-angled** triangle there is one right angle.

An **obtuse-angled** triangle has one obtuse angle.

Iso-sceles means same legs; an **isosceles** triangle has two sides equal in length.

Equi-lateral means equal sides; an **equilateral** triangles has all three sides equal in length.

Exercise 12a (Oral/class discussion)

For each triangle in Fig. 12.2,
(a) measure the sides
(b) measure the angles
(c) find the sum of the angles
(d) name the kind of triangle it is.

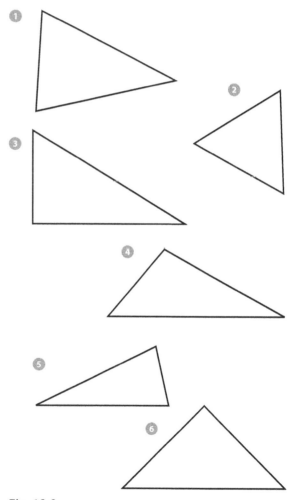

Fig. 12.2

Right-angled triangle

The right-angled triangle is a special triangle. It is used in many situations.

Exercise 12b

Measure and record the lengths of the sides of each right-angled triangle in Fig. 12.3. What do you notice about the position of the longest side in each triangle?

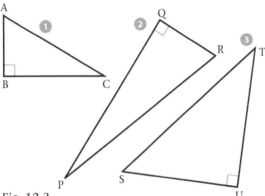

Fig. 12.3

In a right-angled triangle, the longest side is the side opposite the right angle. This longest side is called the **hypotenuse** of the right-angled triangle.

Isosceles triangle

Exercise 12c (Group work)

1. Work in small groups. Make an isosceles triangle by folding and cutting a sheet of paper as shown in Fig. 12.4.

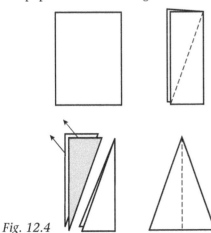

Fig. 12.4

The line of the fold divides the isosceles triangle into two matching parts. This fold is a **line of symmetry**. Discuss whether your isosceles triangle has any other lines of symmetry.

2. Measure all three angles of any isosceles triangle. Discuss your results. What do you notice?

3. In Fig. 12.5, CM is the line of symmetry. On *your* triangle, measure the following: AĈM, BĈM, AM̂C, BM̂C. What do you notice? Also measure the following: AM, BM. Discuss your results. What do you notice?

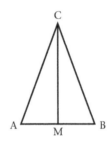   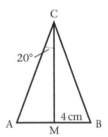

Fig. 12.5   *Fig. 12.6*

4. In Fig. 12.6, △ACB is isosceles, CM is the line of symmetry, AĈM = 20° and BM = 4 cm. Make a sketch of the diagram. Fill in as many other angles and lengths as you can.

5. Why do you think that the triangle in question 4 is named △ACB and not △ABC?

6. In Fig. 12.7, △PQR is a right-angled triangle.

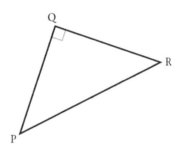

Fig. 12.7

Measure the lengths of PQ and PR. What do you notice? How else can you describe △PQR?

Equilateral triangle

Exercise 12d

1. The diagrams in Fig. 12.8 show a quick way of drawing an equilateral triangle using ruler and compasses.

(a)

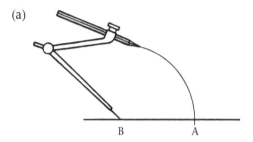

(b)   (c)

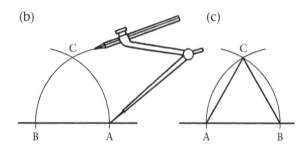

Fig. 12.8

Note that the compasses are kept open at the same position in (a) and (b). Draw some equilateral triangles using this method.

2. Measure all three angles of an equilateral triangle you have drawn. What do you notice?

3. Write down three differences between an equilateral triangle and an isosceles triangle.

4. Draw a large equilateral triangle on a sheet of paper and carefully cut it out. Fold the triangle to find out if it has any lines of symmetry. How many lines of symmetry has an equilateral triangle? What do you notice about the lines of symmetry?

5. Fig. 12.9 shows an equilateral triangle and two of its lines of symmetry.

Fig. 12.9

Make a sketch of the diagram. Fill in as many angles as you can.

6. How many equilateral triangles can you see in each of the diagrams in Fig. 12.10?

(a)   (b)

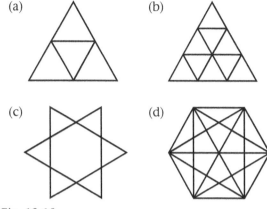

(c)   (d)

Fig. 12.10

Quadrilaterals

Quadri-lateral means four-sided. A **quadrilateral** is a plane shape with four sides and four angles. There are many different kinds of quadrilateral (Fig. 12.11).

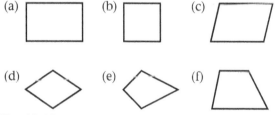

Fig. 12.11
(a) Rectangle   (b) Square   (c) Parallelogram
(d) Rhombus   (e) Kite   (f) Trapezium

Rectangle

Place a matchbox (or other similar small box) on a sheet of paper. Draw round the shape of the face touching the paper. Repeat for two other different faces. Your drawings should look like Fig. 12.12.

Each shape is a **rectangle**.

A rectangle is a quadrilateral in which all the angles are right angles.

Chapter 12

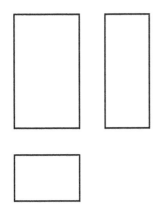

Fig. 12.12

A **diagonal** of a rectangle is a straight line from one corner to the opposite corner. Every rectangle has two diagonals as shown in Fig. 12.13.

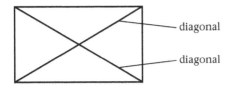

Fig. 12.13

The point where the diagonals cross is the centre of the rectangle.

Exercise 12e (Group work)

Work in small groups and discuss the following in your group.

1. Name at least ten things in your classroom which have a rectangular shape.

2. How many sides has a rectangle?

3. How many angles are in a rectangle?

4. What is the total of the four angles of a rectangle?

5. In each of these rectangles:
   (i) a page out of your exercise book,
   (ii) the shape of the top of your desk,
   (iii) the face of a cuboid (e.g. a chalkbox),
   (a) measure the lengths of both diagonals; what do you notice?
   (b) measure the distance from the centre to each corner; what do you notice?

6. At the centre of each rectangle, where the diagonals cross, there are four angles. How many are acute? What kind of angles are the others?

7. One pair of sides of a rectangle point in the same direction (Fig. 12.14). We say they are **parallel** to each other.
   In pairs, discuss if the other two sides are parallel to each other.

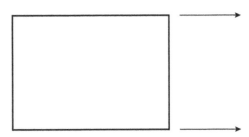

Fig. 12.14

8. Work in small groups. Fold a rectangular sheet of paper so that opposite sides meet. Unfold the paper as in Fig. 12.15.

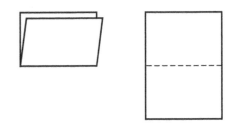

Fig. 12.15

The line of the fold divides the rectangle into two matching parts. This fold is a line of symmetry of the rectangle.

Find out if the rectangle has another line of symmetry. If so, fold the paper another way so that both halves match. How many lines of symmetry has a rectangle? Compare your results with those of another group.

Square

A **square** is a rectangle in which all the sides are of equal length.

Exercise 12f (Group work)

Work in small groups.

① Make a square by folding and cutting a rectangular sheet of paper as shown in Fig. 12.16.

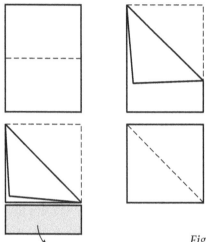

Fig. 12.16

② Think of and list as many things as you can which have square shapes.

③ Measure the lengths of the diagonals of a square. What do you notice?

④ At the centre, where the diagonals meet, there are four angles. Discuss what kind of angles they are.

⑤ Each diagonal meets the corner of a square and makes two angles (Fig. 12.17). Measure these angles. What do you notice?

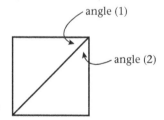

Fig. 12.17

⑥ Discuss how many lines of symmetry a square has. Check by folding.

⑦ In your groups, write down three properties of both a square and a rectangle. Check your results with those of other groups.

⑧ Write down three differences between a square and a rectangle.

Parallelogram

A **parallelogram** is a quadrilateral which has both pairs of opposite sides parallel.

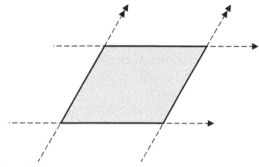

Fig. 12.18

The parallelogram in Fig. 12.18 is formed between two pairs of parallel lines.

Exercise 12g (Pair and group work)

① Work in small groups. Try to name a few things in your home, class or town which have the shape of a parallelogram.

② How many sides has a parallelogram?

③ How many angles has a parallelogram?

④ (a) In pairs, draw a parallelogram with sides at least 5 cm long. Use your rulers of different widths (small and large) to draw a parallelogram (see Fig. 12.19).

(b) In pairs, draw a large parallelogram on a sheet of newspaper. Cut out the parallelogram.

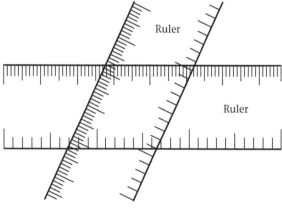

Fig. 12.19

5 Working in your pairs use the parallelograms you made in question 4 to do the following:

(a) Measure the lengths of the sides of the parallelograms. Discuss the results. What do you notice?

(b) Measure the angles of the parallelograms. Discuss the results. What do you notice?

(c) Find the total of the four angles of each parallelogram.

(d) Draw the diagonals of the parallelograms.

(e) Measure the lengths of both diagonals of each parallelogram. What do you notice?

(f) In each parallelogram, measure the distance from the centre to each corner. What do you notice? (see Fig. 12.20)

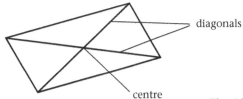

Fig. 12.20

(g) At the centre, where the diagonals cross, there are four angles. How many are acute? What kind of angles are the others?

(h) Try to fold the newspaper parallelogram so that opposite sides meet. Do they meet completely? Do the folded parts of the parallelogram cover each other completely?

(i) Try to fold the newspaper parallelogram so that opposite angles meet. Do this for both pairs of opposite angles. Do the folded parts cover each other completely?

(j) Do your parallelograms have any lines of symmetry? Compare your results with those of another pair of students.

Rhombus

A **rhombus** is a quadrilateral which has all four sides equal.

Fig. 12.21 shows a rhombus; it is sometimes said to be diamond shaped.

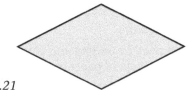

Fig. 12.21

Exercise 12h (Pair and group work)

Work in small groups.

1 Try to name a few things in your home, class or town that have the shape of a rhombus.

2 How many sides has a rhombus?

3 How many angles has a rhombus?

4 In pairs use the following method to make a rhombus from a sheet of paper.

(a) Fold a rectangular sheet of paper so that opposite sides meet.

(b) Fold the paper again. Draw a line as shown in Fig. 12.22(b).

(c) Cut along the line through all four thicknesses of paper.

(d) Unfold the triangular part. This gives a rhombus.

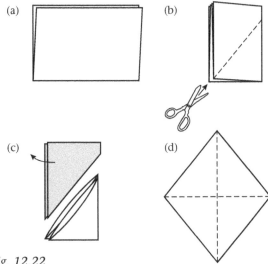

Fig. 12.22

5 Working in your pairs, use the rhombus you made in question 4 and the rhombus in Fig. 12.23 to do the following.

(a) Measure the lengths of the sides of the rhombuses. Discuss your results. What do you notice?

(b) Measure the angles of the rhombuses. Discuss your results. What do you notice?

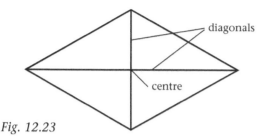

Fig. 12.23

(c) Find the total of the four angles of each rhombus.
(d) Draw the diagonals of each rhombus.
(e) Measure the lengths of both diagonals of each rhombus. What do you notice?
(f) In each rhombus, measure the distance from the centre to each corner. What do you notice?
(g) At the centre, where the diagonals cross, there are four angles. What kind of angles are they?
(h) How many lines of symmetry has a rhombus? Check by folding your paper rhombus
(i) Fig. 12.24 shows angles *a* and *b* between the short diagonal and the sides. In each rhombus, measure the angles *a* and *b*. What do you notice?

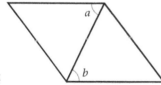

Fig. 12.24

(j) Fig. 12.25 shows angles *x* and *y* between the long diagonal and the sides. In each rhombus, measure the angles *x* and *y*, What do you notice?

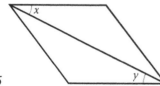

Fig. 12.25

⑥ In your pairs, write down three properties shown by both a rhombus and a parallelogram.

⑦ Write down three differences between a rhombus and a parallelogram.

⑧ What is the difference between a square and a rhombus? Is a square a rhombus?

⑨ Compare your results for questions 6, 7 and 8 with those of another pair.

Kite

A **kite** is a quadrilateral which has two pairs of adjacent sides equal. Fig. 12.26 shows some kites.

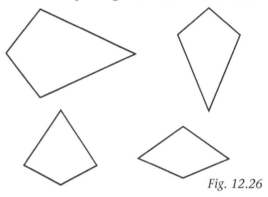

Fig. 12.26

Exercise 12i (Group work)

Work in small groups.

① Try to name a few things in your home, class or town which have the shape of a kite.

② How many sides and angles has a kite?

③ Use the following method to make a kite from a piece of paper.

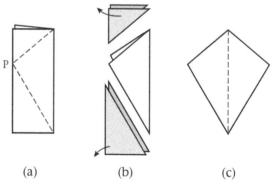

(a)　　　(b)　　　(c)

Fig. 12.27

(a) Fold a rectangular sheet of paper so that opposite sides meet. Mark a point, P, on the long edge such that it is not at the mid-point of the long edge. Draw lines from the point to each end of the fold.

(b) Cut off the four corners along the lines you have drawn.

(c) Unfold the remaining part. This gives a kite shape.

④ Use the kite you made in question 3 (on page 111) and those in Fig. 12.26 to do the following.

(a) Measure the lengths of the sides of the kites. Discuss the results. What do you notice?

(b) Measure the angles of the kites. Discuss the results. What do you notice?

(c) Find the total of the four angles of each kite.

(d) Measure the lengths of both diagonals of each kite. Discuss the results. What do you notice?

(e) At the centre, where the diagonals cross, there are four angles. Discuss together what kind of angles they are.

(f) How many lines of symmetry has a kite?

⑤ Write down two properties that a kite and a rhombus both have.

⑥ Write down three things that are different between a kite and a rhombus. Is a rhombus a kite?

⑦ Check your answers to questions 5 and 6 with those of another group.

Trapezium

A **trapezium** is a quadrilateral which has one pair of parallel sides. Fig. 12.28 shows some trapeziums.

(a)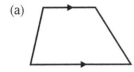

(b)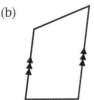

(c)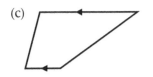

(d)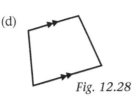

Fig. 12.28

Exercise 12j

① Name four different trapeziums in Fig. 12.29.

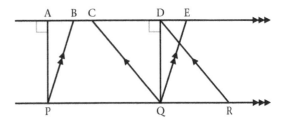

Fig. 12.29

② In Fig. 12.29, what kind of quadrilaterals are the following?

(a) ADQP   (b) BDQP
(c) CDRQ   (d) AEQP
(e) BDRP   (f) BEQP

③ Do any of the trapeziums in Fig. 12.28 have a line of symmetry?

④ Fig. 12.30 shows two trapeziums. Each has a line of symmetry.

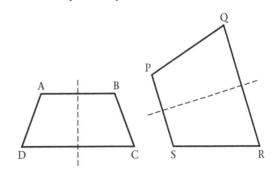

Fig. 12.30

In each trapezium,

(a) name a pair of equal sides,
(b) name two pairs of equal angles.

Polygons

A **polygon** is any closed shape which has three or more straight sides. Hence triangles and quadrilaterals are examples of polygons.

A **regular polygon** is one in which the sides are equal in length and the angles are all equal to each other. For example, the equilateral triangle and square are regular polygons. Fig. 12.31 shows the first six regular polygons.

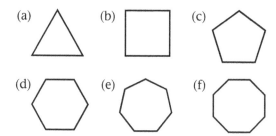

Fig. 12.31   Retgular polygons:
(a) equilateral triangle   (b) square
(c) pentagon   (d) hexagon
(e) heptagon   (f) octagon

Circles

As the number of sides of a regular polygon increases, so their shape approaches that of a **circle**.

The circle is a very common and important shape. Most wheels are circular.

It is nearly impossible to draw a good circle without help. Fig. 12.32 shows some ways of drawing a circle.

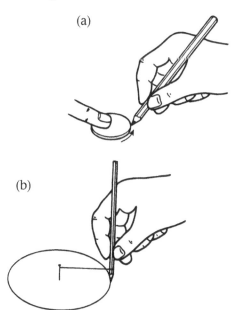

Fig. 12.32   Drawing a circle:
(a) using a circular object such as a coin
(b) using string
(c) using compasses

The parts of a circle

The **centre** is the point at the middle of a circle.

In Fig. 12.33, the **circumference** is the curved boundary of the circle.

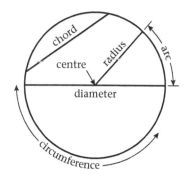

Fig. 12.33

An **arc** is any part of the circumference.

A **radius** is any straight line drawn from the centre to any point on the circumference.

A **chord** is any straight line joining two points on the circumference.

A **diameter** is any chord that passes through the centre of the circle.

In Fig. 12.34, a **sector** is a region bounded by two radii and an arc.

A **semi-circle** is the sector bounded by a diameter and half the circumference, i.e. a half circle.

A **segment** is the region bounded by a chord and an arc.

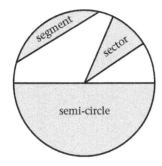

Fig. 12.34

Exercise 12k

Questions 3, 4 and 5 give practice in using compasses.

1. Draw a circle and include the following parts on it: two radii, a sector, a chord, a segment, a diameter, an arc. Label each part and shade any regions.

2. Draw a circle on a piece of paper. Cut out the circle carefully. Check, by folding, to see whether a circle has any lines of symmetry. How many lines of symmetry does a circle have? (Be careful with your answer.)

3. Draw three circles, with radii 3 cm, 4 cm and 5 cm, so that each circle has the same centre.

4. Draw a circle centre O and radius 3 cm. Mark a radius OP (see Fig. 12.35). Measure an angle of 30° at O and with length PQ mark 12 points on the circumference as shown.

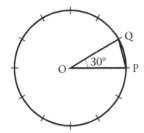

Fig. 12.35

Draw a circle with one of the other points as centre so that the circumference goes through P. Repeat this for all 12 points on the original circle. What shape does your finished pattern look like?

5. (a) How many lines of symmetry does Fig. 12.35 have?
   (b) How many lines of symmetry does the shape you made in question 4 have?

Summary

A **polygon** is *any* closed shape that has three or more straight sides. Polygons are subdivided into groups according to their number of sides:

triangle − 3 sides
quadrilateral − 4 sides
pentagon − 5 sides ..., etc.

These groups are further classified according to the relationship of their sides and angles. A **regular polygon** has equal sides and angles. As the number of sides of a regular polygon increases, the shape approaches that of a **circle**.

Practice exercise P12.1

1. Write True or False for each of the following statements.
   (a) All triangles have three equal sides
   (b) The angles of an isosceles triangle are equal.
   (c) A circle is a polygon
   (d) A polygon has more than three sides
   (e) The sides and angles of a regular polygon are equal
   (f) The set of quadrilaterals includes all shapes bound by four straight sides
   (g) A rhombus is a parallelogram
   (h) A square is a parallelogram.
   (i) A kite is a parallelogram
   (j) A rhombus has four equal sides
   (k) A kite has two pairs of equal sides
   (l) The diameter of a circle is not a chord
   (m) The length of the radius is twice the length of the diameter

(n) The arc of a circle is part of the circumference

2. Name a shape that has
   (a) no line of symmetry
   (b) one line of symmetry
   (c) two lines of symmetry
   (d) three lines of symmetry
   (e) four lines of symmetry
   (f) many lines of symmetry.

3. (a) Draw a circle, centre O, radius 5 cm. Using the same centre draw a circle, radius 8 cm and a circle, radius 10 cm.
   (b) Draw a line that is a chord of the largest circle only.
   (c) Draw a line that is a chord of the two larger circles.
   (d) Draw a line that is a chord of all three circles.
   (e) Draw and shade a sector of the smallest circle only.
   (f) Draw and shade a sector of all three circles.

4. Name the shape or shapes best described by each of the following statements.
   (a) All angles are equal
   (b) All angles and sides are equal
   (c) Two angles and sides are equal
   (d) All angles are equal and has three lines of symmetry
   (e) All sides are equal and has four lines of symmetry

(f) Two parallel sides and no line of symmetry
(g) Two parallel sides and one line of symmetry
(h) Two pairs of parallel sides and no line of symmetry
(i) Four equal sides and two lines of symmetry
(j) Two pairs of equal sides and one line of symmetry
(k) Two pairs of adjacent sides equal
(l) Eight equal sides and angles
(m) Ten equal sides and angles
(n) Three sides of different lengths

5. Copy and complete the table below showing the property of some shapes:

Shape	Equal sides	Equal angles	Parallel sides	Lines of symmetry
Scalene triangle	none			
Isosceles triangle		2		
Equilateral triangle		3		
Square		4		
Rhombus				
Kite				

Chapter 13

Approximation and estimation

Pre-requisites
- the place-value system; decimal fractions; basic units of measurement

Note that the rules in this chapter for approximation and estimation apply to numbers in the base ten place-value system.

Approximation

Rounding off whole numbers

Suppose that on a certain day the population of a town is exactly 18 279 people. What will be the population one week later? It is impossible to say. However, we will have a good idea of the population. It will be 18 000 **approximately**.

We say that 18 279 = 18 000 **to the nearest thousand**. Notice that 18 279 is between 18 000 and 19 000, but is nearer 18 000.

We can also say that 18 279 = 18 300 **to the nearest hundred**. Notice that 279 is between 200 and 300, but is nearer 300.

In each case we have **rounded off** the original number. We may round up or round down. 18 279 is rounded down to 18 000, but it is rounded up to 18 300.

Table 13.1 shows how to round off numbers 630, 631, 632, ..., 639 to the nearest ten.

Table 13.1 Rounding off to the nearest ten

630	no need to round off
631 632 633 634	round *down* to 630 (these numbers are nearer 630 than 640)
635	halfway between 630 and 640: round up to 640*
636 637 638 639	round up to 640 (these numbers are nearer 640 than 630)

*There are other rules for rounding off a 5. However, the above rule will be used in this course

Example 1
Round off 14 505 to the nearest (a) thousand, (b) hundred, (c) ten.

(a) 14 505 = 15 000 to the nearest thousand.
Note: 14 505 is a little nearer 15 000 than 14 000.

(b) 14 505 = 14 500 to the nearest hundred.

(c) 14 505 = 14 510 to the nearest ten.
Note: the last digit of 14 505 is 5; round up.

Exercise 13a (Oral)

1. Round off the following to the nearest
   (i) thousand,   (ii) hundred,   (iii) ten.

   (a) 18 624   (b) 25 246   (c) 32 781
   (d) 7163   (e) 2968   (f) 9476
   (g) 14 939   (h) 26 888   (i) 45 072
   (j) 9895   (k) 30 097   (l) 8350

2. Approximate the following to the nearest ten.

   (a) 345   (b) 375   (c) 695
   (d) 705   (e) 715   (f) 995

Example 2
Approximate 79.65 to the nearest (a) hundred, (b) ten, (c) whole number, (d) tenth.

(a) 79.65 = 100 to the nearest hundred.

(b) 79.65 = 80 to the nearest 10.

(c) 79.65 = 80 to the nearest whole number.
Note: the fraction 0.65 is nearer 1 than 0; round up 79 to the next whole number, 80.

(d) 79.65 = 79.7 to the nearest tenth.
Note: the last digit of 79.65 is 5; round up.

Exercise 13b

1. Round off the following to the nearest whole number.

   (a) 6.9   (b) 12.3   (c) 78.75
   (d) 29.6   (e) 9.5   (f) 99.49

② Approximate the following to the nearest tenth.
(a) 0.71   (b) 0.09   (c) 0.15
(d) 0.45   (e) 0.98   (f) 0.02

③ Round off the following to the nearest hundredth.
(a) 0.164   (b) 0.167   (c) 0.706
(d) 0.702   (e) 0.295   (f) 0.404

④ Approximate the following to the nearest
(i) whole number, (ii) tenth.
(a) 1.38   (b) 4.09   (c) 9.65

⑤ Round off the following to the nearest
(i) tenth, (ii) hundredth.
(a) 0.372   (b) 0.625   (c) 0.155

⑥ Approximate the following to the nearest
(i) ten, (ii) whole number, (iii) tenth.
(a) 26.48   (b) 8.35   (c) 5.84

⑦ Round off 0.798 to the nearest (a) whole number, (b) tenth, (c) hundredth.

⑧ Round off 69.55 to the nearest (a) hundred, (b) ten, (c) whole number, (d) tenth.

We may extend Table 13.1 to find the other whole numbers that are approximated to 640 to the nearest ten.

Table 13.2 Rounding off to the nearest ten

640	no need to round off
641	round
642	down
643	to 640
644	
645	round up to 650
646	
647	round up
648	to 650
649	

From Tables 13.1 and 13.2, we see that for the whole numbers equal to 640 to the nearest ten, the smallest number is 635 and the largest number is 644.

The **range of values** is then 635 to 644, that is, between 635 and 644.

Example 3
Find the range of values approximately equal to
(a) 800 to the nearest hundred
(b) 110 to the nearest ten
(c) 74 to the nearest whole number
(d) 12.2 to the nearest tenth
(e) 3.17 to the nearest hundredth.

(a) The smallest number is 750.
The largest number is 849.
The range of values is 750 to 849.
(b) The smallest number is 105.
The largest number is 114.
The range of values is 105 to 114.
(c) The smallest number is 73.5.
The largest number is 74.4.
The range of values is 73.5 to 74.4.
(d) The smallest number is 12.15.
The largest number is 12.24.
The range of values is 12.15 to 12.24.
(e) The smallest number is 3.165.
The largest number is 3.174.
The range of values is 3.165 to 3.174.

Exercise 13c (Oral)

Find the range of values:

① to the nearest thousand
(a) 2000   (b) 7000
(c) 15 000   (d) 23 000

② to the nearest hundred
(a) 300   (b) 1200
(c) 800   (d) 2700

③ to the nearest ten
(a) 270   (b) 390
(c) 1540   (d) 720

④ to the nearest whole number
(a) 643   (b) 28
(c) 1141   (d) 729

⑤ to the nearest tenth
(a) 4.6   (b) 14.8
(c) 261.7   (d) 325.5

⑥ to the nearest hundredth
(a) 2.31   (b) 4.63
(c) 0.29   (d) 21.46

Decimal places

Decimal places are counted from the decimal point. Zeros after the point are also counted, if they are 'holding' a place that is empty. Digits are rounded up or down as before. Place values must be kept.

Example 4

Read the following examples carefully.

(a) 14.902 8 = 14.9 to 1 decimal place
14.902 8 = 14.90 to 2 d.p.
14.902 8 = 14.903 to 3 d.p.
Note: d.p. is short for decimal places.

(b) 2.397 5 = 2.4 to 1 d.p.
2.397 5 = 2.40 to 2 d.p.
2.397 5 = 2.398 to 3 d.p.

(c) 0.007 2 = 0.0 to 1 d.p.
0.007 2 = 0.01 to 2 d.p.
0.007 2 = 0.007 to 3 d.p.

Exercise 13d (Oral)

Round off the following to (a) 1 d.p., (b) 2 d.p., (c) 3 d.p.

① 12.934 8   ② 24.117 7

③ 5.072 5   ④ 1.937 5

⑤ 1.987 5   ⑥ 0.584 6

⑦ 0.062 5   ⑧ 0.037 5

⑨ 0.900 2   ⑩ 0.008 9

Significant figures

Significant figures begin from the first *non-zero* digit at the left of a number. As before, the digits 5, 6, 7, 8, 9 are rounded up and 1, 2, 3, 4 are rounded down. Digits should be written with their correct place value.

Example 5

Read the following examples and notes carefully.
(a) 546.52 = 500 to 1 significant figure
546.52 = 550 to 2 s.f.
Note: s.f. is short for significant figures.
546.52 = 547 to 3 s.f.
546.52 = 546.5 to 4 s.f.

(b) 8.029 6 = 8 to 1 s.f.
8.029 6 = 8.0 to 2 s.f.
Note: in this case the zero must be given after the decimal point; it is significant.
8.029 6 = 8.03 to 3 s.f.
8.029 6 = 8.030 to 4 s.f.
Note: the 4th significant digit is zero; it must be written down.

(c) 0.009 25 = 0.009 to 1 s.f.
Note: 9 is the first non-zero digit. The two zeros after the decimal point are not significant figures. However, they must be written down to keep the correct place values.
0.009 25 = 0.009 3 to 2 s.f.

Exercise 13e (Oral)

① Round off the following to (i) 1 s.f., (ii) 2 s.f., (iii) 3 s.f.
(a) 7284   (b) 6035   (c) 14 612
(d) 3604   (e) 8009   (f) 5050
(g) 28 336   (h) 9852   (i) 9395
(j) 26 002

② Approximate the following to (i) 1 s.f., (ii) 2 s.f., (iii) 3 s.f., (iv) 4 s.f.
(a) 7.038 4   (b) 18.502   (c) 12.675
(d) 3.799 8   (e) 234.06

③ Round off the following to (i) 1 s.f., (ii) 2 s.f., (iii) 3 s.f.
(a) 0.067 52   (b) 0.305 9
(c) 0.006 307   (d) 0.000 666 6
(e) 0.033 55

Example 6

In the 1996–97 budget it was decided to spend $8 614 160 on land purchase. How might the Finance Minister say this amount in a speech?

He might say:
$8.6 million
or, about $8.6 million
or, just over $8.6 million

He might also say:
$9 million
or, just under $9 million

In the first case the amount was rounded to 2 s.f. In the second case it was rounded to 1 s.f.

The $160 at the end of the $8 614 160 would be
a lot of money for a student to spend. However,
in terms of the budget for a whole country,
$160 is insignificant. The Finance Minister is
not likely to mention it. If a newspaper says,
'Government to spend $8.6 million on land
purchase', we take this to mean the Government
will spend between $8.55 million and $8.64
million.

Exercise 13f

1. (a) There are two '6's in the amount
$8 614 160. Write down the amounts of
money that each 6 represents.
   (b) There are two '1's in the amount
$8 614 160. Write down the amounts of
money that each 1 represents.

2. The actual age of a man is
39 years 8 months 29 days. A doctor asks
the man his age. What will the man say?

3. In 1998 it was proposed to spend
$183 922 343 on Education and Culture.
Give at least three different ways in which a
government Minister might say this amount
in a speech.

4. A newspaper headline says, 'New road to
cost $21.8 million!' Between what amounts
will the road cost?

5. Table 13.3 gives the provisional amounts
spend on school building in a country for
the years 2004–05 and 2005–06.

Table 13.3

| | Capital costs in $'000 | |
	2004–05	2005–06
Secondary Schools	11 360	10 500
Primary Schools	13 895	9160

(a) How might a Finance Minister give the
2004–05 Secondary Schools amount in a
speech?
(b) How might a newspaper print the
2005–06 Primary Schools amount?
(c) Find the total spent in 2004–05 correct
to 2 s.f.

(d) Find the total spent on Secondary
Schools during the 2 years correct to
2 s.f.
(e) Find the total amount spent on school
building over the 2 years correct to 1 s.f.

False accuracy

Consider the following:
(a) A boy measures a line with a ruler. He says
the line is 162.83 mm long. Since it is
impossible to measure 0.83 mm on a ruler,
his answer is an example of **false accuracy**.
It is more realistic to say that the line is
163 mm long.
(b) A report estimates the 2005 population of a
country to be 814 983. Since this number is
an estimate it would have been more realistic
to give the population to 3 s.f.: 815 000.

Most measuring instruments, such as a ruler,
protractor, thermometer, balance, give results
which are correct to only 2 or 3 significant
figures. The answers to calculations using
measurements by such instruments should
likewise be given to 2 or 3 significant figures
only. This is of particular importance when
using electronic calculators.

Some instruments can measure very accurately.
Where possible, ask your science teacher to
show you a micrometer screw gauge, a travelling
microscope, a chemical balance, any Vernier
gauge or other highly accurate measuring
instrument.

Exercise 13g (Oral)

1. Give the measures which will complete
Table 13.4.

Table 13.4

Instrument	Can measure accurately to the nearest
ruler	mm
protractor	
thermometer	
spring balance	
clock	

② Each of the following is an example of false accuracy. Round the given numbers to 3 significant figures.
  (a) The line is 185.39 mm long.
  (b) The woman is 165.778 cm tall.
  (c) The radius of the earth is 6 383.4 km.
  (d) The bank is 985.372 m from the traffic lights.
  (e) Mount Everest is 8 847.73 m high.
  (f) The villages are 14.275 km apart.
  (g) The man has a mass of 67.883 kg.
  (h) Each biscuit has a mass of 6.629 5 g.
  (i) The parcel weighed 2.865 5 kg.
  (j) The lorry carried a load of 7.643 21 tonnes.
  (k) The tin contained 125.108 g of coffee.
  (l) It took 11.372 5 hours to fly to London.
  (m) A full petrol tank holds 40.117 litres.
  (n) The new house cost $51 243.64.
  (o) The temperature was 26.247 °C.

③ Table 13.5 gives some estimated populations. Round each number to a suitable degree of accuracy.

Table 13.5

a country	32 837 541
a large city	1 325 062
a large town	385 460
a small town	7825
a village	683

Estimation

There are many advantages in being able to make **rough calculations**. Consider the following.

(a) A butcher is thinking of buying 22 goats at $37 each. He does a rough calculation first:

$$\$37 \times 22 \simeq \$40 \times 20 = \$800$$

The symbol $\simeq$ means is approximately equal to. $800 is an **estimate** of the cost of the goats. The estimate is not accurate, but it gives the butcher a good idea of the true cost. He may think that $800 is too much money. However, he may think that he can buy at this price. He then does an accurate calculation:

$$\$37 \times 22 = \$814 \qquad \begin{array}{r} 37 \\ \times\ 22 \\ \hline 74 \\ 740 \\ \hline 814 \end{array}$$

Notice that $\$814 \simeq \$800$.

(b) A student does the following problem:
  Calculate the cost of 42 pencils at 86c per pencil. The student gets an answer of $3612. He looks at his answer and thinks, 'It is nearly impossible for 42 pencils to cost $3612'. He does a rough check:

$$86c \times 42 \simeq \$1 \times 40 = \$40$$

This makes him check his working. He sees that he forgot to change cents to dollars:

$$86c \times 42 = 3612 \text{ cents}$$

The correct answer is $36.12. This agrees with his rough check, $40.

It is a good habit always to check calculations by making a rough estimate. A quick estimate can stop you making errors. It can also tell you whether your answer is sensible or not. This is important when using electronic calculators.

When making an estimate, it is usually enough to round off numbers to 1 significant figure or to the nearest whole number.

Example 7
(a) Find the rough value of $4\frac{1}{5} \times 1\frac{7}{8}$.
(b) Find the value of $4\frac{1}{5} \times 1\frac{7}{8}$ accurately.

(a) $4\frac{1}{5} \times 1\frac{7}{8} \simeq 4 \times 2 = 8$
(b) $4\frac{1}{5} \times 1\frac{7}{8} = \frac{21}{5} \times \frac{15}{8} = \frac{63}{8} = 7\frac{7}{8}$

Exercise 13h (Oral)

In questions 1−16 round off each number to 1 s.f. Then estimate each answer.

① $23 + 19$
② $73 - 48$
③ 24×37
④ $572 \div 22$
⑤ $37 + 52$
⑥ $653 - 287$

7. 99×95

8. $171 \div 18$

9. $\$47 + \$61$

10. $\$51 - \$17$

11. $\$9.60 \times 5.8$

12. $43 \, kg \div 8.2$

13. $133 \, g + 452 \, g$

14. $943 \, m - 482 \, m$

15. $\$53 \times 18$

16. $\$672 \div 24$

In questions 17–32 round off each number to the nearest whole number. Then estimate each answer.

17. $6.2 + 3.7$

18. $\$13.09 - \$4.81$

19. 3.4×5.8

20. $14.07 \div 6.7$

21. $3.47 + 12.75$

22. $5\frac{2}{3} - 2\frac{3}{4}$

23. 16.7×1.09

24. $15\frac{3}{4} \div 4\frac{1}{5}$

25. $6\frac{3}{4} + 5\frac{1}{8}$

26. $7.55 - 3.45$

27. $8\frac{1}{2} \times 7\frac{4}{7}$

28. $9.774 \div 3.64$

29. $\frac{5}{6} + 2\frac{3}{8}$

30. $\$4.72 + 89c$

31. $17\frac{3}{4} - 6\frac{7}{8}$

32. $\$2.58 \times 4.5$

In questions 33–40 round off each number to 1 s.f. Then estimate each answer.

33. 0.41×0.92

34. 0.075×0.025

35. 0.333×0.667

36. 0.047×0.023

37. $0.617 \div 0.028$

38. $0.49 \div 0.12$

39. $0.067 \div 0.25$

40. $0.108 \div 0.027$

Exercise 13i

1. Calculate the accurate answers to every fifth question in Exercise 13h, i.e. questions 5, 10, 15, … Use your estimates to check if your answers appear correct.

2. In 2005 the value of a house was $\$238\,000$. Its value rises by about 10% each year. Estimate its value in 2006 correct to the nearest $\$1000$.

3. The mass of a sack of sugar is 10.5 kg. A shopkeeper fills bags each with 320 g of sugar. Estimate the number of bags she can fill from the sack.

4. A farmer has $\$4000$ to spend on cattle. He wants to buy 9 calves. Each calf costs $\$372$ on average. Estimate to check that the farmer has enough money. Find, accurately, how much change he will get after buying the calves.

5. The populations of five towns are 15 600, 17 300, 62 800, 74 000 and 34 400, each to the nearest 100. Find the total population of the five towns to the nearest 1000.
(*Hint*: First find the total population to the nearest hundred as given.)

6. 36 football teams meet at the National Sports Centre. Each team has 12 players. First estimate, then find accurately, the total number of players at the Sports Centre.

7. $x = 0.176 \div 0.32$. By doing a rough calculation, decide which one of the following is the value of x:
(a) 0.18   (b) 0.2   (c) 0.21
(d) 0.3   (e) 0.55

8. A student tries the following problem:
Calculate the cost of 7.8 metres of cloth costing $\$4.15$ per metre.
His answer is $\$3237$.
(a) Is this answer sensible?
(b) Estimate a sensible answer by rounding to the nearest whole numbers.
(c) What error do you think the student made?

9. A shop sells about 340 magazines each week. The selling price of a magazine is $\$4.80$. Estimate the amount of money the shop gets each week from selling magazines.

10. An aeroplane flies 2783 km in $5\frac{3}{4}$ hours. First estimate, then calculate, the average distance it flies in 1 hour.

Body measures

You should know the sizes of parts of your body, such as your hand-span, the length of your foot, your waist measurement and your body mass. You can often use these to estimate other measures.

The following assignments show how to use body measurements to find other distances.

Exercise 13j (Group work)

Work in small groups. You will need a metre rule, a tape measure and a spring balance from the science laboratory.

① (a) Use a metre rule to measure your hand-span in cm (Fig. 13.1).
   (b) Rank the hand-span of all the members of your group from smallest to largest.
   (c) Does anyone in your class have a hand-span less than 15 cm?
   (d) Does anyone in your class have a hand-span greater than 24 cm?

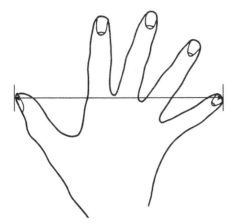

Fig. 13.1 Hand-span: distance between outstretched tips of small finger and thumb

Use your hand-span to estimate:
   (e) the width of your desk to the nearest 10 cm
   (f) the width of the blackboard.
   (g) compare your answer with the other members of your group.
   (h) Use a metre rule to check your estimates in (e) and (f).

② (a) Measure the length of a new pencil.
   (b) Use the pencil to estimate the width of your desk.
   (c) Does your answer agree with that of question 1(e)?

③ Go outside and walk ten paces (Fig. 13.2).
   (a) Use a measuring tape, or a metre rule, to measure the distance you have walked.

←— pace —→

Fig. 13.2 Pace: length of step in normal walking

   (b) Calculate the length of one of your paces. For example, if you walked 6.3 metres, then
   1 pace = 6.3 m ÷ 10
         = 0.63 m
         = 63 cm
   (c) rank the length of paces of all the members of your group from smallest to largest.

④ Measure the following distances in paces:
   (a) from the front of the classroom to the back of the classroom
   (b) from the classroom door to the door of the library
   (c) the length of a football field or netball court
   (d) from the school gate to the Principal's office
   (e) another distance of your own choice.
   Compare your answers with the other members of your group.

⑤ Use your length of pace, calculated in question 3, to estimate the distances in question 4.
   For example, if the football field is 126 paces long, and your pace is 63 cm, then
   length of football field ≈ 0.63 m × 126
   (rounding off)   ≈ 0.6 × 130
   length of football field ≈ 78 m

⑥ If possible, use a tape measure to check some of your estimates in questions 4 and 5.

7 Two students, A and B, walked across the width of a school assembly area. A took 30 paces. B took 36 paces. Which of the following could be reasons why their results are different?
(a) A is lazy
(b) B is a girl
(c) A is quicker than B
(d) B has longer legs than A
(e) A has a longer pace than B
(f) B was carrying a heavier school bag

8 Use a balance from the science laboratory.
(a) Find a stone which has a mass of about 1 kg.
(b) Using your hands as a balance, try to find three things which have the same mass as the stone.
(c) Check your estimate by measuring the three things on the balance.

9 Copy and complete Table 13.6 with your personal statistics.

Table 13.6

Name		
my height		cm
my mass		kg
my hand-span		cm
my pace		cm
my foot		cm

Identify other students who seem to have similar statistics. Compare your estimates.

Summary

In the base ten place-value system, a number may be given correct to approximate values in different ways.

(a) A number may be **rounded off** to a place value, for example, to the nearest hundred, to the nearest whole number, to the nearest tenth, and so on.

The number is **rounded up** if the digit in the next lower place value is 5, 6, 7, 8 or 9 so that the approximate value is greater than the exact value.

The number is **rounded down** if the digit is 1, 2, 3 or 4 so that the approximate value is less than the exact value.

(b) The approximate value may be given to a stated number of **significant figures**. In this method the digit with the highest place value, that is the digit at the left of the number, is the most significant in the number. All other digits to the right, including zero, are significant. Note that in a decimal fraction the most significant digit *cannot* be zero although the zero must be written to indicate the correct place value of the other digits. There are four significant figures in all of the following: 3209, 320.9, 32.09, 3.209, 0.320 9, 0.032 09, 0.003 209

(c) The approximate value may be given to a stated number of **decimal places**. In this method all the digits (including zero) after the decimal point, starting from the tenths place, are counted in the approximation. The following examples are all given correct to three decimal places: 5.042, 0.504, 0.050, 0.005

In (b) and (c), the digit in the lowest place is rounded off according to the rule in (a). Note that the digits in the approximation must keep the correct place value.

Care must be taken to give answers to the same degree of accuracy as that read from the measuring instrument.

When calculating a value, an estimate should be made as a guide to the accuracy of the exact answer by first using approximate values.

Practice exercise P13.1

1 Round these numbers to the nearest whole number.
(a) 23.4   (b) 112.7   (c) 29.35
(d) 18.5   (e) 19.8   (f) 121.09
(g) 94.95   (h) 36.5

② Round these numbers to the nearest ten.
(a) 34   (b) 45   (c) 97   (d) 232

③ Round these numbers to the nearest hundred.
(a) 430   (b) 180   (c) 250   (d) 949

④ Round these numbers to the nearest thousand.
(a) 6200   (b) 3600   (c) 7500   (d) 9470

⑤ Complete the tables.

(a)

Number	Nearest 10	Nearest 100	Nearest 1000
2251			
16 235			
841			
25			

(b)

Country	Natural Gas used – 2003 (terajoules)	Rounded to nearest 10	1000	10 000	1 000 000
Japan	56 653				
Germany	2 104 486				
Italy	1 531 580				
Belgium	361 014				
Denmark	67 497				
U.S.A.	17 028 646				

Practice exercise P13.2

① Use your calculator to find these powers, rounding the answers to 2 s.f.
(a) 50^5 (b) 9^4 (c) 105^3 (d) 2^8

② Use your calculator's root keys to find these roots, rounding the answers to 2 d.p.
(a) $\sqrt[3]{181}$ (b) $\sqrt[3]{25}$ (c) $\sqrt[4]{100}$ (d) $\sqrt[4]{200}$

③ Complete the tables below, rounding your answers to 2 d.p. if necessary.

(a)

Number x	Powers square, x^2	cube, x^3	sixth, x^6
12			
15			
20			

(b)

Number x	Roots square, $\sqrt{}$	cube, $\sqrt[3]{}$	fifth, $\sqrt[5]{}$
12			
15			
20			

④ Round the following numbers to one decimal place.
(a) 3.28   (b) 12.56   (c) 6.24
(d) 8.35   (e) 11.06   (f) 34.55
(g) 12.95   (h) 32.649

⑤ Round the following numbers to two decimal places.
(a) 3.217   (b) 5.625   (c) 14.334
(d) 25.276   (e) 24.295   (f) 0.025 5
(g) 0.005 42   (h) 10.885

⑥ Copy and complete this table.

Element	Atomic Weight	Rounded to Nearest whole number	1 d.p.	2 d.p	3 d.p.
Aluminium	26.981 54				
Chromium	51.996 1				
Helium	4.002 602				
Lanthanum	138.905 5				
Scandium	44.955 91				

⑦ Round the following numbers as indicated.
(a) 58.296 (to 1 s.f)
(b) 239 400 (to 2 s.f)
(c) 0.000 555 2 (to 2 s.f)
(d) 9.000 496 (to 3 s.f)

Practice exercise P13.3

① Frank uses a calculator to work out bank balances. Round each amount to the nearest cent
(a) $124.334 5   (b) $1 269.895 5
(c) $256.375   (d) $78.238 9
(e) $456.905 7   (f) $219.996 75

② Write each number correct to the given degree of accuracy.
(a) 259.648 (to 4 s.f)
(b) 0.005 869 2 (to 3 d.p.)
(c) 95.004 83 (to 3 s.f)
(d) 0.000 300 962 (to 3 s.f)

③ Round these amounts to describe them to a friend. Write the number of significant figures you would use.
(a) A row of houses is made using 328 497 bricks.
(b) A computer processes 546 283 692 518 bytes of data in a year.
(c) A mug has a capacity of 0.449 6 litres.
(d) A bank earns $0.014 938 on every transaction.

④

	Agripro	Insurcom	Caritech
Total annual sales ($)	78 394 206	34 790 000	600 925 409
Rounded to			
% of sales donated to charity	0.1093	0.0413	0.1237
Rounded to			
Number of employees	77	936	1052
Rounded to			
Annual profit ($millions)	6.097 38	11.294 863	743.600 052
Rounded to			

The table shows some statistics of three companies. Round the numbers in each row so the data can easily be compared. Complete the table, stating the number of decimal places (d.p.) or significant figures (s.f.) of the rounded numbers that you have used in your table.

⑤ Each of the following numbers has been rounded to the given degree of accuracy. Write down two possible values for the original number: one smaller and one larger than the rounded number.

(a) 9.6 to 1 d.p.
(b) 3500 to the nearest 100
(c) 0.096 2 to 4 d.p.
(d) 0 to the nearest 10
(e) 490 000 to the nearest 1000
(f) 6 400 000 000 to the nearest 100 000

⑥ Use your calculator to work out these powers and roots to the given degree of accuracy.
(a) 0.49^5 to 3 d.p.
(b) $\sqrt{59\,200\,843}$ to the nearest 100
(c) 346^3 to the nearest 10 000
(d) $\sqrt{0.000\,089}$ to 2 d.p.
(e) 1.68^{20} to the nearest 100

⑦ Round each of these numbers to one significant figure.
(a) 23.7   (b) 17.8   (c) 235.2
(d) 3675   (e) 67.3   (f) 0.025 4
(g) 0.005 79   (h) 9.95

Practice exercise P13.4

① Four of these calculations are wrong. Check them using estimates. Write *correct* or *wrong* for each question.
(a) $13.6 \times 3.4 + 8.2 = 157.76$
(b) $95.3 - 16.2 \times 3.75 = 34.55$
(c) $123.8 + 23.6 \times 5.8 = 260.68$
(d) $233.7 - 37.6 \times 4.3 = 843.23$
(e) $72.7 + 301.35 \div 12.3 = 97.2$
(f) $165.6 - 54.6 \div 2.4 = 46.25$
(g) $(89.4 + 12.3) \times 0.9 = 91.53$
(h) $189.2 + 104.257 \div 6.85 = 42.84$

② Two answers, rounded to 2 d.p., are given for each of these calculations. Use estimated values to select the correct answer. The first problem has been done as an example.
(a) $57.3 + 23.4 \times 16.2$
   (i) 436.38
   (ii) 1307.34.
      $60 + 20 \times 20 = 460$..............(i)....
(b) $249.5 + 3.9 \times 28.7$
   (i) 7272.58
   (ii) 361.43
(c) $76.75 + 39.2 \div 2.4$
   (i) 93.08
   (ii) 48.31

(d) $116.6 - 12.3 \times 0.495$
   (i) 51.63
   (ii) 110.51

(e) $\dfrac{12.27 + 213.4}{24.6 + 31.2}$
   (i) 40.37
   (ii) 4.04

(f) $\dfrac{156.7 - 25.4}{2.5 \times 16.2}$
   (i) 850.82
   (ii) 3.24

2. Estimate the answers to these calculations. Do not use a calculator. Round the numbers to 1 s.f. Show your working.
  (a) Calculate the price of 0.65 m of chain costing $7.23 per metre.
  (b) A rectangular rug has an area of $18.2\,m^2$ and width 3.54 m. Calculate its length.
  (c) A company manufactured 6 365 842 computer disks in a month. Each disk has a capacity of 706 560 000 bytes. What is the total capacity of all the disks in megabytes?
  (d) A train operator sold 185 regular tickets at $17 each and 32 1st class tickets at $37.50 each. What is the total cost of the tickets?
  (e) A lottery prize of $729 543 was divided equally between 46 people. How much did each person receive?

3. (a) Check the answers to these calculations by finding an estimate. Do not use your calculator.
  (b) Do you think the answer is correct? Explain your reasoning.
    (i) $2352 \times 193 = 45\,936$
    (ii) $95\,312 \div 368 = 259$
    (iii) $225^3 = 11\,390\,625$
    (iv) $0.089\,28 \div 0.248 = 0.036$

4. (a) Estimate the percentages of the following amounts by rounding numbers to 1 s.f..
  (b) State the expression with the rounded values.
  (c) Work out the estimated answer
    (i) 86% of $32
    (ii) 6.42 % of 9.69 m
    (iii) 27.3% of 43 950 g
    (iv) 63% of $0.004\,731\,m^2$
    (v) 0.283% of $4 596 820

5. Calculate the following. Round your answers to 3 d.p. where necessary.
  (a) 3^8
  (b) 1.2^{10}
  (c) $\sqrt{37}$
  (d) $\sqrt{0.5}$

Plane shapes (1)
Perimeter

Pre-requisites
- properties of plane shapes; basic units of measurement of length; fractions and decimal fractions; approximation and estimation

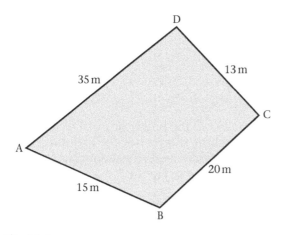

Fig. 14.1

Fig. 14.1 is a diagram of Mr Brown's field. If Mr Brown starts at A and walks along the boundary of his field what distance would he have walked by the time he got back to A? Does it matter in which direction Mr Brown walks?

The distance around the field is

$$15\,m + 20\,m + 13\,m + 35\,m$$

Another name for the distance around is **perimeter**. The perimeter of a shape is the outside boundary or edges of the shape. We often use the word perimeter to mean the length of the boundary of a shape.
The perimeter of Mr Brown's field is 83 m.

Measuring perimeters

The simplest way of finding a perimeter is to measure it with a ruler or tape measure.

Example 1
Measure the perimeter of the quadrilateral ABCD in Fig. 14.2.

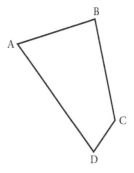

Fig. 14.2

By measurement, AB = 22 mm
BC = 27 mm
CD = 10 mm
DA = 35 mm
Perimeter = 94 mm

Example 2
Find, in cm, the perimeter of the regular hexagon ABCDEF in Fig. 14.3.

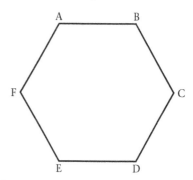

Fig. 14.3

Length of side AB = 2.1 cm.
There are 6 equal sides

Perimeter = 6×2.1 cm
= 12.6 cm

Exercise 14a

① Use a ruler to measure the perimeters of the shapes in Fig. 14.4. Give your answers in mm.

(a)

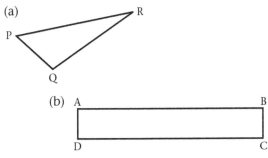

(b) A

(c) W

(d)

Fig. 14.4

② Each of the shapes in Fig. 14.5 is regular. Find the perimeter of each shape by measuring one side and multiplying by the number of sides. Give your answers in cm.

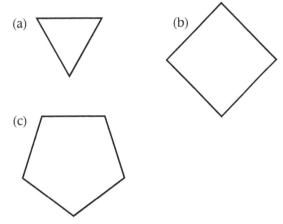

(a)

(b)

(c)

Fig. 14.5

③ Each of the shapes in Fig. 14.6 contains at least one curved edge. Use thread to find the

approximate perimeter of each shape. Give your answers in mm.

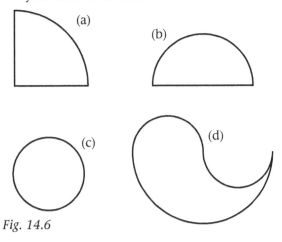

(a)

(b)

(c)

(d)

Fig. 14.6

Perimeter of a rectangle

We usually call the longer side of a rectangle the **length** and the shorter side the **breadth**. The letters l and b are used to stand for the length and the breadth in Fig. 14.7.

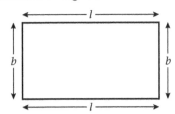

Fig. 14.7

From the diagram:

$$
\begin{aligned}
\text{perimeter of rectangle} &= l + b + l + b \\
&= (l + b) + (l + b) \\
&= 2 \times (l + b) \\
&= 2(l + b)
\end{aligned}
$$

The algebraic expression $2(l + b)$ is called the **formula** for the perimeter of a rectangle. To calculate the perimeter of a rectangle, substitute the numerical value for l and for b.

Example 3

Calculate the perimeter of a football field which measures 80 m by 50 m.

$$
\begin{aligned}
\text{perimeter of field} &= 2(l + b) \\
&= 2 \times (80 + 50)\,\text{m} \\
&= 2 \times 130\,\text{m} = 260\,\text{m}
\end{aligned}
$$

Perimeter of a square

A square is a regular 4-sided shape. If the length of one side of a square is l, then the formula for the perimeter of a square is

$$l \times 4 = 4l$$

The formula for perimeters of rectangles and squares can be useful. However, if you find it difficult to remember formulae, always sketch the given shape and work from that.
Note: *formulae* is the plural of *formula*.

Exercise 14b

① Copy and complete the table for rectangles (Table 14.1).

Table 14.1

	length	breadth	perimeter
(a)	3 cm	2 cm	
(b)	5 cm		18 cm
(c)	16 mm	10 mm	
(d)	$2\frac{1}{2}$ km	1 km	
(e)		$2\frac{1}{2}$ m	17 m
(f)	$8\frac{1}{4}$ cm	$4\frac{1}{2}$ cm	
(g)	5.1 cm	3.2 cm	
(h)	4.3 cm	0.8 cm	
(i)	7.35 m		29 m
(j)		0.9 km	19.8 km

② Copy and complete the table for squares (Table 14.2).

Table 14.2

	length of side	perimeter
(a)	6 km	
(b)	14 km	
(c)	$2\frac{1}{4}$ m	
(d)	8.3 cm	
(e)		40 cm
(f)		360 m
(g)		18 mm
(h)		14.4 cm

③ A rectangular field measures 400 m by 550 m. What is the perimeter of the field (a) in metres, (b) in kilometres?

④ A school compound is made up of a rectangle and a square as in Fig. 14.8. Find the perimeter of the compound.

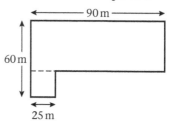

Fig. 14.8

⑤ A woman fences a 3 m by 4 m rectangular plot to keep her chickens in. The fencing costs $1.20 per metre. How much does it cost to fence the plot?

⑥ A man has 36 square tiles. Each tile measures 1 m by 1 m. He lays the tiles in the shape of a rectangle as in Fig. 14.9.

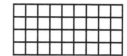

Fig. 14.9

Find the length, breadth and perimeter of this rectangle. Use squared paper to show different rectangles the man can make using all 36 tiles. Find the perimeters of these rectangles. What do you notice?

⑦ Find the perimeter of the parallelograms in Fig. 14.10.

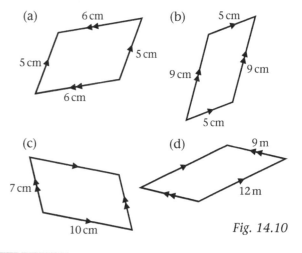

Fig. 14.10

Perimeter of a circle

The perimeter of a circle is called the **circumference**.

Measuring circumference

Here are two ways of finding the circumference of a cylindrical tin can.

(i) Rolling

(a) Make a mark on the circumference of the circular end-face (Fig. 14.11). Make a mark A on a long sheet of paper (e.g. newspaper). Start with the two marks opposite each other.

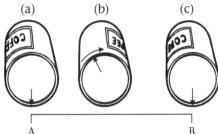

| (a) | (b) | (c) |

Fig. 14.11   A   B

(b) Roll the tin along the paper.

(c) Stop when the mark is against the paper again. Make a mark B on the paper opposite the mark on the tin. Measure the length AB.

The length AB will be the circumference of the circular end-face.

(ii) Winding thread

(a) Wind a piece of thread (or string) once around the tin (Fig. 14.12). Mark the thread at A and B where it crosses.

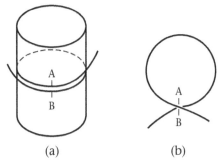

Fig. 14.12   (a)   (b)

(b) Remove the string from the tin. Pull the string straight and measure AB against a ruler.

The length AB will be the circumference of the circle.

Exercise 14c (Group work)

Find three cylindrical objects, e.g. a tube, a tin can and a torch battery. Work in small groups and show your results as in Table 14.3, where c is the length of the circumference and d is the diameter of the circle.

➊ Use both methods, rolling and winding thread, to find the circumference of the circular faces of the three objects.

➋ Mark out the circular face of each object on a sheet of tracing paper. Carefully cut out each circle and fold the circle to find the diameter. Mark and measure the diameters of the three circles.
Compare the diameter of each object with its circumference.

(a) Is the circumference greater than the diameter in each case?

(b) If so, approximately how many times greater?

(c) Is this true for all three objects?

(d) Compare your results with your friends'. Do other people get results like yours?

Table 14.3

	c	d	$\frac{c}{d}$
Battery	80 mm	25 mm	$\frac{80}{25} = 3\frac{1}{5}$
Tin can	31.5 cm	10 cm	$\frac{31.5}{10} = 3.15$

Your results in Exercise 14c may have been as shown in Table 14.3.

In each case, the circumference is just over 3 times the diameter.

For *any* circle it will be found that, approximately, circumference ÷ diameter = 3.1

so that   circumference = 3.1 × diameter

The number 3.1 is not exact. More accurate values are 3.142 or $3\frac{1}{7}$ but even these are not exact. It is impossible to express the number as an exact fraction or decimal. We use the Greek letter π, *pi*, to represent this number.

circumference = π × diameter

or $c = \pi d$

The diameter, d is twice the radius, r, thus:

$$c = 2\pi r$$

This formula is used to calculate the value of the circumference of a circle of radius r. In any question where this formula is used, the value of π will be given, usually 3.14 or $3\frac{1}{7}$ (i.e. $\frac{22}{7}$).

Example 4

Calculate the circumference of a circle of radius $3\frac{1}{2}$ metres. Use the value $3\frac{1}{7}$ for π.

Circumference $= 2\pi r$
$= (2 \times 3\frac{1}{7} \times 3\frac{1}{2})$ m
$= (2 \times \frac{22}{7} \times \frac{7}{2})$ m
$= 22$ m

Example 5

A bicycle wheel has a diameter of 65 cm. During a journey, the wheel makes 1000 complete revolutions.

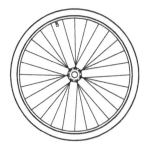

Fig. 14.13

How many metres does the bicycle travel? (Use the value 3.14 for π.)

Distance moved in one revolution of the wheel is equal to the circumference of the wheel.

Circumference $= \pi d$

Distance travelled in one revolution $= 3.14 \times 65$ cm

Distance travelled in 1000 revolutions
$= 3.14 \times 65 \times 1000$ cm
$= \dfrac{3.14 \times 65 \times 1000}{100}$ m
$= 31.4 \times 65$ m
$= 2041$ m

Example 6

Calculate the perimeter of the shape in Fig. 14.14. All lengths are in centimetres. Use the value $\frac{22}{7}$ for π.

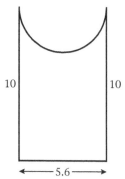

Fig. 14.14

Notice that the perimeter is made up of three sides of a rectangle and a semi-circle. The diameter of the semi-circle is 5.6 cm.

Total length of straight sides
$= 10 + 5.6 + 10$ cm
$= 25.6$ cm

Length of semi-circle
$= \frac{1}{2}$ of $\pi \times 5.6$ cm
$= \frac{1}{2} \times \frac{22}{7} \times \frac{56}{10}$ cm
$= 11 \times \frac{8}{10}$ cm
$= \frac{88}{10}$ cm $= 8.8$ cm

Perimeter of shape
$= 25.6 + 8.8$ cm
$= 34.4$ cm

Exercise 14d

1 Copy and complete Table 14.4 for circles. Use the value $\frac{22}{7}$ for π.

Table 14.4

	Radius	Diameter	Circumference
(a)	7 m		
(b)		7 cm	
(c)	14 mm		
(d)		42 m	
(e)	350 mm		
(f)		5.6 cm	

Chapter 14

② Calculate the perimeter of a circle of radius 70 m. (Use the value $3\frac{1}{7}$ for π.)

③ Find the perimeters of the shapes in Fig. 14.15. All measurements are in cm. Use the value $\frac{22}{7}$ for π.

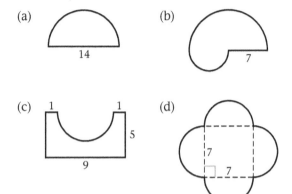

(a) 14

(b) 7

(c) 1 1 5 9

(d) 7 7

Fig. 14.15

④ The diameter of the circular base of a round nut is 14 mm. What is the circumference of the base of the nut? (Use the value $\frac{22}{7}$ for π.)

⑤ The minute hand of a clock is 10.5 cm long. How far does the tip of the hand travel in 1 hour? In $\frac{1}{2}$ hour? In 20 min? (Use the value 3 for π.)

⑥ A record has a diameter of 30 cm and rotates at $33\frac{1}{3}$ revolutions per minute. How far does a point on the edge of the record travel in a minute? Use the value 3.14 for π and give your answer in metres.

⑦ Fig. 14.16 shows a wire circle set inside a wire square of side 7 cm.
(a) Calculate the perimeter of the square.
(b) Calculate the circumference of the circle. (Use $\frac{22}{7}$ for π.)
(c) Calculate the total length of wire used.

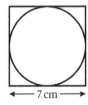

←— 7 cm —→

Fig. 14.16

⑧ A rope is wound 50 times round a cylinder of radius 25 cm (Fig. 14.17). What is the approximate length of the rope? (Use the value 3.14 for π.)
Why is this an approximation?

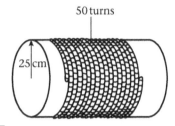

50 turns

25 cm

Fig. 14.17

⑨ A bicycle wheel has a diameter of 63 cm. How many metres does the bicycle travel for 100 revolutions of the wheel? (Use the value $\frac{22}{7}$ for π.)

⑩ An arch of a bridge is made by bending a steel beam into the shape of a semi-circle. See Fig. 14.18.

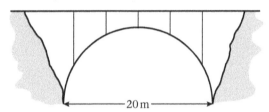

←————— 20 m —————→

Fig. 14.18

If the span (diameter) of the arch is 20 m, use the value 3.14 for π to find the length of the steel beam to the nearest metre.

Summary

The **perimeter** is the outline of a shape. Perimeter is often used, however, to mean the measure of the outline. For each of the shapes below (see Fig. 14.19), the perimeter may be found by using a formula.

Shape	Perimeter
(a) Rectangle	$2 \times (l + b)$
(b) Square	$4 \times l$
(c) Circle	$2\pi r$

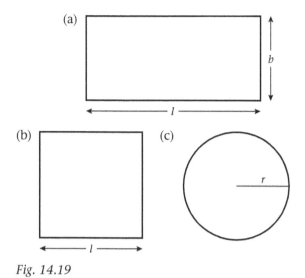

Fig. 14.19

Practice exercise P14.1

Remember
1. *The perimeter of a shape is the distance around the shape.*
2. *The measure is a measure of length.*

1 A jogger jogs around a rectangular playing field measuring 24.5 m by 32 m. What distance does the jogger cover in going
(a) once around the field
(b) 9 times around the field.

2 Calculate the perimeter of the top of a table which is in the shape of a regular hexagon of side 90 cm. Give your answer in cm and metres.

3 A lawn is of the shape as shown below:

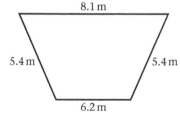

Fig. 14.20
Calculate the perimeter of the lawn.

4 A rectangular table cloth measures 90 cm by 120 cm.
(a) What is the distance around the edge of the table cloth?
(b) How much lace would be needed to go around the tablecloth once allowing 20 cm for corners?

5

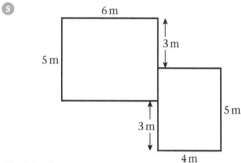

Fig.14.21
Calculate the distance around the shape above.

6 Calculate the circumference of a circle of radius 56 cm. (Use $\pi = \frac{22}{7}$)

7 The floor of a room is covered using square tiles of side 30 cm. The floor is covered by thirteen tiles along the width and twenty tiles along the length.
(a) What is the length of the room?
(b) What is the width of the room?
(c) Calculate the perimeter of the room.
(d) How many tiles are used to cover the room?

8 The diameter of a CD (compact Disc) is approximately 11.8 cm.
(a) What is the radius of the CD?
(b) Calculate the circumference of the CD.
Use $\pi = 3.14$ and give your answer correct to one decimal place.

Chapter 15

Algebraic equations (1)
Simple linear equations, flow charts

Pre-requisites
- basic algebraic processes; simplifying algebraic expressions

Equations

The statement $3x = 18$ is an algebraic sentence. It means: three times an unknown number x is equal to eighteen. A sentence which has a symbol, usually a letter, and an equals sign in this way is called an **equation**. $3x = 18$ is an equation in x. Other examples of equations are: $x + 2 = 5$, $x = 7$, $4 - y = 0$, $3a - 3 = 0$.

The letter in an equation is sometimes called **the unknown**. A sentence like $3x = 18$ may be true or false. It depends on the value of the unknown. For example, $3x = 18$ is false when $x = 2$; it is true when $x = 6$.

Exercise 15a (Oral)

Say whether the following are true or false.

1. $x + 3 = 7$ when $x = 4$
2. $5x = 15$ when $x = 3$
3. $x - 2 = 9$ when $x = 10$
4. $11 + x = 16$ when $x = 7$
5. $\frac{12}{x} = 4$ when $x = 3$
6. $2x = 22$ when $x = 2$
7. $23 - x = 20$ when $x = 3$
8. $\frac{x}{8} = 2$ when $x = 16$
9. $x - 5 = 7$ when $x = 2$
10. $6x = 36$ when $x = 6$
11. $4 = 3 + x$ when $x = 1$
12. $25 = 2\frac{1}{2}x$ when $x = 10$
13. $\frac{24}{x} = 4$ when $x = 4$
14. $9 = 14 - x$ when $x = 5$
15. $6 = \frac{x}{3}$ when $x = 12$
16. $15 = x + 2$ when $x = 17$
17. $19 = 19 + x$ when $x = 0$
18. $1 = \frac{x}{2}$ when $x = 1$
19. $10x = 10$ when $x = 1$
20. $12 = 12 - x$ when $x = 12$

Exercise 15b (Oral)

Each sentence is true. Find the number that each letter stands for.

1. $x = 2 + 7$
2. $x = 3 - 0$
3. $y = 9 + 5$
4. $y = 14 - 4$
5. $12 + 8 = p$
6. $18 - 7 = p$
7. $9 + 7 = q$
8. $17 - 8 = q$
9. $2 + m = 5$
10. $7 - m = 6$
11. $11 + n = 14$
12. $15 - n = 10$
13. $x + 6 = 29$
14. $x - 14 = 0$
15. $c + 11 = 30$
16. $c - 15 = 15$
17. $b + 7 = 21$
18. $b - 17 = 12$
19. $5 = 4 + a$
20. $6 = 7 - a$
21. $17 = 9 + q$
22. $16 = 20 - q$
23. $28 = e + 13$
24. $6 = e - 18$
25. $22 = d + 15$
26. $21 = d - 6$
27. $x + x = 8$
28. $y + y = 12$
29. $18 = b + b$
30. $16 = c + c$
31. $e + e + e = 3$
32. $f + f + f = 12$
33. $24 = j + j + j$
34. $12 - m = m$
35. $p = 28 - p$

Solution of an equation

We can usually find the value of the unknown which makes an equation true. We call this value the **solution** of the equation. $x = 6$ is the solution of $3x = 18$. **To solve** an equation means to find the value of the unknown which makes the equation true.

For example, to solve the equation

$18 - x = 7$

is to find a number which when taken from 18 gives 7.
The number is 11.
$x = 11$ is the solution.

Again, for example, to find the solution of the equation

$\frac{x}{6} = 5$

is to find a number which gives 5, when divided by 6.
The number is 30.
$x = 30$ is the solution.

Exercise 15c (Oral)

Solve the following equations.

①	$x + 8 = 12$	②	$x + 3 = 8$
③	$20 + x = 28$	④	$14 + x = 20$
⑤	$14 - x = 11$	⑥	$16 - x = 13$
⑦	$x - 2 = 15$	⑧	$x - 3 = 8$
⑨	$4x = 20$	⑩	$2x = 50$
⑪	$12 = 3x$	⑫	$72 = 9x$
⑬	$\frac{x}{2} = 5$	⑭	$4 = \frac{x}{9}$
⑮	$\frac{28}{x} = 7$	⑯	$3 = \frac{24}{x}$
⑰	$14 = x + 8$	⑱	$8 = 9 - x$
⑲	$22 = 11 + x$	⑳	$12 = x - 12$
㉑	$7x = 7$	㉒	$\frac{x}{6} = 1$
㉓	$x + 5 = 5$	㉔	$x - 3 = 0$

The balance method of solving equations

Consider the equation $3x = 18$. The $3x$ on the left-hand side (LHS) equals, or **balances**, the 18 on the right-hand side (RHS). We can show $3x$ and 18 balancing on a pair of scales (Fig. 15.1).

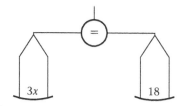

Fig. 15.1

As with real scales, the two sides will balance if we add equal amounts to both sides or if we subtract the same from both sides (Fig. 15.2).

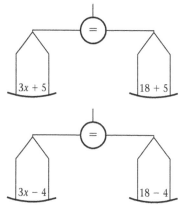

Fig. 15.2

The scales also balance if we multiply or divide by the same amount on both sides (Fig. 15.3).

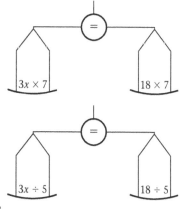

Fig. 15.3

The two sides will stay balanced if we **do the same to both sides**.

Our aim in solving an equation is to find the value of the unknown. Therefore we must choose what to add or subtract, and whether to multiply or divide in such a way that the term in x, and finally x, is by itself on one scale. To solve $3x = 18$ by the balance method, first find which side the unknown, x, is on. In Fig. 15.1, $3x$ is on the scale on the LHS. If $3x$ is divided by 3 the result will be x. If we divide the LHS by 3, we must also divide the RHS by 3 to keep the balance (Fig. 15.4).

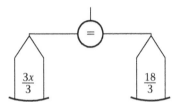

Fig. 15.4

Divide both sides by 3:

$$3x = \frac{18}{3}$$
$$x = 6$$

Example 1

Solve the following equations using the balance method:

(a) $x + 11 = 18$,　(b) $\frac{1}{4}y = 7$,　(c) $8 = x - 5$.

(a) $x + 11 = 18$

Subtract 11 from both sides:
$x + 11 - 11 = 18 - 11$
$x = 7$

(b) $\frac{1}{4}y = 7$

Multiply both sides by 4:
$\frac{1}{4}y \times 4 = 7 \times 4$
$y = 28$

(c) $8 = x - 5$

Add 5 to both sides:
$8 + 5 = x - 5 + 5$
$13 = x$
$x = 13$

Exercise 15d

Use the balance method to solve the following equations. Write down the steps and working as in the above examples.

1.　$3x = 21$
2.　$5x = 20$
3.　$32 = 8x$
4.　$27 = 9x$

5.　$x + 3 = 8$
6.　$x + 5 = 11$
7.　$8 + x = 18$
8.　$17 + x = 23$
9.　$\frac{1}{2}x = 5$
10.　$\frac{1}{4}x = 9$
11.　$2 = \frac{1}{3}x$
12.　$6 = \frac{1}{5}x$
13.　$x - 1 = 20$
14.　$x - 3 = 1$
15.　$13 = x - 7$
16.　$0 = x - 6$
17.　$9 = x + 2$
18.　$20 = x + 14$
19.　$15 = 13 + x$
20.　$7 = 1 + x$

It is possible to solve the equations in the last exercise from knowledge of numbers. However, the balance method is very useful with more difficult equations. For example, the equation $4x + 5 = 17$ is more difficult to solve directly. Figs. 15.5–15.9 show how this equation is solved by the balance method.

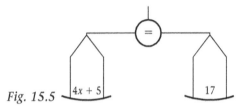

Fig. 15.5

(a) The LHS of $4x + 5 = 17$ contains the unknown. Subtract 5 to leave $4x$ on the LHS. Since 5 is taken from the LHS, 5 must also be taken from the RHS to keep the balance. Subtract 5 from both sides:

$$4x + 5 - 5 = 17 - 5$$

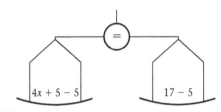

Fig. 15.6

Simplifying:　　$4x = 12$

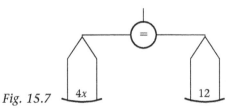

Fig. 15.7

(b) The equation is now easier. Divide the LHS by 4 to leave x. The RHS must also be divided by 4 to keep the balance.

Divide both sides by 4:

$$\frac{4x}{4} = \frac{12}{4}$$

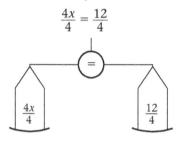

Fig. 15.8

Simplifying: $\qquad x = 3$

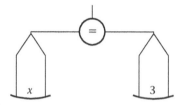

Fig. 15.9

$x = 3$ is the solution of $4x + 5 = 17$

It is a good habit to always check your solution. Substitute the value of the unknown in the equation you solved to see whether the sentence is true.

Check: When $x = 3$,
LHS $= 4 \times 3 + 5 = 12 + 5 = 17 =$ RHS

The following examples are also solved by the balance method.

Example 2

Solve $5x - 6 = 29$.

$5x - 6 = 29$

Add 6 to both sides:
$$5x - 6 + 6 = 29 + 6$$
$$5x = 35$$

Divide both sides by 5:
$$5x \div 5 = 35 \div 5$$
$$x = 7$$

Check: When $x = 7$,
LHS $= 5 \times 7 - 6 = 35 - 6 = 29 =$ RHS

Example 3

Solve $21 = 9 + 2y$

$21 = 9 + 2y$
The unknown is on the RHS.

Subtract 9 from both sides:
$$21 - 9 = 9 + 2y - 9$$
$$12 = 2y$$

Divide both sides by 2:
$$\frac{12}{2} = \frac{2y}{2}$$
$$6 = y$$

Check: When $y = 6$,
RHS $= 9 + 2 \times 6 = 9 + 12 = 21 =$ LHS

Exercise 15e

Solve the following equations by the balance method. Write down every step and show all working. Check each solution.

1. $5y + 6 = 21$
2. $4a + 3 = 15$
3. $3x + 2 = 14$
4. $6p + 2 = 20$
5. $2n - 3 = 5$
6. $3m - 4 = 26$
7. $5t - 2 = 18$
8. $8x - 9 = 7$
9. $6 + 2a = 18$
10. $8 + 5y = 23$
11. $4 + 3d = 25$
12. $1 + 7q = 22$
13. $5 = 7b - 9$
14. $16 = 2a - 4$
15. $3 = 4a - 1$
16. $9 = 5x - 1$
17. $16 = 2a + 4$
18. $19 = 10 + 3x$
19. $8x - 24 = 0$
20. $3x + 7 = 7$

The value of the unknown can be fractional as in the following example.

Example 4

Solve the equation $2x + 7 = 12$.

$2x + 7 = 12$

Subtract 7 from both sides:
$$2x + 7 - 7 = 12 - 7$$
$$2x = 5$$

Divide both sides by 2:
$$\frac{2x}{2} = \frac{5}{2}$$
$$x = 2\tfrac{1}{2}$$

Check: When $x = 2\tfrac{1}{2}$,
LHS $= 2 \times 2\tfrac{1}{2} + 7 = 5 + 7 = 12 =$ RHS

Exercise 15f

Solve the following equations.

1. $3x + 4 = 17$
2. $6x - 5 = 6$
3. $5x + 8 = 11$
4. $4x - 1 = 2$
5. $3 + 2x = 10$
6. $2 = 7x - 4$
7. $19 = 6 + 9x$
8. $8x + 3 = 4$
9. $10x - 3 = 5$
10. $12x + 3 = 21$

Using a flow chart

In Chapter 6, we used flow charts to help us to represent by an algebraic expression information given in words. In the same way we can use flow charts to help us to solve algebraic equations.

To *form* an algebraic equation we use a sequence of arithmetic operations and we do the operations in a given order. To *solve* the equation, that is, to find what we started with, we need to 'undo' what we did. We have to use the inverse arithmetic operations and reverse the order of doing the operations.

Remember that: addition and subtraction are inverse operations; multiplication and division are inverse operations.

The equation $x + 14 = 17$ can be represented by a flow chart, like Fig. 15.10:

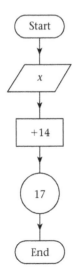

Fig. 15.10

To solve the equation, that is to find the value of x that makes the sentence true, we need to reverse the order and do the inverse operation (Fig. 15.11).

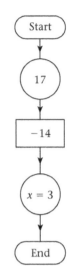

Fig. 15.11

To solve $3x = 18$. In the equation, we multiply the unknown by 3. Therefore to solve the equation, we divide by 3 (Fig. 15.12).

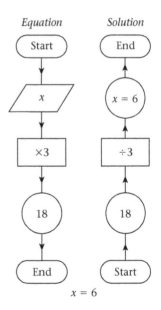

Fig. 15.12

To solve $4x + 5 = 17$.

(b) Using a flow chart

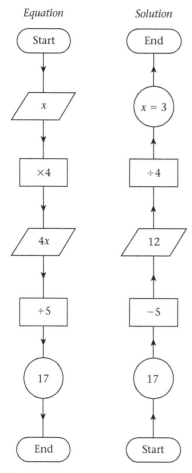

Fig. 15.13

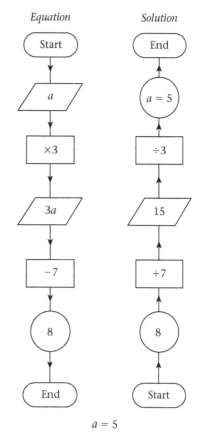

Fig. 15.14

Check: When $a = 5$,
LHS $= 3 \times 5 - 7 = 15 - 7 = 8 =$ RHS

Example 5

Solve the equation $3a - 7 = 8$.

(a) Balance method
$$3a - 7 = 8$$

Add 7 to both sides:
$$3a = 15$$

Divide both sides by 3:
$$a = 5$$

Exercise 15g

Solve the following equations. Write down as many steps as you need. Check each solution.

1. $4n + 3 = 19$
2. $3y + 8 = 41$
3. $9c + 11 = 65$
4. $5b - 12 = 3$
5. $2x - 19 = 5$
6. $7s - 10 = 18$
7. $1 + 2u = 23$
8. $5 + 6m = 29$
9. $3 + 8p = 51$
10. $7 = 3a + 4$
11. $15 = 7t + 1$
12. $29 = 5f + 4$
13. $47 = 10h - 33$
14. $1 = 3y - 14$
15. $0 = 8d - 56$
16. $32 = 6 + 13k$
17. $19 = 7 + 4q$
18. $21 = 5 + 2n$

(19) $2t + 3 = 14$ (20) $5a + 14 = 22$

(21) $6z + 9 = 23$ (22) $4x - 5 = 5$

(23) $9y - 4 = 0$ (24) $3b - 15 = 16$

(25) $9 = 3h + 7$ (26) $11 = 10k + 4$

(27) $5 = 4m + 4$ (28) $0 = 6z - 11$

(29) $8 = 11y - 8$ (30) $19 = 16y - 21$

Summary

When an algebraic expression is put equal to another algebraic expression or equal to a numeric value, the statement is called an **equation**.

In an equation, when we find the value of the letter that makes the statement true, it is called **solving the equation**.

In the **balance method** of solving an equation, the same mathematical operation, that is addition, subtraction, multiplication and/or division, is applied to *both* sides of the equation until only the letter (the unknown, e.g. x, a) is on one side of the equation.

Using a flow chart, doing the inverse operations in the reverse order gives the solution of the equation.

Practice exercise P15.1

Solve the following equations using the balance method

(1) $k + 7 = 13$ (2) $n + 4 = 14$

(3) $v - 6 = 4$ (4) $g - 2 = 13$

(5) $5b = 30$ (6) $\frac{c}{4} = 6$

(7) $b + 14 = 37$ (8) $f - 14 = -8$

(9) $4u = 48$ (10) $5y = 80$

Practice exercise P15.2

For each question, draw flow charts
(a) to show the operation
(b) to find the unknown number, n.

Example $p + 4 = 11$

Answer

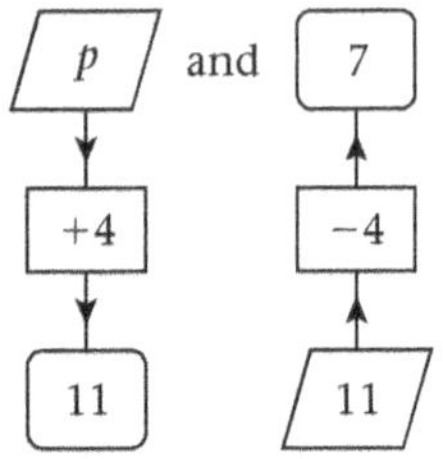

Fig. 15.15

(1) $n + 7 = 11$ (2) $n - 11 = 27$

(3) $n \times 6 = 36$ (4) $n \div 5 = 7$

(5) $2n + 3 = 15$ (6) $3n - 5 = 1$

(7) $7n \times 2 = 56$ (8) $\frac{1}{3}n \div 4 = 3$

Practice exercise P15.3

Solve the following equations

(1) $\frac{n}{5} = 10$ (2) $\frac{s}{12} = 21$

(3) $2b + 9 = 21$ (4) $5a + 6 = 26$

(5) $9y + 11 = 83$ (6) $7b + 11 = 67$

(7) $3u - 5 = 34$ (8) $4l - 12 = 40$

(9) $\frac{m}{7} - 5 = 7$ (10) $\frac{v}{8} - 15 = 3$

Practice exercise P15.4

Solve the following equations. Write down all the steps and check your answers.

(1) $5x + 6 = 1$ (2) $4c = -15$

(3) $\frac{5u}{8} - 3 = \frac{3}{4}$ (4) $4 = 2x + 10$

(5) $b - 4 = -2$ (6) $3(4m - 1) = 0$

(7) $-3 = 5(n + 3)$ (8) $2(2a + 1) = 1$

(9) $-(2y - 2) = 1$ (10) $\frac{m}{5} - 2 = 3$

Revision exercise 4 (Chapter 9)

1. Simplify the following.
   (a) 3.46×1000
   (b) 80×10^2
   (c) $5.15 \div 100$
   (d) $247 \div 10^3$

2. Simplify the following.
   (a) $9.04 + 6.7$
   (b) $9.04 - 6.7$
   (c) 8×0.5
   (d) 0.6×0.04
   (e) $\dfrac{0.42}{7}$
   (f) $36 \div 0.0004$

3. How many cans, each holding $1.8 \, \text{cm}^3$, can be filled from a tank containing $54 \, \text{cm}^3$?

4. A man bought a house for $253 500. After 5 years its value had increased by $33\frac{1}{3}\%$. Calculate its value after 5 years.

5. Express 0.275
   (a) as a fraction in its lowest terms,
   (b) as a percentage.

6. Divide 4.914 by 0.091.

7. During a Physics lesson lasting 1 hour, it took 9 minutes to set up and clear away apparatus. What percentage of the lesson time was used in this way?

8. Express the following percentages as fractions in their lowest terms
   (a) 25% (b) 28% (c) 55% (d) $62\frac{1}{2}\%$

9. A man bought a car for $70 000. He sold it a year later for $56 000. What percentage of his money did he lose?

10. A man works 46.5 hours and is paid $12.00 per hour. What is his total pay?

Revision test 4 (Chapter 9)

1. $1200 \div 0.04 =$
   A 30 000   B 3000   C 300   D 30

2. If $23 \times 54 = 1242$, then $1.242 \div 0.54 =$
   A 0.023   B 0.23   C 2.3   D 23

3. 8.705 m in mm =
   A 0.8705   B 87.05
   C 870.5   D 8705

4. $3\frac{1}{2} + 2\frac{5}{6} =$
   A $5\frac{5}{12}$ B $5\frac{3}{4}$ C $6\frac{1}{3}$ D $6\frac{5}{12}$

5. Four pages of a 16-page newspaper are missing. The percentage missing is
   A $\frac{1}{4}$ B 4   C 16   D 25

6. Express the following fractions as recurring decimals.
   A $\frac{5}{6}$ B $\frac{5}{9}$ C $\frac{5}{11}$ D $\frac{5}{12}$

7. What percentage of the letters in the word *PROTRACTOR* are vowels?

8. Express the first quantity as a fraction of the second. Give your answers in their lowest terms.
   (a) 2 min 15 s,   7 min 30 s
   (b) 35c, $2
   (c) 2.25 kg,   3.75 kg
   (d) 75 mm,   1 m

9. 12.5 kg of flour costs $56.25. What is the cost of 1 kg of flour?

10. Find the quantity of which
    (a) 8c is 20%,
    (b) 12 m is 75%,
    (c) 5 g is $33\frac{1}{3}\%$,
    (d) $9 is 30%,

Revision exercise 5 (Chapters 10, 12)

1. How many degrees are there in
   (a) a right angle,
   (b) $\frac{1}{6}$ of a revolution,
   (c) $2\frac{1}{2}$ revolutions,
   (d) $1\frac{1}{8}$ revolutions?

② State whether the following angles are acute, obtuse or reflex.
(a) 212° (b) 81° (c) 95°
(d) 5° (e) 179° (f) 198°

③ Use a protractor to measure AÔB and BÔC in Fig. R5.

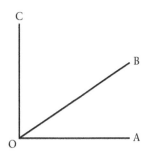

Fig. R5

④ Use a protractor to measure and draw angles of
(a) 40°, (b) 90°, (c) 150°.

⑤ Construct a copy of Fig. R6 such that AÔB = 66°. Measure BÔC.

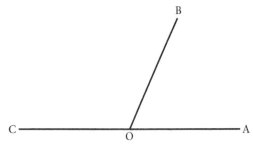

Fig. R6

⑥ Find in degrees the angle between the hour hand and the minute hand of a clock at half past 10.

⑦ How many lines of symmetry do the following have:
(a) a rectangle
(b) a square
(c) an equilateral triangle
(d) an isosceles triangle?

⑧ Two sides of an isosceles triangle are 3 cm and 10 cm. What must be the length of the third side?

⑨ In Fig. R7, ABCD is a parallelogram and ABXY is a rhombus. If BC = 3 cm and CD = 5 cm, find the total distance around the whole shape.

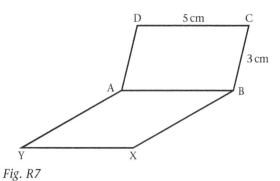

Fig. R7

⑩ In Fig. R8, ABCD and CDEF are squares and CFXY and BCYZ are rectangles. Their diagonals cross at P, Q, R, S respectively. What kind of quadrilateral is

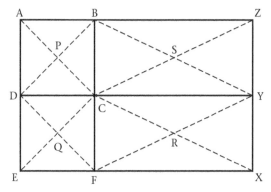

Fig. R8

(a) BPCS, (b) CQFR, (c) PDQC,
(d) SCRY, (e) SCFY, (f) BDFY,
(g) PDEC, (h) PQRS?

Revision test 5 (Chapters 10, 12)

① The angle between the hands of a clock at 2 o'clock is

A 2° B 24° C 30° D 60°

② The number of degrees in $\frac{1}{8}$ of a revolution is

A $12\frac{1}{2}$ B $22\frac{1}{2}$ C 45 D 60

③ The diagonals of one of the following *always* cross at right angles. Which one?

A rectangle   B square
C parallelogram   D regular pentagon

④ In Fig. R9, PQRT and TQRS are parallelograms.

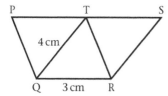

Fig. R9

QR = 3 cm and TQ = 4 cm. What is the length of PS?

A 3 cm   B 5 cm   C 6 cm   D 7 cm

⑤ Which of the following is the reflex angle, in degrees, between the hands of a clock at half past 3?

A 270°   B 75°   C 285°   D 90°

⑥ Use a protractor to measure $\widehat{ABC}$ in Fig. R10.

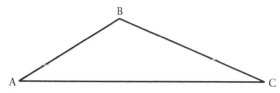

Fig. R10

⑦ Use a protractor to draw an angle of 56°.

⑧ Name four quadrilaterals which have at least one pair of parallel sides.

⑨ In a triangle ABC, AC = BC and $\widehat{CAB}$ = 58°.
(a) What type of triangle is △ABC?
(b) Calculate the two remaining angles.

⑩ Name 3 quadrilaterals whose sides may *not* meet at right angles.

Revision exercise 6 (Chapters 11, 15)

① Simplify the following.
(a) $6a + 3a$ (b) $10x + 4x - 8x$
(c) $3n - 8n + 7n$ (d) $4x \times 5$
(e) $24y \div 3$ (f) $\frac{1}{2}$ of $8c$

② Simplify the following.
(a) $8n - 5 + 2n$
(b) $5 - x - 5x$
(c) $3a + 4b - a + 2b$

③ If $x = 2$, find the value of
(a) $3x$ (b) $3x - 1$
(c) $17 - 5x$ (d) $\frac{14}{x} - 5$

④ Simplify the following.
(a) $2 \times 7 + (-12) + 15 \div 5$
(b) $2 \times (7 + (-12) + 15) \div 5$
(c) $18x \div 6 - 4x + 3 \times 2x$
(d) $3x + 8 \times (3x - 13x) \div 5$

⑤ Say whether each of the following is true or false.
(a) $11 - x = 8$ when $x = 19$
(b) $36 = 24x$ when $x = 1\frac{1}{2}$

⑥ Solve the following.
(a) $5x = 35$ (b) $4x = 18$
(c) $6c - 2c = 12$ (d) $\frac{x}{5} = 20$
(e) $\frac{28}{x} = 7$ (f) $\frac{3}{4}w = 15$

⑦ Solve the following.
(a) $x - 4 = 2$ (b) $8 = \frac{1}{2}x$
(c) $12 - x = 0$ (c) $3a + 5 = 23$
(e) $12 = 5b - 8$ (f) $7 + 2c = 19$

⑧ Solve the following.
(a) $2x - 8 = 1$
(b) $2 = 3x - 2$
(c) $5x + 3 = 10$

⑨ A table top is a square of side l metres. A cloth to cover the table must be a square of side 0.5 m longer than each edge of the table. Write an expression in l for:
(a) the length of each edge of the cloth,
(b) the total length of the four edges of the cloth.

⑩ The total length around the edges of the cloth in Question 9 is 10 m.
(a) Write an equation in l.
(b) Solve the equation to find the value of l.

Revision test 6 (Chapters 11, 15)

1. When $x = 8$, the value of $18 - x$ is
   A 1   B 8   C 10   D 18

2. $9 \times 2 - 12 \div 2 + 2 =$
   A -43 B -18 C 5   D 14

3. $12x^2y \div 3x =$
   A $4xy$ B $8y$ C $9xy$ D $12xy$

4. If $x = 3$, $5x - 2x \times 4 + 12x \div 3 =$
   A 3   B 9   C 24   D 48

5. If $6x + 7 = 55$, then $x =$
   A 7   B 8   C 42   D 48

6. Find the value of the following when $a = 3$, $b = 1$ and $c = 4$.
   (a) $-2b$ (b) $\dfrac{a+b}{c}$ (c) $\dfrac{c}{a-b}$

7. Simplify the following:
   (a) $a - 5a + 8a - 2a$
   (b) $6x - 2y - 5y - 3x$

8. Simplify the following:
   (a) $3xy \times 9y$ (b) $2n \times 5an^2$
   (c) $36a^2b \div 12ab$ (d) $\dfrac{5x^2}{x}$

9. Solve the following:
   (a) $13 - x = 10$ (b) $\dfrac{a}{3} = 3$
   (c) $y + 8 = 20$ (d) $4n - 3 = 17$
   (e) $50 = 7d + 1$ (f) $12x + 8 = 20$

10. In a basket, there are grapefruits, oranges, pineapples and mangoes. There are n grapefruits and twice as many oranges as grapefruits. There are three more mangoes than oranges and half as many pineapples as there are grapefruits.
    (a) Write an expression in n for the total number of fruits.
    (b) If there are 14 fruits, write an equation and solve the equation to find the value of n.
    (c) Find the number of mangoes in the basket.

Revision exercise 7 (Chapters 13, 14)

1. Round off 29 835 to the nearest
   (a) thousand,   (b) hundred,
   (c) ten.

2. Round off 0.845 to the nearest
   (a) tenth,   (b) hundredth,
   (c) whole number.

3. Estimate the following
   (a) 8.6×5.4 (b) $10\frac{1}{3} \times 5\frac{3}{4}$
   (c) $0.82 \div 0.39$

4. A farmer has 385 cattle. The average value of each beast is $518. Estimate the total value of the farmer's cattle.

5. A box contains eight identical record players. If the mass of the box and players is 101.6 kg, estimate the mass of one record player.

6. Divide 2.647 by 0.9 and give the answer correct to 2 d.p.

7. The perimeter of a rectangle is 36 cm. Find the breadth of the rectangle if its length is
   (a) 17 cm,   (b) 12 cm,   (c) 9 cm.

8. A wire ring has a diameter of 1 m. Use 3.14 for π to calculate the length of the wire.

9. A lorry has a wheel of diameter 66 cm. Use the value 3 for π to estimate the number of times the wheel turns when the lorry travels 1 km.

10. (a) The number of people at a lecture is given as 150. If this value was stated to the nearest ten, state the range of values for the number present.
    (b) The length of a box is measured as 10.3 cm to the nearest tenth of a centimetre. State the range of values for the length of the box.

Revision test 7 (Chapters 13, 14)

1. The perimeter of a rectangle is 26 cm. Its breadth is 4 cm. Its length in cm is
   A 9   B 11   C 13   D 17

2. 67.053 to the nearest tenth is
   A 67   B 67.0   C 67.1   D 67.05

③ Which one of the following is most likely to be the correct value of $\$3.90 \times 7.8$?
A $20 B $21.52
C $30.42 D $33.12

④ Which *one* of the following is *not* sensible?
A The woman's hand-span was 20 cm.
B The boy ran 100 km in an hour.
C The height of the tree was 5.8 m.
D The cup held 280 cm³ of tea.

⑤ 0.003 867 to 3 s.f. is
A 0.004 B 0.003 86
C 0.003 87 D 386

⑥ The frame shown in Fig. R11 is a semi-circle joined to a square of side 1.2 m.

Fig. R11

Use $\pi = 3.14$ to calculate the distance around the frame, correct to 2 decimal places.

⑦ Estimate the cost of 20.5 hectares of land at $19 600 per hectare.

⑧ A hotel bill for nine days was $596.40. Estimate the daily cost.

⑨ A man's foot is about 28 cm long. He finds that the width of a room is about 11 of his feet. Find the approximate width of the room in metres.

⑩ A student writes 78 words in eight lines of writing.
(a) Find, to the nearest whole number, the average number of words per line.
(b) Hence estimate how many lines of writing it will take to write a 1 500-word essay.

General revision test B (Chapters 9 – 15)

①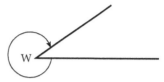

Fig. R12

The angle marked W in Fig. R12 is
A acute B right C obtuse D reflex

② $a(b + c) - 2b(a + c) =$
A $-ab$ B $-ab + 2c$
C $-ab + ac + 2bc$ D $-ab + ac - 2bc$

③ How many lines of symmetry does an isosceles triangle have?
A 0 B 1 C 2 D 3

④ $16 - x = x$ is true when $x =$
A 0 B 8 C 14 D 16

⑤ If $3.4 \times 1.8 = 6.12$, then $61.2 \div 0.18 =$
A 0.34 B 3.4 C 34 D 340

⑥ In 2005 the estimated population of a country was 270 000. The area of the country is 13 900 km². Estimate the 2005 population density (i.e. the number of people per km²).
A 20 B 5 C 0.5 D 0.2

⑦ $x = 23\frac{4}{5} \div 8\frac{1}{2}$.
Use estimation to decide which one of the following is the accurate value of x.
A $1\frac{2}{5}$ B $2\frac{4}{5}$ C $3\frac{9}{10}$ D $4\frac{1}{2}$

⑧ A length of wire is given as 6.8 cm correct to 2 s.f. What is the least possible length of the wire?
A 6.7 cm B 6.74 cm
C 6.75 cm D 6.8 cm

⑨ One pencil costs r cents and one pen costs $\$s$. The cost in $ of b pencils and d pens is
A $\dfrac{br}{100} + ds$ B $(b + d)(r + s)$
C $br + ds$ D 100

⑩ $166\frac{2}{3}\%$ of 0.6 is
A 0.2 B 0.36 C 1.0 D 2.7

⑪ Simplify:
   (a) $2m \times 3mn$ (b) $\frac{1}{3}$ of $12x^2y$
   (c) $\dfrac{4a^3b^4}{18a^2b^4}$ (d) $6p + 6p \div 3$
   (e) $12s \div (4 + 2)3t$

⑫ A thread is wound 100 times round a reel of diameter 3 cm. Use 3.14 for π to calculate the length of the thread.

⑬ When $u = 10$, $a = 4$, $t = 15$, find the value of
   (a) $u + at$
   (b) $ut + \frac{1}{2}at^2$

⑭ What fraction of $1.75 is 77c? Express this fraction as (a) a decimal, (b) a percentage.

⑮ Solve the following:
   (a) $5 + 8a = 37$ (b) $40 = 14a - 30$
   (c) $2a - 1 = 31$ (d) $3x - 4 = 1$
   (e) $7 = 5 + 5x$ (f) $7 + 8x = 9$

⑯ A man walks at the rate of 88 paces to the minute. If each pace is 0.85 m long, how far does he walk in 10 min?

⑰ Ten tomatoes weigh 628 g. A woman buys $2\frac{1}{2}$kg of tomatoes. Approximately how many tomatoes will she get?

⑱ Express 0.504 6 correct to (a) 3 decimal places, (b) 2 significant figures, (c) the nearest tenth.

⑲ A shopkeeper bought a radio for $412.50. The price was increased by 42%. What was the new price?

⑳ Make a drawing like that in Fig. R13 such that $A\widehat{C}D = 114°$. $\widehat{A}$ and $\widehat{B}$ can be any size. Measure $\widehat{A}$ and $\widehat{B}$. Find the sum of $B\widehat{A}C$ and $A\widehat{B}C$.

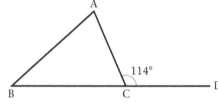

Fig. R13

Chapter 16
Plane shapes (2)

Pre-requisites
■ properties of plane shapes; basic units of measurement of length; approximation and estimation

Area

The **area** of a shape is a measure of its surface.

Exercise 16a (Group work)

Work in pairs. Taking turns, one student does the task and his/her partner checks the results.

① Measure the area of the shapes in Fig. 16.1 by counting the squares that each shape contains. If bits are left over, try to estimate by adding them together to make whole squares.

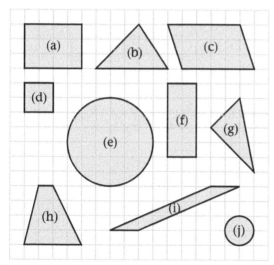

Fig. 16.1

② Find a large leaf. Put the leaf on $1\,cm^2$ squared paper and draw round it. Measure the area of the leaf in cm^2 by counting squares. Estimate the total number of parts of squares as before.

③ Use the method of question 2 to find the area of your hand. (First draw round your hand on graph paper.)

The square is used as the shape for the basic unit of area. A square of side 1 m covers an area of **1 square metre** or **1 m²**. A square of side 1 cm (Fig. 16.2) covers an area of **1 square centimetre** or **1 cm²**.

Fig. 16.2

Area of a rectangle and a square

A rectangle 5 cm long by 3 cm broad can be divided into squares of side 1 cm as shown in Fig. 16.3.

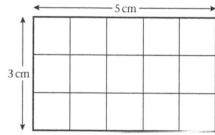

Fig. 16.3

By counting, the area of the rectangle $= 15\,cm^2$. Notice also that $5 \times 3 = 15$; thus, in general:

area of rectangle = length × breadth

Example 1
Calculate the area of a rectangle 6 cm by 3.5 cm.

Area of rectangle $= (6 \times 3.5)\,cm^2 = 21\,cm^2$

A square is a rectangle whose length and breadth are equal, thus:

area of square = (length of side)²

Exercise 16b

1. Copy and complete the table for rectangles (Table 16.1).

   Table 16.1

	Length	Breadth	Area
(a)	3 cm	2 cm	
(b)	5 m	4 m	
(c)	6 m	$2\frac{1}{2}$ m	
(d)	5.2 m	3 m	
(e)	2.1 cm	3.4 cm	
(f)	4 m		18 m²
(g)	5 m		15 m²
(h)		3 m	12 m²

2. Copy and complete the table for squares (Table 16.2).

   Table 16.1

	Length of side	Area
(a)	6 cm	
(b)	9 m	
(c)		49 m²
(d)		16 cm²
(e)	1.2 m	
(f)	$2\frac{1}{2}$ cm	

Example 2

An assembly area is in the shape of a 30 m by 30 m square. Part of the area is a concrete rectangle 25 m by 5 m; the rest is grass. Calculate the area of the grass.

Make a sketch of the assembly area as shown in Fig. 16.4.

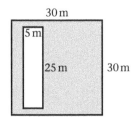

Fig. 16.4

Area of assembly area $= (30 \text{ m})^2 = 900 \text{ m}^2$
Area of concrete $= (25 \times 5) \text{ m}^2$
$= 125 \text{ m}^2$
Area of grass $= 900 \text{ m}^2 - 125 \text{ m}^2$
$= 775 \text{ m}^2$

Example 3

The area of a rectangle is 24 cm² and one side is 6 cm in length. Find its breadth and perimeter.

Breadth of rectangle $= (24 \div 6) \text{ cm} = 4 \text{ cm}$
Perimeter $= 2(6 + 4) \text{ cm} = 2 \times 10 \text{ cm} = 20 \text{ cm}$

Example 4

Calculate the area of the shape in Fig. 16.5. all measurements are in metres and all angles are right angles.

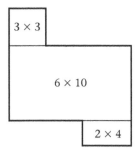

Fig. 16.5

The shape can be split into a 3 by 3 square and 6 by 10 and 2 by 4 rectangles (Fig. 16.6).

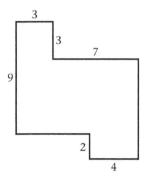

Fig. 16.6

Area = area of square + area of two rectangles
$= (3 \times 3 + 6 \times 10 + 2 \times 4) \text{ m}^2$
$= (9 + 60 + 8) \text{ m}^2 = 77 \text{ m}^2$

Exercise 16c

1. Measure the length and width of this page to the nearest millimetre. Hence find the area of this page to the nearest square centimetre.

2. Calculate the areas of the shapes in Fig. 16.7. All lengths are in metres and all angles are right angles.

(a)

(b)

(c)

(d)

(e)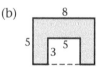

(f)

Fig. 16.7

3. Calculate the *shaded areas* in the diagrams in Fig. 16.8. All lengths are in centimetres and all angles are right angles. (The diagrams are not drawn to scale.)

(a)

(b)

(c)

(d)

(e)

(f)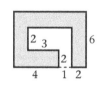

Fig. 16.8

4. A sheet is 2.15 m long and 1.6 m wide. What is the area of the sheet?

5. In a dining hall 25 m by 12 m, an area 8 m square is kept clear for cooking. What area is left over for dining?

6. A 12 m by 12 m square garden has a 1 m wide path through the centre, parallel to one side of the square. This is shown in Fig. 16.9. Calculate the area of garden left over for planting.

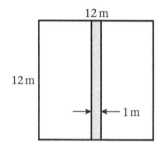

Fig 16.9

7. A cardboard box is 20 cm long, 12 cm wide and 8 cm deep (Fig. 16.10). Calculate the total area of cardboard in the box.

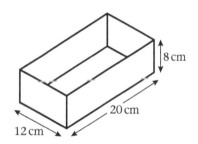

Fig 16.10

8. The box in question 7 is cut down the edges and flattened as shown in Fig. 16.11(a).

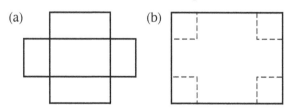

Fig 16.11

Calculate the area of the smallest single rectangle from which the box could be made, i.e. calculate the area of the rectangle in Fig. 16.11(b). What area would be wasted?

9 Calculate the total area of the walls of a room 5 m long, 4 m wide and $2\frac{1}{2}$ m high. (Do not allow for doors and windows.)

10 1 litre of paint covers 15 m². How many litres of paint will be needed to paint the walls of the room in question 9?

11 Calculate the area of the ceiling of the room in question 9. How many $2\frac{1}{2}$ m by 1 m ceiling boards will be needed to cover the ceiling of the room?

12 A boy has 16 matchsticks. He lays them in the shape of a rectangle (Fig. 16.12). Call each matchstick a unit and find the perimeter and area of the rectangle in units and units².
Use matchsticks or squared paper to show the different rectangles which the boy can make using all 16 matchsticks. Find the perimeters and areas of these rectangles. What do you notice?

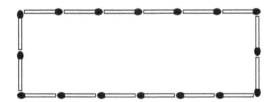

Fig 16.12

Area of a parallelogram

The diagrams in Fig. 16.13 show how any parallelogram, P, can be changed to a rectangle, R, by moving a triangle, T, from one end to the other.

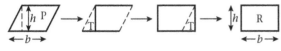

Fig 16.13

Draw a parallelogram on a sheet of graph paper. Mark a triangle as in Fig. 16.13. Cut off the triangle and rearrange the parallelogram to form a rectangle.

What can you say about the area of the parallelogram and the area of the rectangle?

area of parallelogram, P = area of rectangle, R
$$= b \times h$$

In the diagram, the height of the parallelogram is h and its base is b. In general:

area of parallelogram = base × height

A parallelogram can have two bases and two corresponding heights as shown in Fig. 16.14. Note that the height must be **perpendicular** to the base.

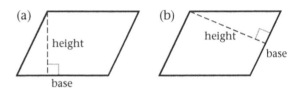

Fig. 16.14 Area = base × height

Example 5

In Fig. 16.15, the base of the parallelogram is 6 cm long and its height is 4 cm. Calculate the area of the parallelogram. If the length of the other side of the parallelogram is 8 cm, calculate its corresponding height, h.

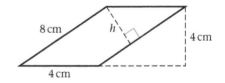

Fig. 16.15

Area of parallelogram $= (6 \times 4)\,\text{cm}^2$
$$= 24\,\text{cm}^2$$
height $=$ area $\div$ base
$$= (24 \div 8)\,\text{cm}$$
$$= 3\,\text{cm}$$

Exercise 16d

Calculate the areas of the parallelograms in Fig. 16.16 (numbers 1 to 5). All dimensions are in cm.

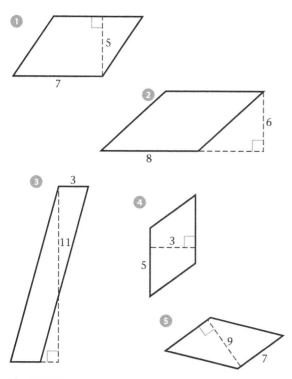

Fig. 16.16

In each of the parallelograms in Fig. 16.17 (numbers 6 to 8), calculate the height, *h*.

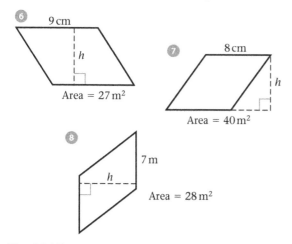

Fig. 16.17

In each of the parallelograms in Fig. 16.18 (numbers 9 to 11), calculate the base, *b*.

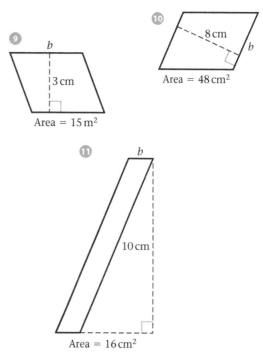

Fig. 16.18

In each of the parallelograms in Fig. 16.19 (numbers 12 to 15), calculate (a) the area, (b) the height *h* or the base *b*.

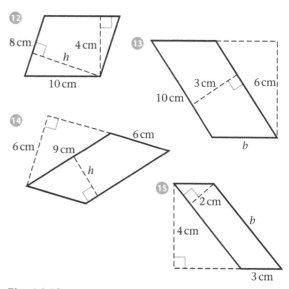

Fig. 16.19

Area of a triangle

Right-angled triangle

Any diagonal of a rectangle divides it into two
equal right-angled triangles (Fig. 16.20). Thus:

area of a right-angled triangle
$$= \tfrac{1}{2} \times \text{length} \times \text{width}$$
$$= \tfrac{1}{2} \times \text{base} \times \text{height}$$

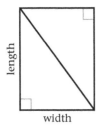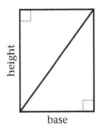

Fig 16.20

Any triangle

Any diagonal of a parallelogram divides it into
two equal triangles (Fig. 16.21).

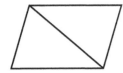

Fig 16.21

Thus the area of each triangle is half the area of
the parallelogram (Fig. 16.22).

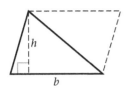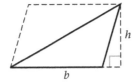

Fig 16.22

Since, area of parallelogram = base × height,

area of triangle = $\tfrac{1}{2}$ × base × height

Notice that any side of a triangle can be taken as
base. Each base has its corresponding height.

Example 6

Calculate the area of the triangle shown in
Fig. 16.23.

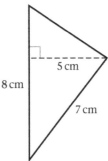

Fig. 16.23

The height is 5 cm. The corresponding base is
8 cm. We do not need the 7 cm side.

Area of triangle $= \tfrac{1}{2} \times \text{base} \times \text{height}$
$$= (\tfrac{1}{2} \times 8 \times 5)\,\text{cm}^2 = 20\,\text{cm}^2$$

Example 7

Calculate the area of the triangle shown in
Fig. 16.24.

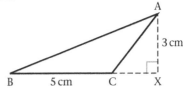

Fig. 16.24

Notice that the height from A to base BC falls
outside the triangle.

Area of $\triangle$ABC $= \tfrac{1}{2} \times \text{base BC} \times \text{height AX}$
$$= (\tfrac{1}{2} \times 5 \times 3)\,\text{cm}^2$$
$$= 7\tfrac{1}{2}\,\text{cm}^2$$

Example 8

Calculate the area of the quadrilateral ABCD in
Fig. 16.25.

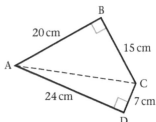

Fig 16.25

Notice that the diagonal AC divides the quadrilateral into two right-angled triangles.

Area of △ABC = ($\frac{1}{2}$ × 20 × 15) cm² = 150 cm²

Area of △ ADC = ($\frac{1}{2}$ × 24 × 7) cm² = 84 cm²

Area of ABCD = (150 + 84) cm² = 234 cm²

Example 9

Calculate the area of the quadrilateral ABCD in Fig. 16.26.

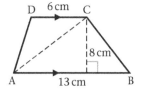

Fig. 16.26

The diagonal AC divides the quadrilateral into two triangles. The height of each triangle is 8 cm.

Area of △ACB = ($\frac{1}{2}$ × 13 × 8) cm² = 52 cm²

Area of △ACD = ($\frac{1}{2}$ × 6 × 8) cm² = 24 cm²

Area of quadrilateral = (52 + 24) cm² = 76 cm²

Exercise 16e

1   Calculate the areas of the triangles in Fig. 16.27. All dimensions are in cm.

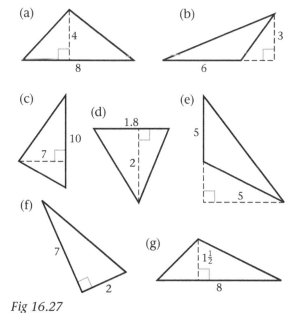

Fig 16.27

2   Calculate the areas of the quadrilaterals in Fig. 16.28. In each case, draw a diagonal to divide the shapes into two triangles. All dimensions are in cm.

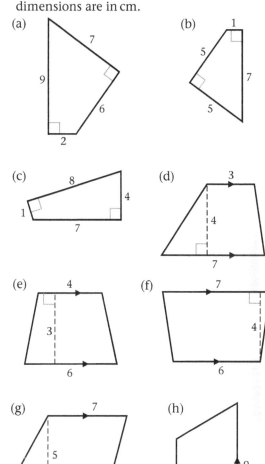

Fig 16.28

Area of a trapezium

In Fig. 16.29, ABCD is a trapezium with AB parallel to DC.

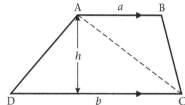

Fig. 16.29

The lengths of AB and DC are a and b respectively. Let their perpendicular distance apart be h. Join AC.

Area of ABCD

$= $ area of $\triangle$ABC $+$ area of $\triangle$ACD

$= \frac{1}{2}ah + \frac{1}{2}bh$

$= \frac{1}{2}h(a + b)$ or $\frac{1}{2}(a + b)h$

The area of a trapezium is the product of the average length of its parallel sides and the perpendicular distance between them.

Area of trapezium =
$\frac{1}{2}$ × (sum of parallel sides) × height

Example 10

Find the area of the trapezium in Fig. 16.30.

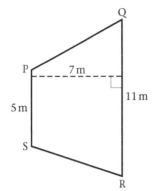

Fig. 16.30

In Fig. 16.30, SP is parallel to RQ.

Area of PQRS $= \frac{1}{2}(5 + 11) \times 7$ m^2

$= (\frac{1}{2} \times 16 \times 7)$ m^2

$= (8 \times 7)$ m$^2 = 56$ m^2

Example 11

If the area of the trapezium in Fig. 16.31 is $40\frac{1}{2}$ cm^2, find the value of x.

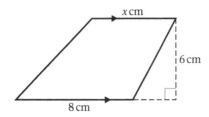

Fig 16.31

Area of trapezium $= \frac{1}{2}(x + 8) \times 6$ cm^2

$= 3(x + 8)$ cm^2

Thus, $3(x + 8) = 40\frac{1}{2}$

$x + 8 = \frac{81}{2} \div 3 = 13\frac{1}{2}$

$x = 13\frac{1}{2} - 8 = 5\frac{1}{2}$

Exercise 16f

① Find the areas of the trapeziums in Fig. 16.32. All dimensions are in cm.

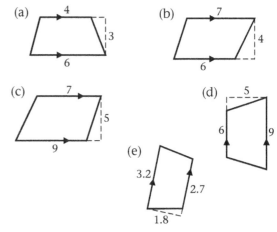

Fig. 16.32

② In each of the trapeziums in Fig. 16.33, find the value of x.

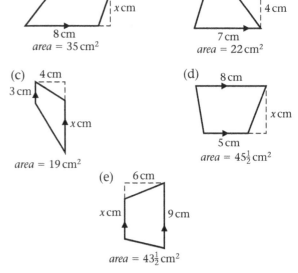

Fig. 16.33

Everyday problems with area

Tiles are often used to cover the floor of a room. The number of tiles needed can be calculated from the dimensions of the room. There is usually some wastage since whole tiles and not fractions or parts are sold.

Example 12

Square tiles, 30 cm × 30 cm, are used to cover a floor. How many tiles are needed for a floor 4.4 m long and 3.8 m wide?

Length of room = 4.4 m = 440 cm

Number of tiles = $\frac{440}{30}$ = $14\frac{2}{3}$

Thus 15 tiles are needed along each length of the room. (The last tile will be cut.)

Width of room = 3.8 m = 380 cm

Number of tiles = $\frac{380}{30}$ = $12\frac{2}{3}$

Thus 13 tiles are needed across each width of the room.

Total number of tiles needed = 15 × 13
$$= 195$$

Exercise 16g

① How many tiles, each 30 cm by 30 cm, will be needed for floors with the following dimensions?
(a) 6 m by 4.2 m   (b) 3.6 m by 3 m
(c) 5 m by 4.2 m   (d) 9 m by 6.2 m
(e) 10 m by 8.4 m   (f) 5.2 m by 4.1 m

② Square polystyrene tiles, 50 cm by 50 cm, are used to cover the ceiling of a classroom measuring 7.4 m by 4.5 m.
(a) Find the number of tiles that are needed.
(b) Find the cost at 65c per tile.

③ An open rectangular box, 1 m long, 70 cm wide and 50 cm deep is painted inside and outside. Find the cost at $1.50 per m².

④ How many paving stones, each 1 m long and 80 cm wide, are needed to cover an area 13.6 m long and 11 m wide?

⑤ Fig. 16.34 is a sketch of a building with a corrugated iron roof.

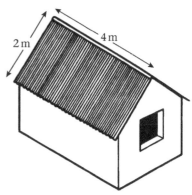

Fig. 16.34

If corrugated iron is sold in sheets measuring 2 m by 60 cm, find the number of sheets that are needed for the building.

⑥ The walls of a bathroom 2.5 m long, 2.05 m wide and 3 m high are to be covered with tiles 15 cm by 15 cm. If a saving of 108 tiles is made on doors and windows, how many tiles will be needed altogether? (Note that some tiles will have to be cut.)

⑦ A rectangular area, 8.55 m long by 5.89 m wide, is to be paved with the largest possible square tiles which will fit in exactly.
(a) Calculate the length of the side of one tile. (b) How many tiles will there be? (*Hint*: express 855 and 589 as products of prime numbers.)

⑧ Using the data of Example 12, divide the area of the floor of the room by the area of one tile. What do you notice?
Can you explain the difference? It might help if you try to cover a graph page 10 cm long by 9 cm wide with square tiles 3 cm × 3 cm.

Area of a circle

Group work

The students form groups of four. In each group, the students work in pairs and compare their results. Each group will need paper, scissors and mathematical instruments.

All the groups will be drawing a circle and dividing it into equal sectors, as in Fig. 16.35.

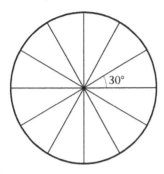

Fig. 16.35

The radius of the circle, r, must be large enough so that the sectors can be cut out and arranged as in Fig. 16.36.

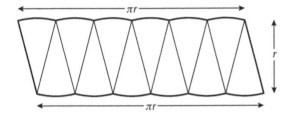

Fig. 16.36

Each group uses a different value for the angle of the sector. The two pairs in each group use the same angle but may decide on different values for the radius of the circle.

The values for the angles, that is: 10°, 12°, 15°, 20°, 24°, 30°, 40°, 45°, 60°, are placed in a box and one student from each group randomly chooses the value for the group. Note that the number of sectors will increase as the size of the angle decreases, so that when the angle of the sector is 60°, there are 6 sectors; and when the angle is 10°, there are 36 sectors. The two pairs in each group compare their final shape in Fig. 16.36 and discuss whether the difference in the values of r was important. The different groups then compare their Fig. 16.36 and discuss how the shape changed as the angle of the sector changed.

The new shape in Fig. 16.36 looks very much like a 'parallelogram'.

The height of the 'parallelogram' is r and its base is of length πr (i.e. half of the circumference of the original circle).

When the circle is divided into more sectors, the sides of the 'parallelogram' become much straighter. (For example, try this activity with 36 sectors, each of 10°.)

Suppose that the shape is a parallelogram, then,

$$\text{area of parallelogram} = \pi r \times r$$
$$= \pi r^2$$

and, hence,

$$\text{area of original circle} = \pi r^2$$

Thus, to calculate the area of any circle of radius r, use the formula:

area of circle $= \pi r^2$

Example 13

Find the area of a circle of radius $3\frac{1}{2}$ metres. Use the value $\frac{22}{7}$ for π.

$$\begin{aligned} \text{Area of circle} &= \pi r^2 \\ &= \tfrac{22}{7} \times (3\tfrac{1}{2})^2 \text{ m}^2 \\ &= \tfrac{22}{7} \times \tfrac{7}{2} \times \tfrac{7}{2} \text{ m}^2 \\ &= \frac{11 \times 7}{2} \text{ m}^2 = 38\tfrac{1}{2} \text{ m}^2 \end{aligned}$$

Example 14

The shape of figure 16.37 is a rectangle with a semi-circle on the open side.

Calculate the area of the shape

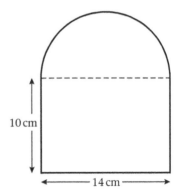

10 cm

14 cm

Fig. 16.37

Area of rectangle $= 14\,\text{cm} \times 10\,\text{cm}$
$= 140\,\text{m}^2$
Diameter of semi-circle $= 14\,\text{cm}$
Radius of semi-circle $= 7\,\text{cm}$
Area of semi-circle $= \frac{1}{2}\pi r^2$
$= \frac{1}{2} \times \frac{22}{7} \times 7 \times 7\,\text{cm}^2$
$= 77\,\text{m}^2$
Area of shape $= 140\,\text{cm}^2 + 77\,\text{cm}^2$
$= 217\,\text{cm}^2$

Exercise 16h

1 Copy and complete Table 17.1 for circles. Use the value $\frac{22}{7}$ for π.

Table 16.3

	Radius	Diameter	Area
(a)	7 m		
(b)		7 cm	
(c)	140 mm		
(d)		28 m	
(e)		$3\frac{1}{2}$ cm	
(f)	2.1 cm		

2 Calculate the area of each shape in Fig. 16.38. All lengths are in cm. Use the value $\frac{22}{7}$ for π. In each case, make a sketch of the shape and put in as many dimensions as possible, especially the radius of any circular parts.

(a)
7

(h)
7

(c)
$1\frac{3}{4}$

Fig. 16.38

3 A circular mat has a diameter of 20 cm. Find its radius and hence calculate the area of the mat. Use the value 3.14 for π.

4 A goat is tied to a peg in the ground. The rope is 3 m long. What area of grass can the goat eat? Use the value 3.1 for π.

5 A protractor is in the shape of a semi-circle of radius 5 cm. Calculate the area of the protractor. Use the value 3.14 for π.

6 There are two circles, one large and one small. The radius of the large circle is three times the radius of the small circle. Find the value of the fraction:

$$\frac{\text{area of small circle}}{\text{area of large circle}}$$

7 A design is made by drawing seven small circles inside one large circle as shown in Fig. 16.39.

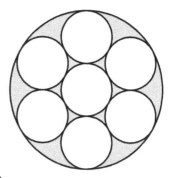

Fig. 16.39

If the diameter of the large circle is 30 cm, calculate:
(a) the radius of the large circle,
(b) the radius of each small circle,
(c) the area shaded in the diagram.
Use the value 3.14 for π.

8 The sports field shown in Fig. 16.40 has a 90 m by 70 m football field with a semi-circular area at each end. A track runs round the perimeter of the sports field.

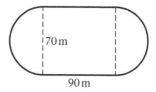
70 m
90 m

Fig. 16.40

Use the value $\frac{22}{7}$ for π to calculate:
(a) the area of the sports field,
(b) the length of one lap of the track.

Summary

The **area** of a shape is a measure of its surface. The following are the areas of some plane shapes:

area of rectangle = length × breadth

area of square = (length of side)^2

area of parallelogram = base × height

area of triangle = $\frac{1}{2}$ base × height

area of trapezium

$= \frac{1}{2}$ (sum of parallel sides)

× perpendicular distance between them

area of circle = πr^2

Practice exercise P16.1

Remember: Unless otherwise stated, use $\pi = \frac{22}{7}$

① Twenty square tiles of side 30 cm completely cover the floor of a room. What is the area of the floor of the room?

② Find the area of the following figures. All measurements are in centimeters.

(a)

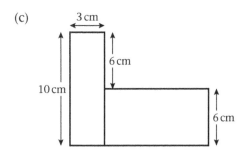

Fig. 16.41

(b)

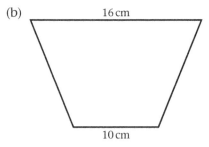

Fig. 16.42

(c)

Fig. 16.43

③ Calculate the area of the following shapes:
   (a) rectangle, length 8.2 cm, width 10 cm
   (b) square of side 13.5 cm
   (c) triangle, height 15 cm, base 9.6 cm.

④ The minute-hand on a clock is 14 cm long. What area does the minute-hand cover in one hour?

⑤ The diameter of an East Caribbean one-dollar coin is 2.8 cm.
   (a) What is the radius of the coin?
   (b) Calculate the area of the surface of the coin.

   Ten of these one-dollar coins are arranged in two rows to just fit in a rectangle.

   (c) What is the length of the rectangle?
   (d) What is the width of the rectangle?
   (e) What is the area of the region covered by the coins?

6 (a) What is the diameter of the largest possible circle that can be cut from a square sheet of cartridge paper of side 12 cm?

(b) What area of paper is left after the circle has been cut off?

7 The top of a stove is a rectangle measuring 72 cm by 50 cm.

(a) What is the area of the stove top?

The rings for the two large burners are of diameter 24.5 cm and the rings for the small burners are 17.5 cm in diameter.

(b) Calculate the total area occupied by the four burners.

(c) What area is not occupied by the burners?

8 Calculate the areas of the following shapes:
(a) a semi-circle of diameter 21 cm
(b) a quarter circle of radius 12 cm
(c) a sector, angle 60°, radius 7 cm
(d) a three-quarter circle, radius 9 cm

9 A circular lawn has a radius of 4.9 m.
(a) What is the area of the lawn?
A walking path of width 0.5 m is cut around and outside the lawn.
(b) What is the diameter of the region covered by the lawn and the walking path?
(c) What is the area of this region?
(d) What is the area of the region covered by the walking path?

Chapter 17

Angles

Angles between lines, angles of polygons

Angles between lines

Exercise 17a revises some work you did in Chapter 10. You will need a protractor, ruler and pencil.

Exercise 17a

Angles on a straight line

1. Make a drawing like that in Fig. 17.1. BCA is a straight line. $\widehat{ACD}$ can be any size.
   (a) Measure $\widehat{ACD}$ and $\widehat{BCD}$.
   (b) Find the sum of $\widehat{ACD}$ and $\widehat{BCD}$.
   (c) Compare your results with other students in your class. What do you notice?

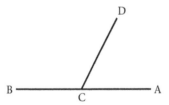

Fig. 17.1

When a straight line stands on another straight line, two adjacent **angles on a straight line** are formed.

In Fig. 17.2, $\widehat{AOB}$ is also adjacent to $\widehat{BOC}$. But AOC is not a straight line.

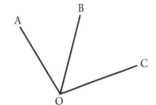

Fig. 17.2

The sum of two adjacent angles on a straight line is 180°. In Fig. 17.1, $\widehat{BCD} + \widehat{DCA} = 180°$.

Vertically opposite angles

2. Draw any two straight lines AB and CD to intersect at a point O (Fig. 17.3).

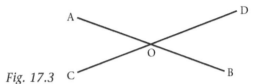

Fig. 17.3

   (a) Measure $\widehat{AOC}$ and $\widehat{DOB}$. What do you notice?
   (b) Measure $\widehat{BOC}$ and $\widehat{AOD}$. What do you notice?

When two straight lines intersect, four angles are formed. The two angles opposite each other are said to be **vertically opposite**. In Fig. 17.3, $\widehat{AOC}$ is vertically opposite $\widehat{BOD}$. $\widehat{AOD}$ is vertically opposite $\widehat{BOC}$. Vertically opposite angles are equal. $\widehat{AOC} = \widehat{BOD}$. $\widehat{AOD} = \widehat{BOC}$.

Angles meeting at a point

3. Mark a point O on your paper. Draw any five lines each starting at O. This will give five angles at O (Fig. 17.4). Mark them a, b, c, d, e. These are **angles at a point**.

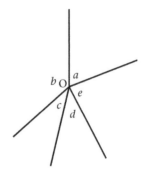

Fig. 17.4

(a) Measure the five angles, *a*, *b*, *c*, *d*, *e*.
(b) Find the sum of the five angles.
(c) Compare your results with other students in your class. What do you notice?

When a number of lines meet at a point they will form the same number of angles. The sum of the angles at a point is 360°.

Example 1

In Fig. 17.5, AOB and COD are straight lines. $\widehat{BOD} = 62°$ and $\widehat{BOE} = 77°$. Calculate the other angles in the figure.

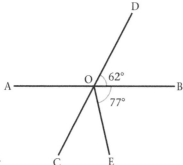

Fig. 17.5

$\widehat{AOC} = 62°$ (vertically opposite to $\widehat{BOD}$)
$\widehat{AOD} = 180° - 62° = 118°$ (adjacent angles on straight line AOB)
$\widehat{COE} = 360° - (62° + 62° + 118° + 77°)$
(sum of angles at O)
$= 360° - 319° = 41°$

Example 2

In Fig. 17.6, $\widehat{APB} = x°$, $\widehat{BPC} = \widehat{APB}$, $\widehat{CPD}$ is twice as big as $\widehat{APB}$ and reflex $\widehat{APD}$ is five times as big as $\widehat{APB}$. Make an equation in *x*. Solve the equation and find the four angles.

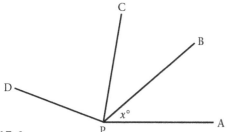

Fig. 17.6

$\widehat{APB} = x°$
$\widehat{BPC} = x°$ $(= \widehat{APB})$
$\widehat{CPD} = 2x°$ $(= 2 \times \widehat{APB})$
$\widehat{APD} = 5x°$ $(= 5 \times \widehat{APB})$
$\widehat{APB} + \widehat{BPC} + \widehat{CPD} + \widehat{APD} = 360°$
(sum of angles at P)
$$x° + x° + 2x° + 5x° = 360°$$
$$9x = 360$$
$$x = \frac{360}{9} = 40$$

The four angles are 40°, 40°, 80° and 200°.

① **(Oral)** Find the size of the lettered angles in Fig. 17.7 (a)−(l). Give reasons.

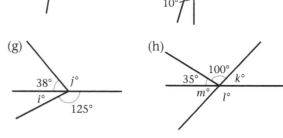

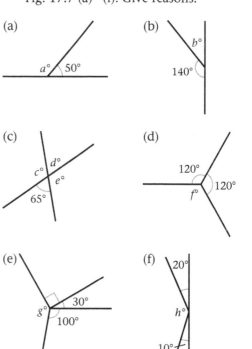

Fig. 17.7(a–h)

(i)
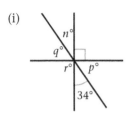

(j)
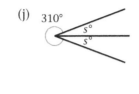

(k)
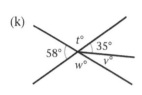

(l)
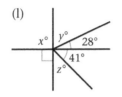

Fig. 17.7(i–l)

② In Fig. 17.8, PÔR = 37°. Calculate the other three angles. Give reasons.

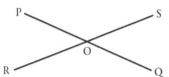

Fig. 17.8

③ In Fig. 17.9, SXP is a straight line. If PX̂Q = 31° and RX̂S = 64°, calculate QX̂R.

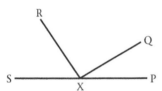

Fig. 17.9

④ In Fig. 17.9, calculate RX̂S, given that PX̂Q = 35° and RX̂Q = 98°.

⑤ In Fig. 17.9, calculate PX̂Q, if QX̂R is a right angle and RX̂S = 68°.

⑥ In Fig. 17.10, BX̂C = 26° and AX̂D = 126°. If BX̂D is a right angle, calculate CX̂D and AX̂B.

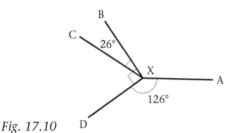

Fig. 17.10

Find the size of AX̂C. In what way could the drawing be made better?

⑦ In Fig. 17.11, AOB is a straight line. The angles marked x° are equal to each other and the angles marked y° are equal to each other. Write an equation using the letters x and y. Hence calculate MÔN.

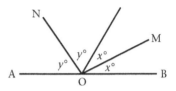

Fig. 17.11

⑧ In Fig. 17.12, EK̂F = x°, FK̂G is twice as big as EK̂F, GK̂H is three times as big as EK̂F and HK̂E is four times as big as EK̂F. Write an equation in x. Solve the equation to find the four angles. Check your answer by finding the sum of the four angles.

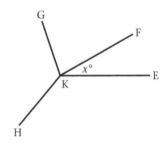

Fig. 17.12

⑨ In Fig. 17.13, VX̂W = 2 × UX̂V and WX̂Y = 3 × VX̂W. Calculate UX̂V. (*Hint*: let UX̂V = x°. Form an equation in x and then solve it.)

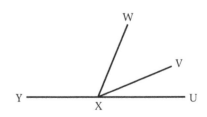

Fig. 17.13

Parallel lines

Usually, if two straight lines are drawn on a plane, they will intersect if the lines are **produced** (i.e. extended) far enough (Fig. 17.14).

Fig. 17.14

If the lines never meet, however far they are produced, we say that they are **parallel**. For example, the lines in your exercise book are parallel to each other. We sometimes show that lines are parallel by drawing arrow heads on them as in Fig. 17.15.

Fig. 17.15

Notice that the distance between a pair of parallel lines is always the same. A line cutting a pair of parallel lines is called a **transversal** (Fig. 17.16).

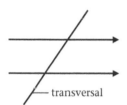

Fig. 17.16 transversal

Exercise 17c

1 Fig. 17.17 represents a view of a hut.

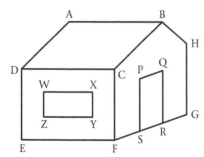

Fig. 17.17

(a) Name as many lines as you can which are parallel to AB.
(b) Name as many lines as you can which are parallel to XY.
(c) Name as many lines as you can which are parallel to PQ.
(d) Is any line parallel to BC?
(e) Is any line parallel to BH?

2 Use the ruled lines in your exercise book to draw a pair of parallel lines as in Fig. 17.18. Draw a transversal in any position.

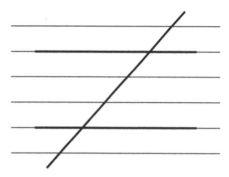

Fig. 17.18

(a) How many angles have been formed?
(b) In Fig. 17.19 the marked angles are called **corresponding angles**. They are in the same, or corresponding, positions at the two intersections.
Measure the sizes of the two corresponding angles on your own diagram. What do you notice?

Fig. 17.19

(c) Fig. 17.20 shows two other pairs of corresponding angles. Draw a sketch to show one more pair of corresponding angles.
Measure each pair on your diagram. What do you notice? Corresponding angles are sometimes called **F angles**. Can you think why?

Fig. 17.20

(d) In Fig. 17.21, the marked angles are called **alternate angles**. Measure the size of this pair of alternate angles on your diagram. What do you notice?

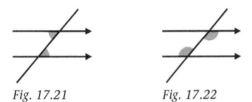

Fig. 17.21   *Fig. 17.22*

(e) There is one other pair of alternate angles (Fig. 17.22). Measure these angles on your diagram. What do you notice? Alternate angles are sometimes called **Z angles**. Can you think why?

③ In Fig. 17.23, name the angles which
(a) correspond to $A\hat{X}P$, $B\hat{X}Y$, $Q\hat{Y}D$, $C\hat{Y}Q$;
(b) are alternate to $B\hat{X}Y$, $X\hat{Y}D$.

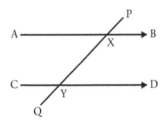

Fig. 17.23

④ In Fig. 17.24, which angle corresponds to
(a) $\hat{r}$,   (b) $\hat{p}$,   (c) $\hat{z}$,   (d) $\hat{j}$?
Which angle is alternate to
(e) $\hat{t}$,   (f) $\hat{m}$,   (g) $\hat{w}$,   (h) $\hat{l}$?

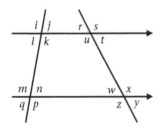

Fig. 17.24

⑤ Fig. 17.25 shows a pair of parallel lines and a transversal intersecting at X and Y. One angle is given as 80°. Sketch a copy of the diagram. Fill in the sizes of all the angles at X. Try to fill in the sizes of the angles at Y.

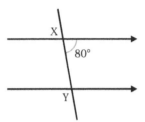

Fig. 17.25

While working through Exercise 17c you may have noticed that

(a) **Corresponding** angles formed on parallel lines are equal.
(b) **Alternate** angles formed on parallel lines are equal.

Exercise 17d

① **(Oral)** Find the sizes of the lettered angles in Fig. 17.26 (a)−(f). Give reasons.

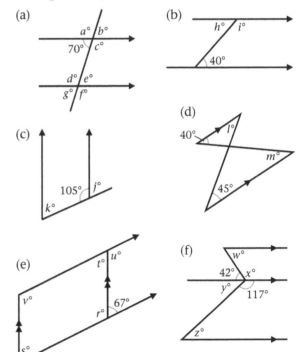

Fig. 17.26

2 Sketch a copy of each diagram in Fig. 17.27. Do not make an accurate drawing. Fill in the sizes of the missing angles.

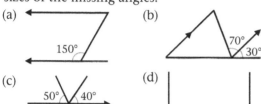

(a)   (b)

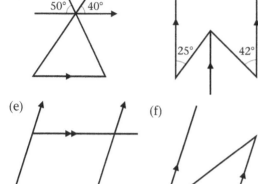

(c)   (d)

(e)   (f)

Fig. 17.27

Angles in a triangle

Exercise 17e

1 (a) Use a protractor to measure the angles in triangles ABC and PQR in Fig. 17.28.
   (b) Find the sum of the angles of △ABC (i.e. find $\widehat{A} + \widehat{B} + \widehat{C}$).
   (c) Find the sum of the angles of △PQR (i.e. find $\widehat{P} + \widehat{Q} + \widehat{R}$).
   (d) What do you notice about your results in (b) and (c)?

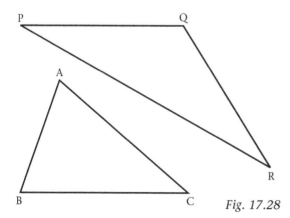

Fig. 17.28

2 Draw two large triangles. One of them should include an obtuse angle.
   (a) Measure the angles in each triangle.
   (b) Find the sum of the angles in each triangle.
   (c) What do you notice about your results in (b)? Do your friends get the same kinds of results?

3 (a) Draw any triangle. Cut it out carefully along its sides.
   (b) Tear off the three angles of your triangle as in Fig. 17.29.

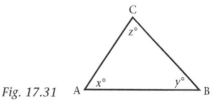

Fig. 17.29   Fig. 17.30

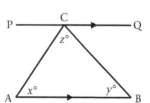

   (c) Take your three angles and arrange them so that they are adjacent to each other as in Fig. 17.30.
   (d) What do you notice? What is the sum of the angles on a straight line?

When working through Exercise 17e you may have noticed that the **sum of the angles of a triangle is 180°**. This is true for any triangle. We can show this in the following way. In Fig. 17.31, ABC is any triangle. Its angles are $x°$, $y°$ and $z°$.

Fig. 17.31

In Fig. 17.32 PCQ is a line through C parallel to AB.

Fig. 17.32

We can use the alternate angles fact, as in Fig. 17.33, to fill in the missing angles at C.

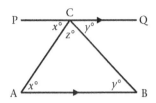

Fig. 17.33

At C, the three angles are adjacent angles on a straight line. Thus $x° + y° + z° = 180°$. But $x°$, $y°$ and $z°$ are also the sizes of the angles of the △ABC. Thus the sum of the angles of any triangle is 180°.

Use this fact in Exercise 17f.

Exercise 17f

1 **(Oral)** State the sizes of the lettered angles in Fig. 17.34 (a)−(i). Give reasons.

Note: in any diagram, lines marked with a small line are equal in length.

(a)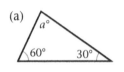

(b)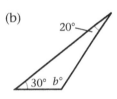

(c)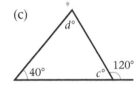

(d)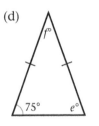

(e)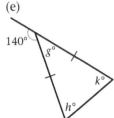

(f)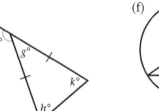

Fig. 17.34(a–f)

(g)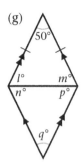

(h)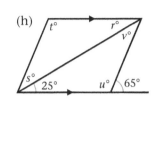

(i)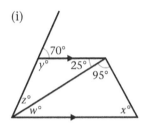

Fig. 17.34(g–i)

2 Calculate the third angle of a triangle in which two of the angles are as follows.

(a) 47° and 65°   (b) 24° and 77°
(c) 56° and 18°   (d) 39° and 21°
(e) each 58°   (f) 103° and 42°
(g) 38° and 71°   (h) 43° and 94°
(i) 60° and 60°   (j) 35° and 55°

3 *Table 17.1*

	$\hat{ABC}$	$\hat{BAC}$	$\hat{ACB}$	$\hat{ACD}$
(a)	58°	47°		
(b)	65°			118°
(c)		59°		126°
(d)	46°		35°	
(e)	43°	94°		
(f)		60°	60°	
(g)	35°			125°
(h)	58°		25°	

The angles in Table 17.1 refer to the diagram in Fig. 17.35. Calculate the missing angles in each row.

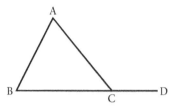

Fig. 17.35

④ In Fig. 17.36, $A\widehat{B}C = x°$, $B\widehat{A}C$ is twice as big as $A\widehat{B}C$ and $A\widehat{C}B$ is three times as big as $A\widehat{B}C$. Make an equation in x. Solve the equation. Find the three angles of the triangle.

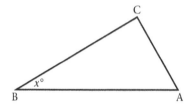

Fig. 17.36

⑤ In Fig. 17.37, ABCDE is a regular pentagon. Its centre, O, is joined to each vertex.

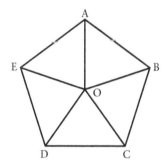

Fig. 17.37

(a) Calculate the size of each angle at O.
(b) What kind of triangle is △AOB?
(c) Calculate the angles of △AOB.

⑥ Sketch a copy of each diagram in Fig. 17.38. Fill in the missing angles.

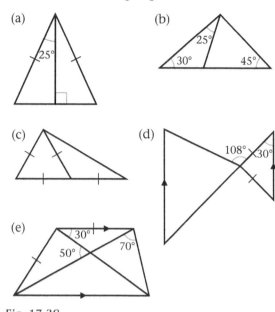

Fig. 17.38

Exterior angles of a triangle

In △ABC (Fig. 17.39) side BC is **produced** to X. (To produce a line means to make it longer.) $A\widehat{C}X$ is called the **exterior angle** of the triangle. $A\widehat{B}C$ and $B\widehat{A}C$ are called the **opposite interior angles**.

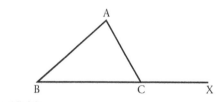

Fig. 17.39

Each side of the triangle can be produced in two directions. In Fig. 17.40, CB is produced to Y.

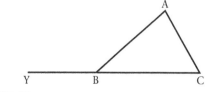

Fig. 17.40

Every triangle has three exterior angles. These are labelled x, y and z in Fig. 17.41.

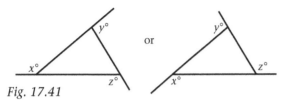

Fig. 17.41

Exercise 17g (Oral/Discussion)

① Calculate the sizes of the lettered angles in Fig. 17.42.

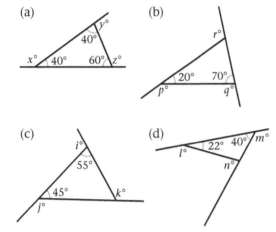

Fig. 17.42

② Calculate the sizes of the interior angles in Fig. 17.43.

③ For each exterior angle in Figs 17.42 and 17.43 find the sum of the two opposite interior angles.

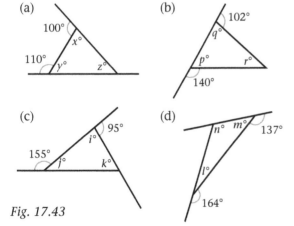

Fig. 17.43

④ Find the sum of the exterior angles of each triangle in Figs 17.42 and 17.43.

The results of Exercise 17g demonstrate the following facts:

1   **The exterior angle of a triangle is equal to the sum of the opposite interior angles.**

2   **The sum of the exterior angles of a triangle is 360°.**
    Use these facts in Exercise 17h.

Exercise 17h

① **(Oral)** State the sizes of the lettered angles in Fig. 17.44(a)−(f). Give reasons.

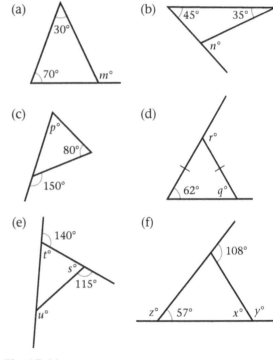

Fig. 17.44

② In Fig. 17.39 (on page 167), if $B\widehat{A}C = 60°$ and $A\widehat{C}X = 100°$, calculate $A\widehat{B}C$.

3 Calculate the missing angles in Fig. 17.45.

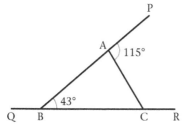

Fig. 17.45

4 Calculate the lettered angles in Fig. 17.46.

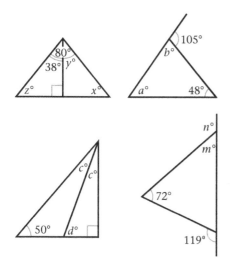

Fig. 17.46

5 Calculate the values of *w, x, y* and *z* in Fig. 17.47.

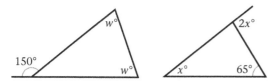

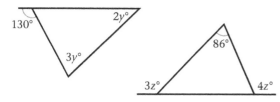

Fig. 17.47

Summary

When lines meet or intersect, angles are formed. Some special features of these angles are given below.

angles on a straight line add up to 180° (a straight angle), i.e. $a + b = 180°$

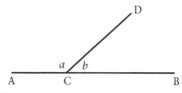

Fig. 17.48

angles at a point add up to 360°, i.e. $a + b + c + d = 360°$

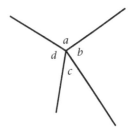

Fig. 17.49

vertically opposite angles are equal

Fig. 17.50

alternate angles on parallel lines are equal

Fig. 17.51

corresponding angles on parallel lines are equal

Fig. 17.52

the sum of the angles of a triangle is 180°, i.e.
$a + b + c = 180°$

Fig. 17.53

Practice exercise P17.1

1 Which of the following angles are equal?

(a) vertically opposite angles
(b) adjacent angles on a straight line
(c) corresponding angles on parallel lines
(d) angles of a square
(e) opposite angles of a rhombus
(f) angles of a regular polygon
(g) alternate angles of an equilateral triangle
(h) opposite angles of a parallelogram
(i) angles of an isosceles triangle

2 In each of the following examples, the sizes of two angles of a triangle are given. Calculate the value of x.

(a) 25°, 74°, $x°$
(b) 51° 104°, $x°$
(c) 96.5°, 48.5°, $x°$
(d) 59°, 104°, $x°$
(e) 28.4°, 83.7°, $x°$

3 The sizes of two exterior angles of a triangle are given in each example below. Calculate the value of p, the third exterior angle,

(a) 88°, 104°, $p°$
(b) 90°, 119°, $p°$
(c) 103°, 140°, $p°$
(d) 113.5°, 107.5°, $p°$

4 Write TRUE or FALSE for each of the following statements. (Write the correct statement for each one that is false.)

(a) Exterior angles of a triangle are always equal.
(b) Exterior angles of an equilateral triangle are equal.
(c) Adjacent angles on a straight angle add up to 90°.
(d) Alternate angles on parallel lines are always equal.
(e) Corresponding angles on parallel lines are equal.
(f) The sum of the angles of a triangle is 180°.
(g) The sum of the exterior angles of a triangle is 180°.

Statistics (1)
Data collection and presentation

Pre-requisites
- collecting information; number line; fractions; angles

Statistical information

Suppose a stranger asks you for **information** about yourself. You could tell him a lot of things. For example: your name; the town you live in; the school you go to; what you ate last night; the things you like; the things you don't like; etc.

You could also give him other information. For example: I am 15 years old; I have 4 brothers and 2 sisters; I am 171 cm tall and my mass is 48 kg; I wear size 8 shoes; my house is 5 km from the school; etc. You will notice that here you are using numbers.

When numbers are used, the information is called **statistics**. The first statistics were tables of births and deaths. Today, we use the word 'statistics' for tables of numbers giving information about many things.

Some statistics about two football teams are given in Table 18.1.

Table 18.1

	Games played	Won	Lost	Drawn	Goals for	Goals against
Giantkillers	18	10	5	3	60	21
Locomotives	15	2	8	5	19	37

The statistics above give a lot of information about the two teams. Giantkillers seem to be more successful than Locomotives. A good player, looking at the statistics, might prefer to play for Giantkillers than to play for Locomotives. Thus statistics can help when making decisions.

Exercise 18a (Oral)

1. Refer to the statistics in Table 18.1 about the Giantkillers and Locomotives.
   (a) Which team has played more games?
   (b) Which team has drawn more games?
   (c) Do the games won, lost and drawn add up to the games played?
   (d) How many goals have the two teams scored altogether?
   (e) How many goals have been scored against the two teams altogether?
   (f) How many games have the two teams lost altogether?
   (g) For every game that Giantkillers have lost, how many have they won?
   (h) For every game that Locomotives have lost, how many have they won?
   (i) For every goal that Locomotives have scored, approximately how many goals have been scored against them?
   (j) If 'goal average' means 'goals for' divided by 'goals against', find, to the nearest whole number, the goal average for Giantkillers.
   (k) If a team gets 2 points for a win and 1 point for a draw, how many points do Giantkillers have?
   (l) Similarly, how many points do Locomotives have?

② Table 18.2 gives the statistics for the numbers of students at a Secondary School for the years 2001 to 2006.

Table 18.2

	2001	2002	2003	2004	2005	2006
Number of students	901	904	939	975	1010	1046

(a) Is the school growing in size?
(b) What is the difference in the number of students in 2006 and 2001?
(c) If there are about 36 students in each class, how many classes did the school have in 2001 and 2002?
(d) In one year the school started a new Form 1 class. Which year?
(e) In which year will that Form 1 class be a Form 5 class?
(f) Estimate the number of students that the school would have in 2007.

③ A girl made a note of the first 100 vehicles that passed her on a road. The numbers of each type of vehicle are given in Table 18.3.

Table 18.3

	Lorry	Bus	Car	Taxi	Motor bike	Bicycle
Number	7	0	28	2	15	48

(a) Nearly half of the vehicles were of one kind. What were they?
(b) How many vehicles had only two wheels?
(c) How many cars were there for every one lorry?
(d) Which was the third most common type of vehicle?
(e) How many buses did the girl see?
(f) Is it true to say that buses never go on the road?
(g) Is this road more likely to be in a big city or in a small village? Give reasons for your answer.

④ Table 18.4 shows how much a family spends on food, rent and entertainment for each of four months.

Table 18.4

	Month 1	Month 2	Month 3	Month 4
Food	$589	$575	$602	$584
Rent	$800	$800	$800	$800
Entertainment	$236	$240	$229	$234

(a) Which item always costs the same?
(b) Which item is always the most expensive?
(c) What is the total money spent on rent?
(d) What is the total money spent on entertainment?
(e) What is the total money spent on food?
(f) During which month does the family spend most money?
(g) During which month does the family spend least money?
(h) The total income of the family is $2680 per month. Approximately what fraction of this is spent on entertainment over the 4 months?

⑤ Table 18.5 shows the numbers of people killed and injured in road accidents in a country in two years.

Table 18.5

	Killed	Injured	Total
1st year	800	3023	
2nd year	925	2985	

(a) Find the totals for each year.
(b) Which year seemed safer?
(c) How many people were killed in the two years?
(d) One of the four numbers looks like an estimate. Which one?

6 Table 18.6 shows the height of a boy between age 10 years and 15 years.

Table 18.6

Age (yr)	10	11	12	13	14	15
Height (cm)	121	127	134	140	152	162

(a) How often was the boy's height measured?
(b) Did the boy's height increase by the same amount each year?
(c) In which year did he grow the most?

Collecting data

It would be impossible to give statistics unless **data** were collected beforehand. 'Data' means basic information, usually in number form. To collect data you need to count and write down, or **record**, the information clearly.

The examples in Figs 18.1 and 18.2 show the same data collected in two different ways by two students. The first student tried to write down every vehicle as it came by. When two bicycles came by he did not have time to write them down properly. It is easy to make mistakes when counting this student's totals (Fig. 18.1).

The second student spent some time before beginning to record. She wrote down all the vehicles she could think of in a column. When a vehicle came by she made a tally. It is easy to count her totals (Fig. 18.2).

(a) bus, car, car, car, lorry, bicycle,
   bicycle car, car, lorry, bicycle, car

Fig. 18.1

Vehicle	Tally	Total
Car	‖‖	6
Bus		1
Lorry	‖	2
Taxi		
Bicycle	‖‖	3
Motorbike		

Fig. 18.2

Work through the assignments in Exercise 18b; they show how to collect data clearly.

Exercise 18b (Group work)

Keep the data you collect in the following assignments. You will use the data in Exercise 18d.

1 **Class statistics**
Make a large chart, like the one in Fig. 18.3, showing the full name of everyone in your class. The chart should contain the columns given in Fig. 18.3.
Pin the chart on the classroom wall. Find your name and enter your personal statistics under the column headings.

Class:	Date:				
Name	Age (years)	Height (cm)	Mass (kg)	Number of brothers	Number of sisters
Alice Mendes	14	169	47	1	4

Fig. 18.3

2 Traffic survey: Types of vehicle

Work in small groups. Make a table in your exercise book as in Table 18.7.

Go to a place where traffic passes. Each member of the group decides which type of vehicle(s) they are looking for and makes a tally of those kinds of vehicle that pass in 1 hour. (The time may be longer or shorter. Try to get between 50 and 100 vehicles.) At the end put all your results together to make the table.

Table 18.7

Name:		
Place:	Date:	
Direction of traffic: towards . . .		
Vehicle	Tally	Totals
Car		
Bicycle		
Motorbike		
Lorry		
Taxi		
Bus		
Others		

Collect the information required for questions 3 and 4 and make it into a table.

3 The shoe size worn by each member of your class.

4 The number of times each letter of the alphabet appears on a particular page in your English textbook.

Presentation of data

Data should be **presented** clearly. Good presentation makes statistical data easy to read and understand. The following example is used to show different kinds of presentation:

An English teacher gave an essay to 15 students. She graded the essays from A (very good), through B, C, D, E to F (very poor). The grades of the students were:

B, C, A, B, A, D, F, E, C, C, A, B, B, E, B

Rank order list

Rank order means in order from highest to lowest. The 15 grades of the students are given in rank order below:

A, A, A, B, B, B, B, B, C, C, C, D, E, E, F

Notice that *all* the grades are put in the list even though most of them appear more than once.

The advantages of the rank order list, in this case, are that we can easily find the following: the highest and lowest grades; the number of students who got each grade; the most common grade; the number of students above and below each grade; and so on.

Frequency table

Frequency means the number of times something happens. For example, 3 students got grade A. The frequency of Grade A is 3. Table 18.8, a **frequency table**, gives the frequency of each grade.

Table 18.8

Grade	A	B	C	D	E	F
Frequency	3	5	3	1	2	1

In most cases, a picture will show the meaning of statistical data more clearly than a list or table. The following methods of presentation give the data of the example in picture, or **graph**, form.

Pictogram

A **pictogram** uses pictures or drawings to give a quick and easy meaning to statistical data. In the pictogram in Fig. 18.4, each pin figure represents a student who gets the grade shown.

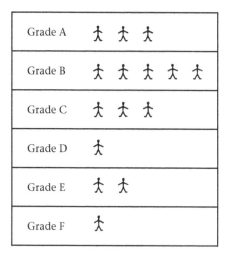

Fig. 18.4

Bar chart

A **bar chart** is very like a pictogram. The number of students who get each grade is represented by a bar instead of a picture. The height of each bar in Fig. 18.5 represents the frequency of that grade.

The scale at the left-hand side of the bar chart shows the frequency. Each bar is the same width.

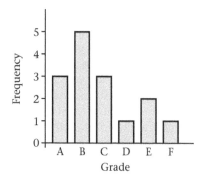

Fig. 18.5

Pie chart

A **pie chart** is a graph in the shape of a circular 'pie'. In the example in Fig. 18.6, the total number of students makes up the whole pie. Each piece of the pie is a sector of the circle. The size of each sector represents the number of students who get the grade shown in that sector.

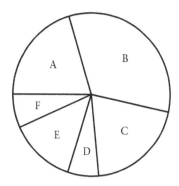

Fig. 18.6

Usually there are *no* numbers on a pie chart. The sizes of the sectors give a quick comparison between the numbers of students getting each grade.

Exercise 18c (Oral)

1 The following is a rank order list of an exam result:

> 87, 82, 78, 76, 75, 70, 66, 64, 59, 59, 59, 51, 49, 48, 41

(a) How many students took the exam?
(b) What was the highest mark?
(c) What was the lowest mark?
(d) What is the mark of the student who came 6th?
(e) What is the position of the student who got 76 marks?
(f) Three students got 59 marks. What is their position?
(g) What is the position of the student who got 51 marks?
(h) How many students got less than 75 marks?
(i) How many students got more than 45 marks?

(j) If 45 is the pass mark, how many students failed?

(k) What is the mark of the student in the middle of the rank order?

② Frequency Table 18.9 shows a tally of types of vehicles that were wrecked in serious accidents during a month on a busy road.

Table 18.9

Vehicle	car	lorry	bus	taxi	others
Frequency	卌	卌	I	II	II

(a) Write the frequencies as numbers instead of tallies.

(b) Which type of vehicle had the biggest number of serious accidents?

(c) How many vehicles were wrecked altogether?

(d) Name some types of vehicles that might be included in 'others'.

(e) Is it true that cars and lorries together had nearly 3 times as many serious accidents as all the other vehicles?

③ The bar chart in Fig. 18.7 shows the average monthly rainfall in mm of Trinidad in a year.

(a) Which month had most rainfall?

(b) How many mm of rain fell in the wettest month?

(c) Which month had least rainfall?

(d) Which months had less than 100 mm of rain?

(e) List the three wettest months in rank order.

(f) Find, in mm, the total rainfall for the year.

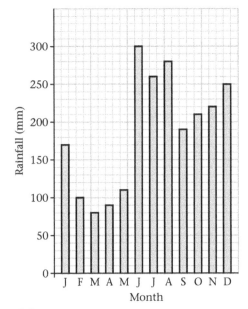

Fig. 18.7

④ The pie chart in Fig. 18.8 shows the amount of money that a government spends on Universities, Teacher Training Colleges, Secondary Schools and Primary Schools.

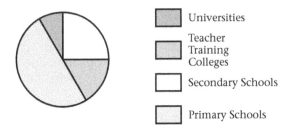

Fig. 18.8

(a) Does the pie chart tell you how much money the government spent?

(b) Which of the four gets least money?

(c) Which of the four gets most money?

(d) Can you think of reasons for your answers to (b) and (c)?

(e) What fraction of the money is spent on Primary Schools?

(f) What fraction of the money is spent on Secondary Schools?

(g) Approximately what fraction of the money is spent on Teacher Training Colleges?

Exercise 18d

Use squared paper to draw pictograms and bar charts.

1. A History test was graded from A to D. The results of ten students are given below:

    B, A, C, C, B, C, D, B, A, B

    (a) List the grades in rank order.
    (b) Which grade did most students get?
    (c) How many students got above grade C?
    (d) Make a frequency table of the results.
    (e) Draw a pictogram to show the results of the test.

2. The dress sizes of 20 women are given in the frequency table below.

Table 18.10

Size	10	12	14	16	18
Frequency	1	6	8	4	1

   (a) Draw a bar chart to show the frequencies of the dress sizes.
   (b) If you were a trader selling dresses, which three sizes of dress would you order most of?

3. A transport company has 6 lorries, 4 vans and 2 cars.
   (a) How many vehicles does the transport company have?
   (b) Draw a pie chart to show how the vehicles are divided.

4. 15 people were asked to name their favourite colour. Their answers are shown in the frequency table below.

Table 18.11

Colour	blue	red	green	yellow	black
Frequency	3	4	5	1	2

   (a) Which is the most popular colour?
   (b) Draw a bar chart to show the results in the table. If possible, use the given colours to colour the bars.

5. Every 800 g of dried fish contains about 300 g of water, 100 g of fats, 300 g of protein and 100 g of other substances.
   (a) What is 100 g as a fraction of 800 g?
   (b) What is 300 g as a fraction of 800 g?
   (c) What is $\frac{1}{8}$ of 360°?
   (d) What is $\frac{3}{8}$ of 360°?
   (e) Draw a pie chart to show the contents of dried fish.

6. Use the data you collected for Exercise 18b (class statistics), question 1.
   (a) Copy and complete the frequency table (Table 18.12) of ages of students.

Table 18.12

Age in years	less than 10	11	12	13	more than 13
Frequency					

   (b) What is the most common age?
   (c) Draw a bar chart to show the frequency of ages of students in your class.

7. Work in small groups. Use the data you collected for Exercise 18b (types of vehicle), question 2. Either:
   (a) draw a pictogram to represent the types and numbers of vehicles, *or*
   (b) draw a bar chart to show the types and frequencies of vehicles.

8. Table 18.13 gives the average monthly rainfall in Barbados.

Table 18.13

Months	J	F	M	A	M	J
Rainfall (mm)	89	57	51	64	83	140
Month	J	A	S	O	N	D
Rainfall (mm)	159	184	191	197	191	121

   Draw a bar chart to show the average monthly rainfall in Barbados.

9 Fig. 18.9 shows the distribution of public expenditure in a Five Year Development Plan of an island.

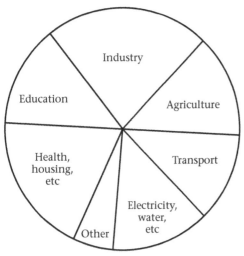

Fig. 18.9

(a) What is most money being spent on under the Five Year Development Plan?
(b) Approximately what fraction of the whole is this?
(c) Which is receiving the larger share, Agriculture or Education?
(d) Can we say how much each is receiving?
(e) What are the advantages and disadvantages of this method of showing data?

10 The following are the sizes of shoes worn by 20 people:

    7, 9, 6, 10, 8, 8, 9, 11, 8, 7,
    9, 6, 8, 10, 9, 8, 7, 7, 8, 9.

Table 18.14

Size	Frequency
6	
7	
8	
9	
10	
11	

(a) Copy and complete the frequency table (Table 18.14).
(b) Draw a bar chart showing the frequency of shoe size.
(c) A trader sells shoes. Which sizes do you think he sells most of?

Summary

Statistical information or **data** can be presented in a **frequency table**, **bar chart**, **pictogram** or **pie chart**.

A frequency table shows the **frequency** of an event (i.e. the number of times the event occurs).

Bar charts, pictograms and pie charts are pictorial, or graphical, forms of showing the information.

Practice exercise P18.1

1 Twenty students were asked to name their favourite team in the 2006 World Cup tournament. Their answers are shown in the frequency table below.

Table 18.15

Team	Brazil	Ghana	Italy	France
Frequency	8	4	3	5

(a) Which is the most popular team?
(b) Draw a bar chart to show the results in the table.

2 The frequency table (Table 18.16) shows the grades in Mathematics for all the students in a class.

Table 18.16

Grade	A	B	C	D	E	F
Frequency	6	9	12	8	4	1

(a) How many students are in the class?
(b) How many students got grades A–C?
(c) Which grade did most students get?

3 Choose a page in your History book and complete Table 18.17 to show the number of times each vowel appears on the page.

Table 18.17

Vowel	a	e	i	o	u
Frequency					

(a) Which vowel was used most often?
(b) Which vowel was used least number of times?
(c) Arrange the vowels in order of number of times used (highest to lowest) on the page.

4 The marks earned by students in a History test are shown below:

25 34 15 21 24 39 12 38 25 28
18 33 29 30 26 32 27 34 19 26
34 28 20 34 27 21 20 28 26 25
32 22 21 19 34

(a) How many students sat the test?
(b) Rank the marks highest to lowest.
(c) Which mark was earned by most students?
(d) How many students earned the highest mark?
(e) How many students earned the lowest mark?
(f) What mark is in the middle position?
(g) Do you think this was a difficult test? Give a reason for your answer.

5 Four students decided to collect ten-cent coins to help a member of their class to go on a school trip. The number of coins collected by each student is shown in table 18.18.

Table 18.18

Student	John	Jamie	Marlon	Najier
Frequency	35	36	45	68

(a) How many coins were collected altogether?
(b) What was the total amount of money collected?
(c) Who collected the most coins?
(d) If the cost of the trip was twenty dollars, how much more was needed?

6 The pictograph shows the number of cars rented by a car rental agency for the months January to December in a year.

Cars rented

Fig. 18.10

⊗ = 30 cars

(a) What was the total number of cars rented over the period?
(b) In which months were most cars rented?
(c) Can you tell from this information how many cars the agency owned?
(d) Draw a bar chart to show the information in the pictograph.

7 Four friends Jan, Sydney, Renee and Kim collected whiz cards. The number of cards they collected are shown on the tally chart below:

Table 18.19

Names of girls	Tally of cards collected
Jan	卅 卅 卅 ‖
Reneé	卅 卅 ‖‖
Sydney	卅 卅 ‖‖‖
Kim	卅 卅 卅 卅 ‖‖‖

(a) Draw a table showing in numbers, how many whiz cards each girl collected.
(b) What is the total number of whiz cards collected?
(c) Who collected most whiz cards?
(d) If all the girls had collected the same number of cards, how many would Kim have collected?

Chapter 19

Relations

Pre-requisites
■ whole numbers; factors; multiples

Here are the names of the members of the Jones family:

    Oswald – father
    Lola   – mother
    Robert – son
    Dale   – son
    Mary – daughter

We can show the relationship between the members of this family in a number of ways, as shown in Figs 19.1–19.3.

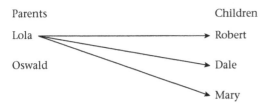

Relation: 'is the mother of'

Fig. 19.1 Relation 'is mother of'

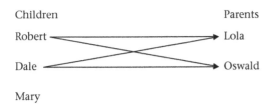

Relation: 'is the son of'

Fig. 19.2 Relation 'is the son of'

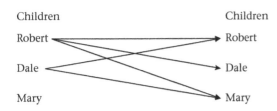

Relation: 'is the brother of'

Fig. 19.3 Relation 'is the brother of'

Draw a similar diagram to show the relationship 'is the sister of'.

The type of diagram shown in Figs. 19.1–19.3 is called an **arrow diagram**. The direction of the arrow shows the direction of the **relation**, or the rule that connects the two members of the family. The arrow starts at one member and 'goes to' another member of the family.

In Fig. 19.1, arrows start at Lola and 'go to' Robert, Dale and Mary.
Fig 19.1 gives the relation 'is the mother of' so that you read

    Lola   ——————→   Robert
as   Lola is the mother of Robert.

In Fig 19.2, arrows start at Robert and 'go to' Lola and Oswald; start at Dale and 'go to' Lola and Oswald.

What does this arrow diagram show?
Note that no arrow goes from Mary to Lola and Oswald. Why?

Note the differences between Fig 19.1 and Fig 19.2.

In Fig 19.1, the three arrows start at one member, Lola, but each arrow goes to a different member of the family. This type of relation is called a **one-to-many** relation.

In Fig 19.2, two arrows start at each member of the family in the relation and go to different members of the family. This type of relation is called a **many-to-many** relation.

In Fig 19.3, arrows start at Robert and 'go to' Dale and Mary; start at Dale and 'go to' Robert and Mary.
Again no arrow goes from Mary – why?
Is this a one-to-many or a many-to-many relation? Why do you think so?

In the examples above, the relations were between different members of one set, the Jones family. We can also have relations between members of more than one set as shown in Example 1.

Example 1

Draw an arrow diagram for the Jones children and the subjects they like at school:

> Robert likes English, Mathematics and Science
> Dale likes English and History
> Mary likes English, Mathematics and Spanish

The children form the first set, {Robert, Dale, Mary}, and the subjects form the second set, {Mathematics, English, History, Science and Spanish}. The relationship between the children and the subjects they like is given in Fig. 19.4.

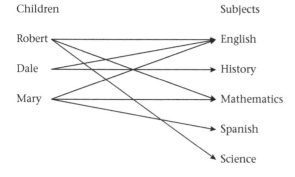

Relation: 'likes'

Fig. 19.4 Relation 'likes'

Is this a one-to-many or a many-to-many relation? Why do you think so?

Notice that:
> each subject is written only once
> more than one arrow starts at each child and more than one arrow goes to each subject.

To show the favourite subject of the Jones children, we can draw the arrow diagram in Fig. 19.5 for the relation 'likes best'.

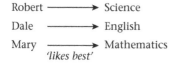

'likes best'

Fig 19.5 Relation 'likes best'

The arrow diagram in Fig. 19.6 shows their favourite day of the week. The first set is {Robert, Dale, Mary} and the second set is {Sunday, Monday, ..., Saturday}.

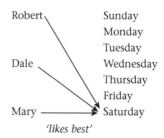

Fig 19. 6   'likes best'

Note the differences between Fig 19.5 and Fig.19.6.

In Fig 19.5, each arrow starts at one member of the first set and goes to a different member of the second set. This type of relation is called a **one-to-one** relation.

In Fig 19.6, each arrow starts at one member of the first set but all go to the same member of the second set. This type of relation is called a **many-to-one** relation. In a **many-to-one** relation, all the arrows may not go to one member only in the second set but only one arrow leaves each member of the first set as shown in Fig. 19.7.

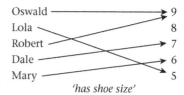

Fig 19.7   'has shoe size'

An arrow diagram can also be used to show relationships between numbers.

Example 2

What is the relationship between the numbers in Fig. 19.8?

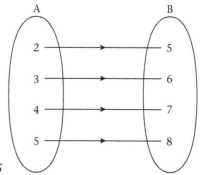

Fig. 19.5

Each number in A is 3 less than the number it goes to in B. The relation is '3 less than' and is a one-to-one relation.

Exercise 19a

1. Draw an arrow diagram to show the relation
   (a) 'is 4 more than',
   (b) 'is a multiple of'.

2. Record the day of the week on which each of 10 members of your class was born. Draw an arrow diagram to show the relation 'was born on'.

3. Record the favourite colours of the members of your family. Draw an arrow diagram to show the relation 'likes the colours'.

4. Complete the arrow diagram (Fig. 19.9) to show the relation 'is paired with'.

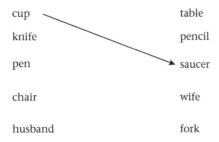

Fig. 19.9

Example 3

What is the relationship between the numbers in the arrow diagram in Fig. 19.10?

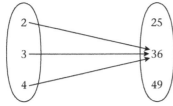

Fig. 19.10

Check carefully that 2, 3, and 4 are all factors of 36 but are not factors of factors of 25 and 49, therefore the relation is 'is a factor of'.

Note that the relation is many-to-one since only one arrow starts at each number in the first set and all three arrows go to the same number, 36, in the second set.

Example 4

Complete the arrow diagram in Fig. 19.11 to show the relation 'is a factor of'.

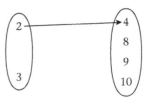

Fig. 19.11

The completed arrow diagram is shown below.

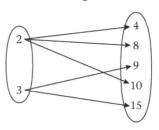

Fig. 19.12

Which type of relation is this?

Compare Fig.19.10 and Fig.19.12.

Note that the type of relation in this example is one-to-many although in both Example 3 and Example 4, the relation is 'is a factor of'.

Exercise 19b

1. State which of the arrow diagrams in Exercise 19a show
   (a) one-to-one,   (b) many-to-one,
   (c) one-to-many,   (d) many-to-many.

2. Draw two arrow diagrams showing a one-to-one relation. State the relation.

3. Draw two arrow diagrams showing a one-to-many relation. State the relation.

4. Draw two arrow diagrams showing a many-to-one relation. State the relation.

5. Draw two arrow diagrams showing a many-to-many relation. State the relation.

6 State which of the diagrams below show
(a) many-to-many (b) one-to-many
(c) many-to-one (d) one-to-one
relations. State the relations.

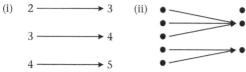

(i) 2 ⟶ 3 (ii)

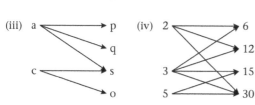

(iii) a (iv) 2 ... 6, 12, 15, 30

Fig. 19.13

Summary

A **relation** is the connection between the members of two sets. An **arrow diagram** is used to show the members of the sets that have the given relation. A relation may be **one-to-one**, **one-to-many**, **many-to-one**, or **many-to-many**.

Practice exercise P19.1

For each of the following diagrams, state the type of relation

1 A B **2** C D

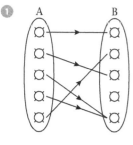 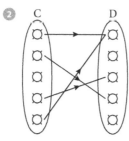

3 G H **4** J K

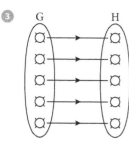 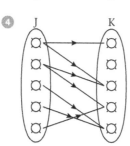

5 L M **6** V W

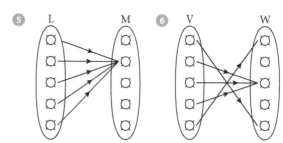

Practice exercise P 19.2

For each of the following,
(a) write down the type of relation
(b) state the relation.

1

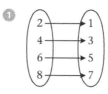

2 ⟶ 1
4 ⟶ 3
6 ⟶ 5
8 ⟶ 7

2

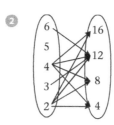

6, 5, 4, 3, 2 → 16, 12, 8, 4

Practice exercise P 19. 3

Complete each of the following arrow diagrams for the given relations.

1 $x \longrightarrow 2x - 1$

−1
0
2
3

2 $x \longrightarrow x^3$

−3
−1
0
1

3 $x \longrightarrow 3x - 2$

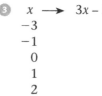

−3
−1
0
1
2

Practice exercise P 19.4

For each of the given domains and ranges, state the relation
(i) in words
(ii) in the format, $x \longrightarrow$

❶
4	1
8	2
12	3
16	4
$x \longrightarrow$	

❷
4	3
9	8
13	12
21	20
$x \longrightarrow$	

❸
−3	18
−1	2
0	0
2	8
4	32
$x \longrightarrow$	

Practice exercise P 19.5

(a) In a family, Verna and Desmond are the mother and father respectively of five children. There are three boys named Richard, Andrew and Brian, and two girls Alanna and Genna.
  (i) Identify the universal set and three subsets.
  (ii) Name and list the members of the sets.
  (iii) Draw a Venn diagram to represent the information.
(b) Given the following relations:
  (1) *is the mother of*
  (2) *is the son of*
  (3) *is the brother of*
  (4) *is the sister of*
  for **each relation**
  (i) draw an arrow diagram
  (ii) state whether the relations are one-to-one, one-to-many, many-to-one, or many-to-many.

Practice exercise P19.6

In each of the following questions,
(a) show the relation between the members of the sets by drawing an arrow diagram. (*Note: in* ❶ *and* ❹, *use only the first letters of the names*)
(b) state whether the relations are one-to-one, one-to-many, many-to-one or many-to-many.

❶ Andy and Rob *like* cycling, football, basketball
Nivan, Dan and Pam *like* basketball, tennis
Dan, and Lauren *like* cycling, football

❷ N = {0, 1, 2, 3, 4, 5, 6}
    is 3 more than

❸
x	3	0	−1	0	3	8
y	−2	−1	0	1	2	3

x is one less than the square of y

❹ {3, 4, 5, 6} and {hexagon, square, kite, triangle, decagon}
is the number of sides in

Practice exercise P 19.7

In each of the following questions, you are given the members of the first set and the relation.
(a) Copy and complete the table, using set notation as in the example, to write the members of the second set of numbers.

Domain	Relation	Range
{15, 12, 9, 6, 3}	$x \longrightarrow \frac{x}{3}$	{5, 4, 3, 2, 1}
❶ {1, 3, 5, 9}	$x \longrightarrow x - 5$	
❷ {−3, −2, 0, 4}	$x \longrightarrow x^2 - 5$	
❸ {−3, −1, 1, 3}	$x \longrightarrow 2x^2$	
❹ {0, 2, 4, 6}	$x \longrightarrow -x$	
❺ {−5, −1, 3, 2}	$x \longrightarrow 4$	

(b) Draw an arrow diagram to represent each of the relations and write the type of each relation.

Ratio and rate

Pre-requisites
- fractions; basic units of measurement

Ratio

Ratios compare quantities of the *same* kind, for example 4 kg : 7 kg. The given units may be different, e.g. 1 cm and 5 km, but the quantities are of the same kind for it is possible to express km in cm.

Suppose the prices of two chairs are $60 and $80. The **ratio** of the prices is 60 : 80, 'sixty to eighty'.

Ratios behave in the same way as fractions. For example,

$$\frac{60}{80} = \frac{30}{40} = \frac{120}{160} = \frac{3}{4}$$

and 60 : 80 = 30 : 40 = 120 : 160 = 3 : 4

Hence, both parts of a ratio may be multiplied or divided by the same number. It is usual to express ratios as fractions in their lowest terms.

Example 1

Express the ratio of 8 cm to 3.5 cm as simply as possible.

$$8 \text{ cm to } 3.5 \text{ cm} = 8 : 3.5$$
$$= 2 \times 8 : 2 \times 3.5$$
$$= 16 : 7$$

Notice that we do not give units in a ratio.

Example 2

Express the ratio of 96c to $1.20 as simply as possible.

Express both sums of money in cents.

$$96\text{c to } \$1.20 = \frac{96\text{c}}{120\text{c}} = \frac{96}{120} = \frac{96 \div 24}{120 \div 24} = \frac{4}{5}$$

The ratio is 4 : 5.

Notice that quantities must be in the same units before they can be given as a ratio.

Exercise 20a

Express the following ratios as simply as possible.

1. 3 : 6
2. 20 to 30
3. 15 : 12
4. 28 to 21
5. $10 to $15
6. 1 h to 15 min
7. 1 mm to 1 cm
8. 400 kg : 1 tonne
9. 80 s to 2 min
10. 18 boys to 12 boys
11. $200 to $150
12. $1.25 : 75c
13. 2 days : 1 week
14. 3 days to 3 weeks
15. 1 h 30 min : 2 h

Example 3

Fill the gap in the ratio 2 : 7 = _____ : 28.

Let the missing number be a.

Then $2 : 7 = a : 28$

or $\dfrac{2}{7} = \dfrac{a}{28} = \dfrac{a}{7 \times 4}$

$a = 2 \times 4 = 8$

The missing number is 8.

Exercise 20b

Fill the gaps in the following.

1. $\dfrac{4}{5} = \dfrac{}{10}$
2. $4 : 5 = $ _____ $: 10$
3. $\dfrac{3}{8} = \dfrac{9}{}$
4. $3 : 8 = 9 : $ _____
5. $6 : 9 = $ _____ $: 15$
6. $10 : 12 = 25 : $ _____
7. $8 : 12 = $ _____ $: 9$
8. _____ $: 9 = 16 : 24$
9. $120 : $ _____ $= 84 : 56$
10. $22 : 18 = 33 : $ _____

Sharing

To divide a quantity into two parts which are in the ratio 2 : 5, first divide it into 7 equal **shares** (since 2 + 5 = 7). The required parts will then be 2 shares and 5 shares, i.e. $\frac{2}{7}$ and $\frac{5}{7}$ of the given quantity.

Example 4

Two boys share 35 oranges in the ratio 2 : 3. How many oranges does each boy get?

One boy gets 2 shares and the other gets 3 shares. There are 5 shares altogether.

Number of oranges in 5 shares $= 35$

Number of oranges in 1 share $= \dfrac{35}{5} = 7$

Number of oranges in 2 shares $= 2 \times 7 = 14$

Number of oranges in 3 shares $= 3 \times 7 = 21$

One boy gets 14 oranges and the other gets 21 oranges.

Check: $14 + 21 = 35$

Example 5

Divide $1.17 between Clive and Rudy so that their shares are in the ratio 8 : 5.

$8 + 5 = 13$

$\frac{1}{13}$ of $1.17 $= \dfrac{117}{13}c = 9c$

Clive's share $= 8 \times 9c = 72c$

Rudy's share $= 5 \times 9c = 45c$

Check: $72c + 45c = \$1.17$

Exercise 20c (Oral)

Share the following quantities in the given ratios.

(a) $15 in the ratio 1 : 2

(b) 26 kg in the ratio 5 : 8

(c) 20 cm in the ratio 4 : 1

(d) $30 in the ratio 2 : 3

(e) 80 ml in the ratio 3 : 7

(f) 22 oranges in the ratio 9 : 2

Exercise 20d

① Divide the following quantities in the given ratios.

(a) 98c   5 : 9

(b) 56 kg $4\frac{1}{3} : 2\frac{3}{5}$

(c) 10.8 kg   11 : 7

(d) 22.95 m   5 : 12

② Two women share 5 dozen eggs in the ratio 1 : 2. How many eggs does each receive?

③ Abel is 12 years old and Gino is 8 years old. They share 15 mangoes in the ratio of their ages. How many does each get?

④ Two builders share 11 tonnes of bricks so that one gets $1\frac{1}{2}$ times as much as the other. The total cost is $227.50. How much does each pay?

⑤ A and B contribute $1400 and $1800 respectively to a business partnership. Of the profit, A receives 20% as manager. The rest is shared in the ratio of their investments. Find the ratio of A's total share of the profits to B's.

Example 6

What is the result of increasing $3.50 in the ratio 6 : 5?

Let the new amount be $n.

Then $\dfrac{n}{3.50} = \dfrac{6}{5}$

$\dfrac{n}{3.50} \times 3.50 = \dfrac{6}{5} \times 3.50$

New amount $= \$4.20$

Example 7

Decrease 3 litres in the ratio 5 : 12.

$\dfrac{\text{new amount}}{3 \text{ litres}} = \dfrac{5}{12}$

new amount $= \dfrac{5}{12} \times 3 \text{ litres}$

$= 1.25 \text{ litres}$

Exercise 20e

Find the result of

① increasing 20 in the ratio 3 : 2.

② decreasing 15 in the ratio 2 : 5.

③ increasing $2 in the ratio 5 : 4.

④ decreasing $2 in the ratio 4 : 5.

⑤ decreasing 1 kg in the ratio 3 : 4.

⑥ increasing 100 m in the ratio 7 : 5.

⑦ increasing 2 km in the ratio 7 : 4.

⑧ decreasing 7.5 litres in the ratio 3 : 5.

⑨ increasing $2\frac{1}{2}$ h in the ratio 6 : 5.

⑩ decreasing $1.80 in the ratio 7 : 12.

Example 8

In what ratio is 20.25 m changed if it is decreased to 11.25 m?

The ratio, $\dfrac{\text{new length}}{\text{old length}} = \dfrac{11.25\,\text{m}}{20.25\,\text{m}} = \dfrac{1125}{2025}$

$$= \dfrac{125}{225} = \dfrac{5}{9}$$

The length is decreased in the ratio 5 : 9.

Exercise 20f

Find the ratio in which the first quantity must be changed to make the second. State whether it is an increase or a decrease.

① $10; $15   ② $10; $7.50

③ 40 girls; 25 girls   ④ 6 g; 16 g

⑤ 5 g; $3\frac{1}{2}$ g   ⑥ 150 m; 120 m

⑦ $2.40; $3.30   ⑧ 6 km; $4\frac{1}{2}$ km

⑨ $2\frac{1}{4}$ m; 3 m   ⑩ 1 h 30 min; 1 h 15 min

Many problems can be solved by using ratios. Read Examples 9, 10 and 11 carefully.

Example 9

If 7 clocks cost $131.25, find the cost of 9 clocks.

Since the number of clocks is increased in the ratio 9 : 7, the cost will also be increased in the ratio 9 : 7.

7 clocks cost $131.25

9 clocks cost $131.25 \times \dfrac{9}{7}$

$$= \$18.75 \times 9$$
$$= \$168.75$$

This is an example of **direct ratio**, or direct proportion.

Example 10

If 5 men dig a drain in 14 days, how long would 7 men take?

The number of men is *increased* in the ratio 7 : 5. The time taken is *decreased* in the ratio 5 : 7. (More men take less time.)

Time taken by 5 men = 14 days

Time taken by 7 men = $14 \times \dfrac{5}{7}$ days

$$= 10 \text{ days}$$

This is an example of **inverse ratio**, or inverse proportion.

Example 11

A 900 g jar of jam costs $11.40. How much jam can be bought for $57?

Money is *increased* in the ratio 57 : 11.40

$11.40 buys 900 g

$57 buys $\left(900 \times \dfrac{57}{11.40}\right)$ g $= 900 \times \dfrac{570}{114}$ g

$$= 900 \times 5\,\text{g}$$
$$= 4500\,\text{g}$$
$$= 4\tfrac{1}{2}\,\text{kg}$$

Exercise 20g

① If 10 ballpoint pens cost $18.00 how much will 3 pens cost?

② A shelf holds 21 books each 5 cm thick. How many books, each 7 cm thick, will the shelf hold?

③ If 13 bars of soap cost $35.10, how much will 17 bars of soap cost?

④ It takes 15 monthly instalments of $96 to pay for a television set. How many monthly instalments of $60 will it take?

⑤ It costs $403.20 to stay at a hotel for 7 days. How much does it cost to stay for 3 days?

⑥ If 5 radios cost $197, how many can be bought for $315.20?

⑦ A rope can be cut into 18 lengths of 17.5 cm. How many lengths of 15.75 cm can it be cut into?

⑧ A motorist does a journey in 2 h 24 min at an average speed of 55 km/h. How long would the journey take at an average speed of 60 km/h?

⑨ A woman has enough money to buy $15\frac{3}{4}$ m of cloth at $5.40 per metre. How many metres at $4.20 could she buy?

⑩ The rent for 126 ha of land is $229.50. Find the rent for 196 ha (of the same land).

Rate

Quantities of *different* kinds may be connected in the form of a **rate**. The following are some examples of rates.

(a) A workman is paid $72.00 for an 8-hour day. His rate of pay is $9.00 per hour.
(b) A cyclist travels 28 km in 2 hours. His rate is 14 km per hour. In this case the rate is called **speed**.
(c) A piece of metal has a volume of 20 cm³ and a mass of 180 g. Its **density** is 9 g per cm³. The density of gases, liquids and solids is a rate giving the mass per unit volume.
(d) A town of 32 000 people has an area of 40 km². The population density of the town is 800 people/km². **Population density** is a rate giving the average number of people per unit area.

Notice that km/h or km h^{-1} is short for kilometres per hour.

Unitary method for solving problems

Note the following:
(a) each line of working is a complete sentence
(b) the quantity to be found comes last in each sentence
(c) the first sentence states the given facts
(d) the second sentence gives the value for a *unit*
(e) the following sentences give other calculations as required.

Example 12

Find, in km/h, the rate at which a car travels if it goes $38\frac{1}{2}$ km *in* 35 min.

In 35 min the car goes $38\frac{1}{2}$ km.

In 1 min the car goes $\dfrac{38\frac{1}{2}}{35}$ km.

In 60 min the car goes $\dfrac{38\frac{1}{2}}{35} \times 60$ km

$= \dfrac{77 \times 60}{2 \times 35}$ km

$= \dfrac{11 \times 60}{2 \times 5}$ km

$= 66$ km

The rate (speed) of the car is 66 km/h.

Exercise 20h

① A workman is paid $380 for a 40-hour week. Calculate his rate of pay per hour.

② A car travels 153 km in $2\frac{1}{4}$ h. Calculate its average speed in km/h.

③ A steel beam is 5.2 m long and has a mass of 137.8 kg. Find its average mass in kg/m.

④ 42 cm³ of sea water has a mass of 43.26 g. Find its density in g/cm³.

⑤ A town has an area of 24 km² and a population of about 31 000. Find the population density of the town per km² correct to 2 s.f.

⑥ A shop reduces all its prices at the rate of 15c in the $. Find the new price of an article which was $7.40.

⑦ A car uses petrol at the rate of 1 litre for every 6.5 km travelled. How many litres does it use when travelling 117 km?

⑧ A village is roughly in the shape of a rectangle $1\frac{1}{3}$ km by $1\frac{1}{4}$ km. What is its population if the average density is 570 people/km²?

⑨ When I travel at 60 km/h it takes me 2 hours for a certain journey. (a) How long is the journey? (b) How long does it take me when I travel at 50 km/h?

10 A bridge is 220 m long and has a mass of 11 220 tonnes. Find its average mass in t/m.

11 A man gets $150.80 for working $14\frac{1}{2}$ hours. Find his rate of pay per hour.

12 The estimated population of Grenada is 110 000 and the island covers an area of 340 km². Find the population density of Grenada to 1 s.f.

13 Each week a typist works from 8.00 a.m. to 12.30 p.m. on six days and from 2.00 p.m. to 5.30 p.m. on four days. Her rate of pay is $9.60/hour. What is her total wage?

14 How long will it take me to cycle a distance of 12 km at an average rate of 5 m/s?

15 A car uses petrol at the rate of 1 litre for every 11 km. If petrol costs 95c per litre, find the cost of the petrol for a journey of 891 km.

Example 13

A man is paid $180 for 10 days of work. Find his pay for (a) 3 days, (b) 24 days, (c) x days.

We are to find money. Money comes last in every line of working.
For 10 days the man gets $180
For 1 day the man gets $180 ÷ 10 = $18.
(a) For 3 days the man gets 3 × $18 = $54
(b) For 24 days the man gets 24 × $18 = $432
(c) For x days the man gets x × $18 = $18x
This is an example of direct proportion. The less the man works (3 days) the less he is paid ($54). The more the man works (24 days) the more he is paid ($432).

Example 14

7 workmen dig a piece of ground in 10 days. How long would 5 workmen take?

We are to find time. Time comes last in every line of working.

7 workmen take 10 days

1 workman takes 10 × 7 days = 70 days

5 workmen take 70 ÷ 5 days = 14 days

Notice the following:
(a) We assume all men work at the same rate.
(b) The second sentence gives the time for 1 unit, a workman.

This is an example of **inverse proportion**, or indirect proportion. Fewer men (5) take a longer time (14 days). More men (7) take a shorter time (10 days).

Compare the method using inverse ratio in Example 10.

When solving problems by the unitary method, always:
(a) write sentences with the quantity to be found at the end;
(b) decide whether the problem is an example of direct or inverse proportion;
(c) find the rate for 1 unit before answering the problem. This is where the **unitary method** gets its name from.

Many problems cannot be answered by the unitary method. For example, 'If a girl is 1 metre tall when she is 6 years old, how tall will she be when she is 18?' We cannot say. She will certainly *not* be 3 metres tall!

Exercise 20i (Oral)

In this exercise (i) give the first sentence of the working, (ii) state whether the problem is an example of *direct proportion* or *inverse proportion* or *neither*, (iii) give the complete answer for part (a) only.

1 A man is paid $100 for 5 days. Find his pay for (a) 1 day, (b) 2 days, (c) 22 days.

2 6 notebooks cost 90c. Find the cost of (a) 1 notebook, (b) 5 notebooks, (c) 44 notebooks.

3 5 men build a wall in 10 days. How long will it take (a) 1 man, (b) 10 men, (c) 25 men?

4 A 2-year-old boy has 6 sisters. (a) How many sisters did he have when he was 1 year old? How many sisters will he have when he is (b) 3 years old, (c) 21 years old?

5 A car travels 84 km in 2 hours. At the same rate, how far does it travel in (a) 1 hour, (b) 3 hours, (c) *x* hours?

6 A piece of land has enough grass to feed 15 cows for 4 days. How long would it last (a) 1 cow, (b) 6 cows, (c) *y* cows?

7 The temperature of 6 litres of liquid is 30 °C. Find the temperature of (a) 1 litre, (b) 8 litres, (c) 400 ml.

8 9 equal bottles hold $4\frac{1}{2}$ litres of water altogether. How much water does (a) 1 bottle, (b) 5 bottles, (c) *x* bottles hold?

9 A container has enough water to last 9 people for 4 days. How long would it last (a) 1 person, (b) 8 people, (c) *z* people?

10 A car uses 10 litres of petrol in 80 km. How far will it go on (a) 1 litre, (b) 5 litres, (c) *x* litres?

Exercise 20j

1 Where possible, work out parts (b) and (c) of the questions in Exercise 20i.

2 A man gets $80 for 5 days' work. How much does he get for 14 days' work?

3 It takes 4 men 3 days to dig a small field. How long would it take 3 men to do the same work?

4 A boy buys 7 pens for 77c. How much would 10 pens cost?

5 A 3-tonne lorry makes 10 journeys to move a pile of earth. How many journeys would a 5-tonne lorry make?

6 A 14-year-old boy runs 100 m in 14 seconds. How long would it take a 6-year-old boy?

7 A bag of corn can feed 100 chickens for 12 days. For how long would the same bag feed 80 chickens?

8 It takes 21 9-litre buckets to fill a drum with water. How many 7-litre buckets would it take?

9 A car travels at 60 km/h and takes 5 hours for a journey. How long would the car take if it travelled at 90 km/h?

10 A 480-page book is 2.4 cm thick (not counting the thickness of the covers). Find the thickness of 380 pages of the book.

Summary

Quantities of the same kind are compared in terms of a **ratio**, or a **proportion**.

Quantities of different kinds may be connected in the form of a **rate**. Speed and density are examples of rates.

Practice exercise P20.1

1 Complete the following ratios .
   (a) $3:8 = ?:24$ (b) $4:? = 16:12$

2 Write the following ratios in their lowest terms.
   (a) $12:20$ (b) $12.5:10$

3 Divide these amounts into the given number of equal parts.
   (a) 2.8 litres   (i) 2 parts   (ii) 5 parts
   (b) $25.83   (i) 3 parts   (ii) 7 parts

4 Divide $90 in the following ratios.
   (a) $4:1$ (b) $2:3:1$ (c) $2.5:3.5:2$

5 Divide each of the following in the given ratio:
   (a) $250, $2:3$ (b) 15 m, $7:3$
   (c) 5 *l*, $2:2:1$ (d) 30 kg, $1:1.5:2.5$

Practice exercise P20.2

Simplify the following ratios and then write the ratios in the form $n:1$ or $1:n$.

1 $350:210$ 2 $17:85$
3 5 cm to 1 km   4 0.5 *l* to 200 ml
5 3 km to 15 m   6 $12 to 40 cents
7 $2\frac{3}{4}:1\frac{5}{6}$ 8 1.5 kg to 600 g

⑨ 56 : 72

⑩ 91 : 18.2

⑪ 30 cm to 1 m 10 cm

⑫ 40 cents to $4

⑬ 500 cc to 3 l

⑭ $2\frac{1}{10} : \frac{3}{4}$

Practice exercise P20.3

For each of the following, complete the ratio

① 300 m to ? km = 4 : 5

② 3 : 4 : 2 = ? : 12 : ?

③ 3 : ? : 2 = $6 : $22 : ?

④ 15 : 6 = ? : 1

Practice exercise P20.4

For each of the following, find x using the given ratio

① $x : 24\,\text{cm}^2$, 4 : 3

② $x : \$210$, 5 : 12

③ $x : \$3224$, 5 : 8

④ $x : 12\,h$, 4 : 3

⑤ $x : 4\,l$, 3 : 2

⑥ $x : 900\,\text{cm}$, 2 : 5

Practice exercise P20.5

① If 12 pens cost $75, how many pens can be bought for $100?

② 30 cows eat the grass in a field in 5 days. Find the number of days that the grass in a field of equal size would last 3 cows eating at the same rate.

③ A computer uses 3 ink cartridges to make 80 copies of a document. Work out
(a) the number of copies that 10 cartridges will make
(b) the number of cartridges required to make 500 copies.

④ A bus is moving at a constant speed. The bus goes 72 km in 90 min
Calculate
(a) the distance covered in 40 min.
(b) the time taken to move 300 km.

⑤ The distance d between the legs of a ladder and the height h of the ladder are in a constant ratio.

If $h = 0.64$ m when $d = 0.4$ m, calculate
(a) d when $h = 2$ m (b) h when $d = 0.6$ m

⑥ $3960 is shared among Alan, Bill and Carl in the ratio 3 : 2 : 4.
(a) Calculate the amount that each person receives.
Increase Alan's share in the ratio 5 : 3, and decrease Carl's share in the ratio 3 : 4.
(b) Write the new ratio of the shares for Alan, Bill and Carl.
(c) Calculate the new amount that each person receives.

⑦ Tim studied Mathematics, English and Science for 50 minutes, 30 minutes and 45 minutes respectively.
Ann studied Mathematics, English and Science for 40 minutes, 40 minutes and 45 minutes respectively.
For each student
(a) find the ratio of the times spent on the three subjects
(b) work out the fraction of the total time spent on each subject.
(c) compare the proportional times spent on each subject by each student.

⑧ Cement, sand and gravel are mixed to make concrete. The ratio used is
cement : sand : gravel = 2 : 3 : 1.
(a) Find the mass of each material in 33 kg of concrete.
The ratio of the mixture is changed. The sand is decreased in the ratio 2 : 3, and the gravel is increased in the ratio 3 : 2.
(b) Find the ratio of the materials in the new mixture.

⑨ The profits of a business are divided so that J's share : K's share = 5 : 3.
(a) If J received $2000, find K's share.
(b) Find the total amount of the profits.
K's share was changed in the ratio 7 : 5.
Calculate
(c) the new ratio of J's share : K's share
(d) the amount each received if the profits did not change.

Chapter 21

Symmetry

Line symmetry

Draw a square, a rectangle and an isosceles triangle on a sheet of paper. Cut these shapes out very carefully. Now fold each shape along as many lines as possible so that one part of the shape fits exactly over the other. Each line along which you folded the shape is a **line of symmetry**.

A line of symmetry divides a shape into two parts, each part a reflection of the other. A shape which has one or more lines of symmetry is said to be **symmetrical**. Fig. 21.1 shows some symmetrical shapes.

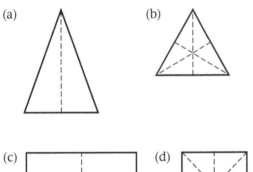

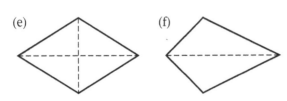

Fig. 21.1   *Lines of symmetry:*
(a) isosceles triangle   (b) equilateral triangle
(c) rectangle   (d) square   (e) rhombus   (f) kite

Fig. 21.2 shows how to make a symmetrical shape by folding and cutting a piece of paper.

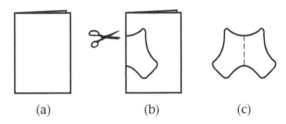

Fig. 21.2   (a) *Fold a sheet of paper*
(b) *Cut out any shape through both thicknesses of paper*
(c) *Unfold the part which has been cut out: the result is symmetrical*

The symmetrical shapes in Fig. 21.3 were made by folding pieces of paper which had wet paint on them.

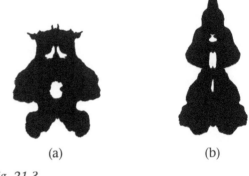

Fig. 21.3

Since one side of a symmetrical shape is a reflection of the other, line symmetry is often called **mirror symmetry** or **bi-lateral symmetry**. *Bi-lateral* means *two-sided*.

Exercise 21a

1. Use the methods shown in Figs 21.2 and 21.3 to make some symmetrical shapes and designs.

Mathematics for Caribbean Schools

2 Fold a sheet of paper twice as in Fig. 21.4. Cut through all four thicknesses of paper.

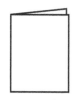

Fig. 21.4

Unfold the shape you have cut out. How many lines of symmetry does it have?

3 Fold a sheet of paper as in Fig. 21.4 then again diagonally. How would you cut this to give the shapes in Fig. 21.5? (Experiment by folding and cutting.)

(a) 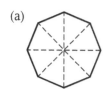   (b)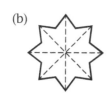

Fig. 21.5 (a) *Regular octagon*
              (b) *Eight-pointed star*

4 Which of the shapes in Fig. 21.6 have lines of symmetry?

(a)   (b)   (c)

(d) 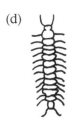 (e) 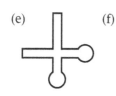 (f)

(g)   (h)   (i)

Fig. 21.6

5 Which of the mathematical instruments in Fig. 21.7 have bi-lateral symmetry? (Neglect any writing on the instruments.)

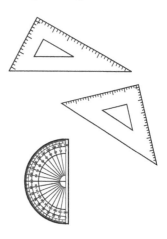

Fig. 21.7

6 Make a copy of each shape in Fig. 21.8. Draw any lines of symmetry.

(a)   (b)  (c)

(d)   (e)  (f)

(g)   (h)   (i)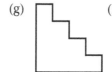

(j)  (k)   (l)

Fig. 21.8

7 Fold a sheet of paper three times as in Fig. 21.9.

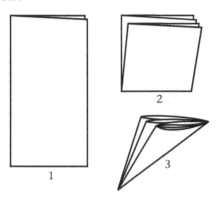

Fig. 21.9

Cut through all eight thicknesses of paper. Unfold the shape you have cut out. How many lines of symmetry does it have?

8 Copy each of the half shapes in Fig. 21.10 onto graph paper. Complete the shapes for the lines of symmetry given by dashed lines.

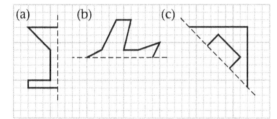

Fig. 21.10

Point symmetry

Fig. 21.11 shows a rectangular sheet of paper pinned to a notice board. The pin is at the centre of the rectangle.

Fig. 21.11

If the rectangle is given a half-turn it will take up the same position as before (Fig. 21.12).

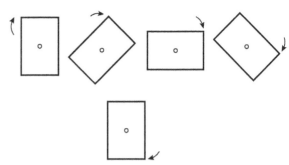

Fig. 21.12

This is an example of **point symmetry** or **rotational symmetry**. If the rectangle is now given another half-turn it will return to its original position. We say that the rectangle has point symmetry of **order** 2 about its centre, i.e. if the rectangle is given one complete revolution about its centre there are 2 positions in which it will appear as it started: one after turning through 180°, the other after turning through 360°.

Fig. 21.13 shows that a parallelogram also has point symmetry of order 2 about its centre.

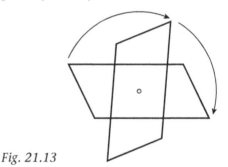

Fig. 21.13

The equilateral triangle in Fig. 21.14 has point symmetry of order 3 about its 'centre'.

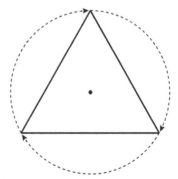

Fig. 21.14

Fig. 21.15 shows that any shape will return to its original position after 1 revolution about a point.

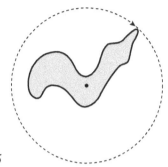

Fig. 21.15

Hence plane shapes are only considered to have point symmetry in cases where the order is 2 or more.

There is often a connection between bi-lateral (line) symmetry and rotational (point) symmetry. However, as Table 21.1 shows, some shapes, such as the parallelogram, may have point symmetry without having line symmetry.

Table 21.1

Shape		Number of lines of symmetry	Order of point symmetry
Rectangle		2	2
Parallelogram		0	2
Equilateral triangle		3	3

Exercise 21b

1 Which of the shapes in Fig. 21.16 have point symmetry?

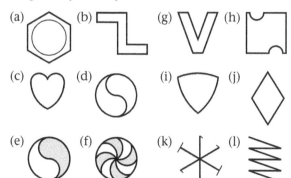

Fig. 21.16

2 Make a table like that of Table 21.1 for the shapes in Fig. 21.16.

3 Complete Table 21.2 by making a tick in the correct column.

Table 21.2

Shape	Has point symmetry only	Has line symmetry only	Both	Neither
Isosceles △				
Equilateral △				
Scalene △				
Rectangle				
Square				
Parallelogram				
Rhombus				
Kite				
Trapezium				
Circle				

4 Do any of the mathematical instruments in Fig. 21.7 (on page 193) have rotational symmetry?

5 Each shape in Fig. 21.17 is part of a shape that has point symmetry. Copy the shapes and complete the symmetry using the given orders and points.

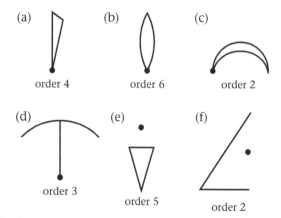

Fig. 21.17

Mirror symmetry of solids

Any cuboid can be cut through the middle to give two parts, each one the same as the other (Fig. 21.18).

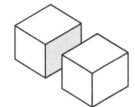

Fig. 21.18

A cuboid is a **symmetrical** solid. It has **mirror symmetry**. If one part is placed against a mirror, as in Fig. 21.19, its reflection looks like the other part.

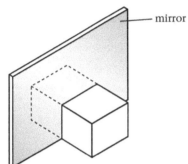

Fig. 21.19

The mirror in Fig. 21.19 is a **plane of symmetry** of the cuboid. Every cuboid has at least three planes of symmetry (Figs. 21.19, 21.20).

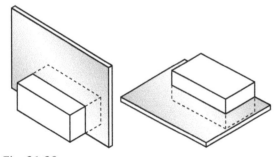

Fig. 21.20

Exercise 21c (Oral)

1. Which of the following have mirror symmetry?
   (a) A lamp
   (b) A chair
   (c) A radio
   (d) A car
   (e) A school bag
   (f) A torch
   (g) A clock
   (h) An aeroplane

2. How many planes of symmetry do the solids in Fig. 21.21 have?

(a) 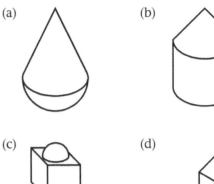 (b)

(c) 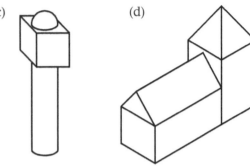 (d)

Fig. 21.21

3. Make a list of ten objects around you that have mirror symmetry.

4. For each object in question 3, write down how many planes of symmetry it has.

5. How many planes of symmetry has a cube?

Summary

A **line of symmetry** divides a plane shape so that one part is a reflection of the other.

A plane shape has **point (rotational) symmetry** if, when it is rotated about a point, it fits upon itself in *more than one* position.

The **order** of point symmetry is the number of times the shape fits upon itself when rotated about a point.

A solid has **mirror symmetry** when the reflection of one part looks like the other part.

Practice exercise P21.1

1 Which of the following letters of the alphabet have
   (a) line symmetry
   (b) point symmetry
   (c) both

Fig. 21.22

2 (i) How many times does each of the following shapes fit upon itself as it is rotated about a centre?
   (ii) State the order of point symmetry for each shape.

(a) a rectangle

(b) a square

(c) an equilateral triangle

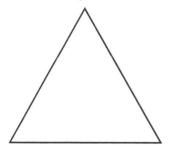

Fig. 21.23

3 (a) State the number of lines of symmetry for each of the following shapes:

(i)   (i)

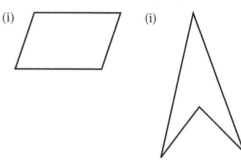

(iii)   (iv)

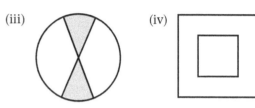

Fig. 21.24

(b) State the order of point symmetry for each of the following shapes:

(i)   (ii)

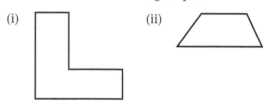

(iii)   (iv)

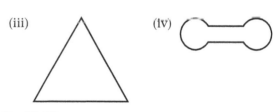

Fig. 21.25

4 Name two letters of the alphabet that have
   (a) 2 lines of symmetry
   (b) point symmetry of order 2.

Chapter 22

Consumer arithmetic (1)
Profit and loss, taxes

Cost price and selling price

The **cost price** of an article is the amount of money it costs. The **selling price** is the amount of money for which it is sold. In any transaction, the cost price is 100%. The selling price may be more or less than 100% since it is calculated in relation to the cost price.

Profit and loss

If the selling price of an article is greater than the cost price, there is a **profit** (or gain). If the selling price is less than the cost price, there is a **loss**.

Example 1
A trader buys a kettle for $80 and sells it at a profit of 15% of the cost price. Find his actual profit and the selling price.

$$\text{Profit} = 15\% \text{ of } \$80 = \tfrac{15}{100} \times \$80$$
$$= \frac{\$1200}{100}$$
$$= \$12$$
$$\text{Selling price} = \$80 + \$12$$
$$= \$92$$

Example 2
A hat is bought for $25.00 and sold for $22.00. What is the loss as a percentage of the cost price?

$$\text{Actual loss} = \$25.00 - \$22.00 = \$3$$
$$\text{The ratio, loss : cost} = \$3 : \$25$$
$$= 3 : 25 = \tfrac{3}{25}$$
Thus the loss is $\tfrac{3}{25}$ of the cost price.
$$\text{Percentage loss} = \tfrac{3}{25} \times 100\%$$
$$= 12\%$$

Generally, the loss (or profit) is calculated as a percentage of the cost price.

Example 3
By selling goods for $5.35 a trader makes a profit of 7%. He reduces his prices to $5.15. What is his percentage profit now?

$$\text{Selling price} = \$5.35$$
$$\text{Profit} = 7\%$$
$$\text{So, } 107\% = \$5.35$$
$$\text{And} \qquad 100\% = \$\left(5.35 \times \frac{100}{107}\right)$$
$$\text{i.e. cost price} = \$5.00$$
$$\text{New selling price} = \$5.15$$
$$\text{Actual profit} = \$5.15 - 5.00$$
$$= 15 \text{ cents}$$

Percentage profit now
$$= \left(\frac{15}{500} \times 100\right)\%$$
$$= 3\%$$

Exercise 22a

1. Find (i) the profit or loss, and (ii) the selling price for the following cost prices.
(a) $4	profit 20%
(b) $10	profit 15%
(c) $3.75	loss 8%
(d) $1.44	loss $12\tfrac{1}{2}$%
(e) 75c	profit 60%

2. Find (i) the actual profit or loss,
   (ii) the percentage profit or loss for the following cost and selling prices.
	cost price	selling price
(a)	$3	$3.45
(b)	$1.80	$2.25
(c)	$360	$324
(d)	$96	$132
(e)	$4.20	$3.57

3. A farmer buys a cow for $1600 and sells it for $1320. What is the percentage loss?

④ A trader bought some hats for $19 each. She sold them at a 30% profit. What was the selling price?

⑤ A man bought some wood for $9. He used the wood to make a box which he sold for $30. What percentage profit was this?

⑥ Brian bought a hat for $25. He sold it to Ned at a 20% profit. What did Ned pay?

⑦ Ned (in question 6) was short of money. He sold the hat to Daniel at a loss of 20%. What did Daniel pay?

⑧ A car which cost $336 000 was sold at a loss of $17\frac{1}{2}$%. What was the selling price?

⑨ A woman buys a car for $350 000 and sells it for $374 500. Her son buys a bicycle for $8000 and sells it for $8600. Which one makes the greater profit per cent?

⑩ An article is bought for $22.50 and sold for $25.20. Find the profit per cent. Find the price at which it should be sold to make a profit of 16%.

⑪ A woman sells an article for $21.75 and makes a profit of 16%. What did the article cost? Find how much she should have sold it for to make a profit of 28%.

⑫ A trader buys pens at $44 a dozen and sells them at $4.40 each. Find her profit per cent.

Discount buying

A **discount** is a reduction in price. Discounts are often given for paying in cash.

Example 4

A television set costs $54 000. A $12\frac{1}{2}$% discount is given for cash. What is the cash price?

Either,

$$\text{discount} = 12\frac{1}{2}\% \text{ of } \$54\,000 = \frac{12\frac{1}{2}}{100} \times \$54\,000$$

$$= \frac{1}{8} \times \$54\,000$$

$$= \$6750$$

cash price $= \$54\,000 - \$6750 = \$47\,250$

or,

$$\text{cash price} = (100\% - 12\frac{1}{2}\%) \text{ of } \$54\,000$$

$$= 87\frac{1}{2}\% \text{ of } \$54\,000$$

$$= \frac{87\frac{1}{2}}{100} \times \$54\,000 = \frac{7}{8} \times \$54\,000$$

$$= \$47\,250$$

Discounts are often given for buying in bulk.

Example 5

A trader sells ballpoint pens at $1.40 each or 4 for $4.40. How much is saved by buying 4 pens at once instead of 4 pens separately?

Normal cost of 4 pens $= 4 \times \$1.40 = \$5.60$
Discount price of 4 pens $= \$4.40$
Saving $\qquad\qquad\qquad = \$5.60 - \$4.40$
$\qquad\qquad\qquad\qquad = \$1.20$

Exercise 22b

① Find the discount price if a discount of
(a) 10% is given on a cost price of $430
(b) $12\frac{1}{2}$% is given on a cost price of $280
(c) 8% is given on a cost price of $1080
(d) 25% is given on a marked price of $92
(e) 20% is given on a marked price of $29.95

② The selling price of a table is $14 000. The trader gives a 25% discount for cash. What is the cash price?

③ During a sale a shop takes 12c in the dollar off all marked prices. What would be the sale price of a bicycle marked $1445?

④ A trader sells eggs at 50c each or $2.80 for 6. How much is saved by buying 3 dozen eggs in sixes instead of separately?

⑤ A 250 g bag of salt costs $4.50. A 20 kg sack of salt costs $280. Calculate the saving per kg by buying the 20 kg sack of salt.

⑥ A market trader asks $150 for some cloth. A woman offers $80. After bargaining, they agree on a price half-way between the two starting prices. How much does the woman pay? What discount did she get by bargaining?

Instalment buying

An **instalment** is a part payment. Most people find it easier to buy expensive items by paying instalments.

Example 6

The cost of a hi-fi set is either $68 000 in cash or a deposit of $8 000 and 12 monthly payments of $5 500. Find the difference between the instalment price and the cash price.

Instalment price = deposit + instalments
$$= \$8000 + 12 \times \$5500$$
$$= \$8000 + \$66\,000$$
$$= \$74\,000$$
Price difference = $74 000 − $68 000 = $6000

Buying by instalment is called **hire purchase**. The buyer hires the use of an item before paying for it completely. He has to pay for the hire of the item. This is why hire purchase is more costly than paying in cash.

Exercise 22c

1. The hire purchase price of a motor bike is $49 680. This is spread over 12 equal monthly instalments. Calculate each instalment.

2. The hire purchase price of a table and chairs is $42 000. 25% is paid as a deposit. The rest is spread over 12 equal monthly instalments.
   (a) Calculate the amount of the deposit.
   (b) Calculate the remainder to be paid.
   (c) Find the amount of each monthly instalment.

3. To buy a suit, a man can either pay $562.50 cash or he can pay 16 weekly instalments of $40.25.
   (a) Find the cost of the suit when paying by instalments.
   (b) Find the difference between the cash and hire purchase prices.

4. A colour television set costs either $4940 cash or 52 weekly payments of $114.25 each week. How much more does the television set cost when paid for weekly?

5. The cash price of a car is $269 118. The hire purchase payments require a deposit equal to 10% of the cash price and then 36 monthly payments of $7326. Calculate the saving when paying in cash.

6. A refrigerator costs $8350. A 5% discount is given for a cash payment. Alternatively, it can be paid by hire purchase. In this case the price is raised by 11%. Calculate the difference between paying cash and paying by hire purchase.

Sales tax

A proportion of the money paid for goods is given to the Government. A part which may be given to the Government is called **sales tax**. This tax is called Value Added Tax (VAT) in Trinidad and Tobago and in Barbados. In Jamaica it is called GCT, which refers to General Consumption Tax. The Government uses this tax and other taxes to pay for services such as defence, education, health and transport.

The sales tax varies with the selling price of the goods and with the type of goods. Sales tax is usually given as a percentage of the selling price.

Example 7

An advertisement for a table says that its price is

$890 plus 18% sales tax

How much does the customer pay?

Either:
Amount paid by customer.
$$= 118\% \text{ of } \$890$$
$$= \$890 \times \frac{118}{100}$$
$$= \$1050.20$$

Note: The difference between $1050.20 and $890 is $160.20. The Government receives $160.20 sales tax.

or:
Sales tax = 18% of $890
$$= \$\frac{18}{100} \times 890$$
$$= \$160.20$$

Amount paid by customer
$$= \$890 + \$160.20$$
$$= \$1050.20$$

Example 8

A bed costs $2242 including tax at 18%. How much tax does the Government receive?

Basic price of bed = 100%
Price including tax = (100 + 18)% of basic price
118% of basic price = $2242
1% of basic price = $\dfrac{\$2242}{118}$

Sales tax = 18% of basic price
$$= \dfrac{\$2242}{118} \times 18$$
$$= \$342$$

The government receives $342 tax.

Exercise 22d

1 Find out how much customers pay for each item in the following advertisement.

ARAR FASHIONS
(a) Shirts $80 + 15% VAT
(b) Dresses $240 + 15% VAT
(c) Jackets $290 + 15% VAT
(d) Trousers $190 + 15% VAT
(e) Shoes $220 + 15% VAT

Fig. 22.1

2 Find the amount of tax that the Government receives on each item in the following advertisement.

MASLAND FASHIONS
(a) Coffee tables $590
(b) Mattresses $885
(c) Beds $1032.50
(d) 3-piece suites $1980
(e) Kitchen tables $690
* All prices include VAT of 15%

Fig. 22.2

Commission

Commission is payment for selling an item. For example, an insurance agent gets commission for selling insurance. The more insurance he sells, the more commission he gets. Factories often employ sales representatives to sell their goods to shops and traders. The sales representatives often receive a proportion of the value of the goods they sell. This proportion is their commission.

Example 9

A sales representative works for an electric fan company. He gets a commission of 14c in the dollar. In one week he sells 4 large fans at $1050 each and 9 small fans at $540 each. Calculate his commission.

Total sales = 4 × $1050 + 9 × $540
= $4200 + $4860
= $9060

He gets 14c for every dollar
Commission = 9060 × 14c
= 126 840c
= $1268.40

Exercise 22e

1 An estate agent gets 2% commission for selling a house. How much money does he get for selling a house for $599 000?

2 A car salesman gets 1c in the dollar commission. Calculate his commission if he sells $523 800 worth of cars in a month.

3 A man sells tickets for a pop concert. He gets $10 for every 5 tickets he sells. How much will he get for selling 285 tickets?

4 A woman is an agent for a mail order company. She gets 10% commission on all monthly payments. How much commission does she get for a monthly payment of $253.80?

5 An insurance agent sells $284 000 worth of insurance. His commission is 20%. How much money does he get?

6 A rent collector's commission is $4\frac{1}{2}$% of his takings. In one month he collects $84 280 in rent. How much money does he get?

7 A cinema manager gets 24c commission on every ticket he sells. How much money does he get during a week when he sells 1318 tickets?

8 An electrical goods salesman gets 14c in the dollar commission. How much commission does he get if he sells 20 radios at $480 each and 2 television sets at $6150 each?

Income tax

Most people pay part of their income to the Government. The part they pay is called **income tax**.

There are many ways of calculating income tax. The following method is a simplified version of a tax system.

1 Tax is paid each year on taxable income. The rates of taxable income for married and single people are given in Table 22.1.

Table 22.1

Annual taxable income (within given ranges of income)	Rate of tax	
	Married	Single
first $1000	10%	14%
second $1000	12%	16%
third $1000	14%	18%
fourth $1000	16%	20%
fifth $1000	18%	22%
sixth $1000	20%	24%
seventh $1000	22%	26%
.	increasing in	
.	steps of 2%	
.	as far as …	
fifteenth $1000	38%	42.5%
sixteenth $1000	40%	45%
seventeenth $1000	42.5%	45%
over $17 000	45%	45%

2 Allowances are given to help to meet the cost of personal and family commitments. Allowances are as follows:

personal
- single: $1800
- married: $3000

children: $500 for each child
dependents: $400

The maximum personal allowances are $6000 for a married person and $3600 for a single person. Allowances may include a percentage of the payments for life insurance premiums, mortgage interest, education and health bills, and so on.

A person's total allowances are called the non-taxable income.

3 Tax is calculated as follows:
Taxable income
= total income − total allowances
Total tax payable
= chargeable income tax + 10% surcharge.

Example 10

A man has a total income of $9500. He has a wife and 3 children. He claims an allowance for a dependent relative. Calculate the amount of tax he pays.

first:
Total income = $9500

second:
Allowances
married allowance $3000
3 children $1500
dependent $400
Total allowances $4900

third:
Taxable income = $9500 − 4900
= $4600
Chargeable income tax
= tax paid on $4600
= $100 + $120 + $140 + $160 + 18% of $600
= $628
Surcharge = 10% of $628
= $62.80
Total tax paid = $628 + $62.80
= $690.80

Exercise 22f

In this exercise use the tax system described above.

1. Calculate the total income tax payable on each of the following annual salaries when earned by (i) married, (ii) single people.
   (a) $4000   (b) $7000   (c) $5500
   (d) $12 000 (e) $3800   (f) $8422

2. A single woman earns an annual salary of $6500 and claims $400 for her mother. Calculate the amount of tax she pays.

3. A married man earns $9100 per annum. He has 5 children.
   (a) Calculate his allowances.
   (b) Calculate his chargeable income tax (i.e. tax before surcharge).
   (c) Calculate the tax he pays.
   (d) Calculate his income after tax is paid.

4. A married man has 4 children and a dependent mother. His salary is $7500 per annum.
   (a) Calculate his allowances.
   (b) Calculate the amount of tax he pays altogether.
   (c) He is paid monthly and tax is taken from his pay in equal monthly instalments. Calculate his monthly 'take home' pay.

5. A husband and wife are both teachers. They each get an annual salary of $10 400. They have 3 children. Calculate how much tax they pay altogether.
   Note: Add the two salaries together but allow only one set of allowances.

Summary

Profit (or **loss**) is the difference between the **cost price**, that is, the price paid by a person as a buyer of an item and the **selling price**, that is, the price charged by that person as a seller of the same item. The profit (or loss) may be expressed as a percentage of the cost price or of the selling price.

A **discount** is a reduction in price; discounts are usually expressed as a percentage of the price.

An **instalment** is a part payment; buying by instalment is called **hire purchase**, since the buyer hires the use of the item before buying it completely.

Sales tax is a part of the money paid for an item; this tax is given to Government and is usually expressed as a percentage of the cost of the item.

Commission is payment for selling an item; it is usually expressed as a proportion or percentage of the selling price.

People pay part of their income to Government – this is called **income tax**. Tax systems vary from country to country; the tax rates and non-taxable allowances granted by different governments vary.

Practice exercise P22.1

1. Fill in the blank spaces in the table below. In the profit and loss columns, put an ✗ in the space which does not apply and write in the amount in the other column.

	Cost price	Selling price	Profit	Loss
(a)	$2100		$320	
(b)		$44.95		$18.50
(c)	$120.00	$99.95		
(d)		$3290.00	$688.20	
(e)	$84			$28

2. A trader buys a washing machine for $2500. She sells it for a 25% profit. Calculate
   (a) the profit she makes
   (b) the selling price of the washing machine.

3. A computer software package was bought for $400 and sold for $516. Calculate the profit and the per cent profit made on the transaction.

④ A profit of 40% was made on a second-hand car that was bought for $15 000. How much money was made on the sale and what was the selling price of the car?

⑤ A set of books costing $480 was sold at a garage sale for $120. What was the loss percentage on the set of books?

⑥ The firm Infinity Hardware made a profit of $720 on a set of machine parts. This amount was a profit per cent of 18%. What was the cost price of the machine parts?

⑦ A grocer buys 10 dozen eggs at $5.70 per dozen. Five per cent of the eggs break. How much per dozen must the grocer sell the eggs to get a profit of 15% on the cost price?

⑧ Office Supplies bought 75 chairs at $3.60 each. They paid $1367.50 freight on the entire shipment, and allowed $35.10 for overhead expenses *on each chair*. The chairs were all sold at $400 each.
  (a) What was the total cost to Office Supplies to obtain the chairs?
  (b) How much were the chairs sold for?
  (c) What profit/loss was made? Express this as a per cent of the cost price.

⑨ Courtney sells his vacuum cleaner for $86.25. He makes a 15% profit. How much did he buy it for?

⑩ Vital Supplies bought 1000 mangoes for $1200. Five per cent of the mangoes were spoiled and had to be discarded. 400 mangoes were sold at $2.50 each; 300 were sold at $1.00 each; and the remainder were sold at 60 cents each.
  (a) How many mangoes were spoiled?
  (b) How much money was obtained from selling the mangoes?
  (c) What was the profit/loss? Express this as a percentage of the cost price.

⑪ A grocer buys apples at $12 per kilogram. Unfortunately he has too many so he sells them for $10.50 per kilogram.
  (a) Calculate the percentage loss.
  His total loss was $300.

(b) How many kilograms of apples did he buy?

⑫ Jon sold his car for $10 370, getting a 22% profit. Lucinda sold her car for $10 560 for a 20% profit. Calculate whose car cost more originally.

Practice exercise P22.2

① The regular price of a dishwasher at PRICE BUSTA is $1800. What is the sale price?

Clearance **SALE** on
ALL DISHWASHERS
40%OFF
at **PRICE BUSTA**

Fig. 22.3

②

SUPER STEREO SYSTEM	
was	$2750
now	**$1999**

Fig. 22.4
  (a) What is the percentage discount on the sale of a super stereo system?
  (b) VAT is calculated at 12% of the selling price *before the discount* is given. What must the customer actually pay to obtain a super stereo system?

③ A clothes shop is having a '75% off' sale. How much does Brian pay for the following items: (a) trousers: marked price $152 (b) a tie: marked price $95

④ Copy and complete the table. VAT is calculated on the marked price. Calculate all monetary values to the nearest cent.

	Marked price	Selling price	Discount	VAT (17.5%)
(a)	$175	$155.75		
(b)	$1998		27%	
(c)		$110.98	8%	
(d)			14%	$89.60

⑤ *Comfy Clothes* sells shoes at a 25% discount. The selling price is $34.50 and they make a 20% profit. What percentage profit would they have made on the marked price?

⑥ Copy and complete the following table. The first one is done for you; note that tax is calculated on the Marked Price.

	Marked price	Per cent discount	Amount of discount	Sales tax/VAT	Amount of Tax	Selling price
(a)	$150	20	$30	7%	$10.50	$130.50
(b)	$200	15			$16	
(c)	$1500		$450			$1170
(d)	$400		$25		$25	
(e)		10		15%	$36	

⑦ Katharine wants to travel to the Caribbean. She purchases an airline ticket which includes a VAT of 20%. She paid $462. What was the price of the ticket?

⑧ Accommodation for a weekend at Ideal Vacation Club is advertised as $500. The actual amount to be paid includes a service charge of 10% and a government accommodation tax of 8%. What is the actual cost of a weekend stay at Ideal Vacation Club?

⑨ Ryan spent three nights at Bestview Hotel. The price of his room was $160 per night. His meals cost $165 before a 5% consumption tax. The hotel's service charge is 10% on the total bill. What was the total cost of Ryan's stay?

⑩ Hard Top Ltd., a manufacturer of school desks, gives special discounts to the purchaser, depending on the quantity bought. He quotes prices for desks as follows:

The cost of one desk is $180 but discounts are offered when ten or more desks are bought. The amount of the discount depends on the number of desks bought.

From ten to 50 desks, the discount is 15%. For more than 50 desks, up to 100, the discount is 25%. Over 100 desks, the discount is 30%.

Calculate the total cost of

(a) 9 desks   (b) 10 desks
(c) 11 desks   (d) 50 desks
(e) 51 desks   (f) 57 desks
(g) 100 desks   (h) 106 desks
(i) 107 desks

Explain the answers to (a), (b) and (c). What do the answers to (d), (e) and (f) tell you?

Practice exercise P22.3

① Freddy needs a new oven. He chooses one that needs a $50 down-payment and $50 per month for one year. Alternatively he can pay cash and get a 15% discount. What is the cash price of the oven?

② A bank has two loan schemes.

	Deposit	Instalments
Scheme A	$5000	300 monthly instalments of 0.36%
Scheme B	5%	240 monthly instalments of 0.5%

A couple needs $150 000 to buy a car and takes a loan from the bank. For each scheme,
what percentage profit would the bank make
what is the total amount paid in instalments to the bank?

③ The cash price of a refrigerator is $8399. The hire purchase price is a down-payment of $2000 and payments of $75 every week for 2 years. Calculate

(a) the total price paid by hire purchase
(b) the difference between the cash price and the hire purchase price
(c) the difference as a percentage of the cash price.

④ The cash price of a television is $3500. The down payment under a hire purchase agreement is $350. 30% interest is charged on the balance.

(a) Calculate the total amount (balance and interest) due.

The total amount due must be paid in 25 equal monthly payments.

(b) Calculate the amount of each payment.

⑤ A bank charges a 90% mortgage for 20 years on a house valued at $300 000. The monthly instalment to be paid to the bank is $2808. Calculate

(a) the down-payment
(b) the mortgage loan
(c) the amount paid in installments to the bank
(d) the total amount paid for the house.

Practice exercise P22.4

① Mrs. Ace charges $35 per hour for her services when she works normal hours. Her normal hours are 9am to 5pm Monday to Friday.

If she works at any other time on a weekday, she gets time-and-a-half.

If she works at the weekend, she gets double time.

Calculate Mrs Ace's earnings during each of the following weeks:

(a) she does a normal week
(b) she does a normal week but works until 8 p.m. on Thursday and does four hours on Sunday
(c) she does a normal week but leaves work two hours early on Wednesday, and works two hours late on Thursday
(d) she gets Monday off, but in addition to her normal hours, starts at 7:30 a.m. on Tuesday and does six-and-a-half hours on Saturday.

② Use the table below to calculate the income tax to be paid in the following cases.

Tax-free allowances (per year)	Tax rates (per year)
Personal allowance = $8000	10% on first $3000
Spousal allowance = $5000 (if not working)	25% on next $8000
Per child (at school) = $900	40% on next $24 000
Per child (at university) = $1300	50% on the remaining
Mortgage interest = all	
National insurance = first $1500 contributions	

(a) Brian Young earns $27 379 every year. He is unmarried with no children. He pays mortgage interest of $1700 a year, and pays national insurance of $90 a month.

(b) Marsha earns $63 419 per year. She is single and has two children at school. Her mortgage interest repayments are $8250 per year and national insurance contributions are $4700 per year.

③ Cherise is offered two jobs: one in Barbados and one in Trinidad and Tobago.
The job in Barbados pays Bds$30 000 but she must pay 15% in taxes.
In Trinidad and Tobago she could earn TT$100 000 with tax at 19%.
In which country will she take home more, and by how much?
(Assume Bds$1 = TT$3).

Revision exercise 8 (Chapters 16, 17, 21)

1. Find the perimeter and area of a rectangle which measures 8 cm by 10 cm.

2. Calculate the area of a parallelogram of height 6 cm and base 9 cm.

3. Two sides of a triangle are 7 cm and 4 cm and the angle between them is a right angle. Calculate the area of the triangle.

4. Two angles of a triangle are 63° and 45°. Calculate the size of the third angle.

5. In Fig. R14, find a, b, c. State reasons for your answers.

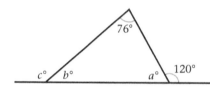

Fig. R14

6. In Fig. R15, find a, b, c, d, e. State reasons for your answers.

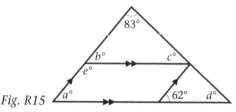

Fig. R15

Use Fig. R16 to answer questions 7–10.

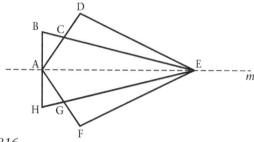

Fig. R16

In Fig. R16, m is a line of symmetry of shape ABCDEFGH.

7. (a) What is the image of point C in m?
   (b) What is the image of point E in m?
   (c) What is the image of line HG in m?
   (d) What is the image of △CDE in m?

8. (a) Name two points which are equidistant from m.
   (b) Name two right-angled triangles.
   (c) Name an isosceles triangle.
   (d) Name two kites.

9. (a) If BÊA = 26°, what is AĤE?
   (b) If HÂG = 31°, what is FÂD?

10. If AB = 3.8 cm and EH = 6.3 cm, what is the perimeter of △EHB?

Revision test 8 (Chapters 16, 17, 21)

1. The area of a floor 3 metres square is
   A 3 m²   B 6 m²
   C 9 m²   D 300 m²

2. A triangle and a parallelogram have the same base and same area. If the height of the triangle is 5 cm, the height of the parallelogram is
   A 1.25 cm   B 2.5 cm
   C 5 cm   D 10 cm

3. In Fig. R17, POQ and SOR are straight lines. If PÔS = 40° and TÔQ = 65°, TÔR =
   A 35°   B 30°   C 25°   D 15°

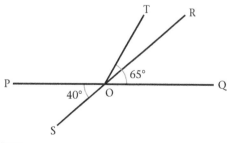

Fig. R17

④ A sheet of paper is folded 3 times so that the folded edges are placed together. All 8 thicknesses of paper are cut through to give a shape. If the paper is unfolded how many lines of symmetry does it have?
A 2   B 3   C 4   D 8

⑤ Which of the following has (have) point symmetry of order 2?
I kite,   II rhombus,   III square
A I only   B II only
C III only   D I and II only

⑥ Calculate the area of the triangle in Fig. R18. Calculate the height *h* shown in the diagram.

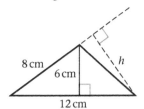

Fig. R18

⑦ In Fig. R19, find *a*, *b*, *c*. State reasons for your answers.

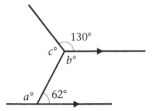

Fig. R19

⑧ In Fig. R20, find *a*, *b*, *c*, *d*. State reasons for your answers.

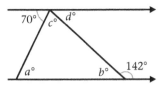

Fig. R20

⑨ In Fig. R21, O is the centre of the circle and OMXN is a square of side 7 cm.

Calculate the area of △XYZ.

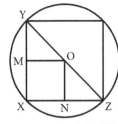

Fig. R21

⑩ Use angle and symmetry properties to find the sizes of all the angles in Fig. R22.

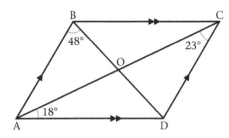

Fig. R22

Revision exercise 9 (Chapters 18, 19)

① (a) Place the following distances in rank order:
5 km, 1 km, 6 km, 4 km, 9 km, 2 km, 15 km, 8 km, 10 km.
(b) Find the fifth greatest distance.

② In one year rain fell on 80 days. There was no rainfall on any of the other days. Taking a year to be 360 days, draw a pie chart to show this information.

③ At a meeting, half of the people are women, one-third are men and the rest are children. Draw a pie chart to show this information.

④ If there were 60 people at the meeting in question 3, draw a pictogram to show the information of the question. (Let one symbol represent 10 people.)

⑤ 100 people were asked their ages; the results are given in Table R2.

Table R2

Age	under 15	15–29	30–44
Frequency	43	32	17

Age	45–59	60 and over
Frequency	5	3

(a) What fraction of the people were under 30?
(b) What percentage of the people were between 45 and 59?

⑥ Draw a bar chart to show the information in question 5.

⑦ Cislyn, Pearl, Rohan and Roland are children of the same parents. Draw an arrow diagram showing the relation 'is the sister of'.

⑧ For the set of students in the same row as yourself, draw an arrow diagram showing the relation 'travel to school by'.

⑨ Consider the set of numbers 1, 2, 3, ..., 9. Draw an arrow diagram showing the relation 'is the square root of'.

⑩ Robert, Andrew and Richard are brothers. Robert likes playing tennis and football; Andrew likes playing tennis and cricket. Richard likes cricket. Draw an arrow diagram to show the relation 'likes playing'.

Revision test 9 (Chapters 18, 19)

① When recording data, the tally marks ЖЖ ЖЖ ЖЖ ||| represent the number
A 13   B 15   C 18   D 33

② Which of the following is *not* a graph?
A pictogram   B frequency
C bar chart   D pie chart

Use Fig. R23 to answer questions 3, 4, 5.

Fig. R23 is a bar chart showing the numbers of students who got grades *A*, *B*, *C*, *D* in an essay.

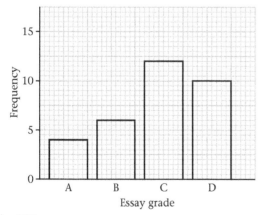

Fig. R23

③ Which grade did most students get?
A *A*   B *B*   C *C*   D *D*

④ How many students got grade *B*?
A 4   B 6   C 10   D 12

⑤ How many students wrote the essay?
A 4   B 12   C 22   D 32

⑥ A book has 120 pages of drawings, 72 pages of photographs and 168 pages of writing. Show this information on a pie chart.

⑦ Show the information of question 6 on a pictogram. (Let each symbol represent 24 pages.)

⑧ Choose any six Caribbean islands. Draw an arrow diagram to show the relation *'is the capital of'*.

⑨ Using any six numbers, draw an arrow diagram to show the relation *'is twice'*.

⑩ Draw an arrow diagram to show the relation *'number of pages'* for four of your textbooks.

Revision exercise 10 (Chapters 20, 22)

① A car uses 20 litres of petrol for a journey of 180 km. How many litres will it use for a journey of 108 km?

② The number of boys in a school is 120. If the ratio of boys to girls is 2 : 3, find the total number of students in the school.

③ (a) Express 15 as a percentage of 40.
(b) Increase 80 in the ratio 6 : 5.
(c) The price of an article decreased from $125 to $100. Express this as a ratio.

④ A factory made 1500 refrigerators in a 40-hour week. Find its production rate in refrigerators/hour.

⑤ A history book costs $42.00. The writers of the book get 10% of the price of each book that is sold. How much will they get if it sells 15 628 copies in one year?

⑥ A furniture salesman gets an 8% commission. How much will he get for selling 44 chairs at $265 each, 11 tables at $1015 each and 5 beds at $770 each?

⑦ Use the tax table on page 202 to answer this question. A man has a wife and 5 children. Calculate his tax allowances. If his income is $11 320, calculate the amount of tax he pays.

⑧ A car cost $15 200. 9% discount is given for paying in cash. The car can also be bought by paying 24 monthly instalments of $728. Find the cost of the car
  (a) when cash is paid,
  (b) when it is paid for by instalments.
  (c) Find the difference between paying in cash and paying by instalments.

⑨ A dealer sells a gold ring for $960 and makes a profit of 60%. Find the selling price if the dealer is to make a profit of 95%.

⑩ A bookseller sells $4812 worth of books in a week. His commission is 4c in the $. How much money does he get?

Revision test 10 (Chapters 20, 22)

① In 1996 a new machine cost $8400. In 2006 the same model cost $27 600. The 1996 to 2006 costs as a ratio in its simplest terms is
  A 1 : 4   B 7 : 23
  C 21 : 69   D 84 : 276

② $30 was shared between Dan and Ken so that Ken's share : Dan's share was 2 : 3. Ken's share was
  A $18   B $15
  C $12   D $6

③ A car dealer gains $600 on a sale which is equivalent to a profit of 8%. What was the cost price?
  A $8100   B $7500
  C $5400   D $4800

④ The price of a chair is $53.10 when a sales tax of 18% is included. The actual sales tax on the chair is
  A $4.38   B $5.31
  C $6.93   D $8.10

⑤ Oranges are sold at 80 cents for one orange or $3.25 for a bag of 5 oranges. The total discount if 50 oranges are bought in bags of 5 is
  A $7.50   B $8.00
  C $15.00   D $15.50

⑥ I travelled at 60 km/h and took 2 h for a certain journey. How long would it have taken me if I had travelled at 50 km/h?

⑦ A shop assistant is paid 8c in the $ on the amount of sales she makes. Find how much she gets if she sells 3 dresses at $390 each, 5 skirts at $180 each and 2 pairs of shoes at $220 per pair.

⑧ A new candle is 15 cm long. It burns at the rate of 1.2 cm per hour. Calculate the length of the candle after it has been burning for 2 h 36 min.

⑨ A married man earns $48 000 per year. He has two children. Tax-free allowances per year are:
  $800 for each adult,
  $300 for each child.
  10% on approved national investments.
  If in 1 year he invested $6000 in an approved national investment, calculate
  (a) the total tax-free allowances,
  (b) the man's taxable income.

Taxable income	Rate of tax
For every dollar of the first $30 000	10c
For every dollar of the taxable income that exceeds $30 000	30c

  (c) Calculate the amount of tax the man paid in that year.

⑩ The hire purchase price of a motor bike is $20 400. $12\frac{1}{2}$% is paid as a deposit. The remainder is spread over 12 equal monthly instalments.
  (a) Calculate the amount of the deposit.
  (b) Calculate the remainder to be paid.
  (c) Find the amount of each monthly instalment to the nearest cent.

General revision test C (Chapters 16–22)

① If 5 men can do a piece of work in 6 days, how many days, to the nearest whole number, should 8 men take to do it? Assume that each man works at the same rate.

   A 4   B 5   C 7   D 9

②

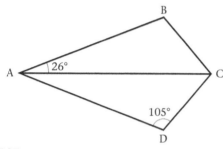

Fig. R24

In Fig. R24, $a =$

   A 12   B 30   C 42   D 54

③

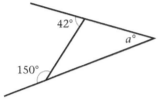

Fig. R25

In Fig. R25, ABCD is a kite, $B\widehat{A}C = 26°$ and $A\widehat{D}C = 105°$. $A\widehat{C}D =$

   A 26°   B 41°   C 49°   D 64°

④ A trader bought a pair of shoes for $124. He sold them at a profit of 55%. The selling price of the shoes was

   A $168.00   B $179.00
   C $192.20   D $204.00

⑤ Four of the five angles in a pie chart are 30°, 50°, 25° and 65°. The size of the fifth angle is

   A 10°   B 90°   C 170°   D 190°

⑥ Two angles of △ABC are 46° and 67°. Hence, △ABC is

   A equilateral   B isosceles
   C right-angled   D obtuse-angled

Use Fig. R26 to answer questions 7–9.

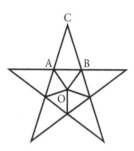

Fig. R26

Fig. R26 is a star shape developed from a regular pentagon.

⑦ What is the order of rotational symmetry of the shape in Fig. R26?

   A 3   B 5   C 10   D 15

⑧ The size of each angle at O is

   A 54°   B 60°   C 72°   D 108°

⑨ $O\widehat{A}B =$

   A 36°   B 54°   C 60°   D 90°

⑩ If $\dfrac{x}{72} = \dfrac{16}{96}$, $x =$

   A 6   B 8   C 12   D 18

⑪ In one year a man spent $650 on tobacco and $240 on sweets. Next year he reduced his tobacco spending by 40% and increased his spending on sweets in the ratio 5 : 3. How much did he save?

⑫

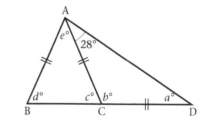

Fig. R27

In Fig. R27, AB = AC = CD and $C\widehat{A}D = 28°$. If BCD is a straight line, find the sizes of angles a, b, c, d, e. State reasons for your answers.

13

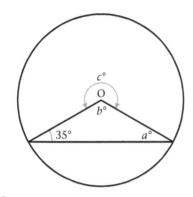

Fig. R28

In Fig. R28, O is the centre of the circle. Find *a*, *b*, *c*. State reasons for your answers.

14 A salesman gets a commission of $8\frac{1}{3}$% of the value of the things he sells. Find his commission for selling 2 guitars at $118 each, 5 tennis rackets at $33 each and 14 kettles at $8.50 each.

15 A traffic survey gave the results shown in Table R3.

Table R3

Vehicle	car	lorry	bus	bicycle
Frequency	12	10	5	23

(a) How many lorries were there for every one bus?
(b) What percentage of the vehicles were bicycles?
(c) Represent the data in Table R3 on a bar chart.

16 First set: 24, 25, 26, … 30
Second set: 2, 3, 4, … 8
Draw an arrow diagram showing the relation 'is a multiple of'.

17 Harry, George, John and Samuel are four boys in a class. Each boy chooses a girl whose name starts with the same letter as his own to team up with him in a table tennis competition.
Draw an arrow diagram to show the teams.

18 The diagram in Fig. R29 shows an arrow diagram for two sets A and B.
(a) What is the relation shown in the diagram?
(b) Why is there no arrow from 1 in set A?

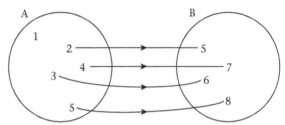

Fig. R29

19 S = {2, 3, 4, 5, 6}   T = {2, 4, 6, 8}
Draw arrow diagrams from S to T to show
(a) 'is a multiple of'
(b) 'is a factor of'
(c) 'is greater than'

20 Make a list of the names of the members of two families you know. Draw arrow diagrams for both families to show the relation 'is a parent of'.

Practice examination

Paper 1

Answer ALL the questions. In each question choose **one** *of the letters* **A, B, C,** *or* **D** *which corresponds to the correct answer.*

1 The prime factors of 18 are
A 2, 3   B 2, 3, 6
C 2, 3, 6, 9   D 2, 3, 6, 9, 18

2 $(-3) + (5) + (-7) =$
A -15 B -5 C $+1$ D $+15$

3 $89.6 \div 0.35 =$
A 0.256   B 2.56   C 25.6   D 256

4 When $m = -5$, $m - 14 =$
A -19 B -9 C 9   D 19

5 The LCM of 4, 5, 6, 8 is
A 120   B 240   C 480   D 960

6 Which of the following could represent the face of a solid?
A cube   B cone
C square   D prism

7 0.05078 to 3 significant figures $=$
A 0.05   B 0.05007
C 0.05008   D 0.051

8 $(3) + (-5) \times (-2) + (4) \times (-1) =$
A -8 B 0   C $+4$ D $+9$

9 Given S = {c, h, a, r}, which of the following sets are equal to S?
A P = {h, a, r, m}
B Q = {e, a, c, h}
C R = {c, r, a, m}
D T = {a, r, c, h}

10 N is an even number divisible by 3 and 5. A possible value of N is
A 18   B 50   C 60   D 75

11 The number of lines of symmetry in an equilateral triangle is
A 0   B 1   C 2   D 3

12 103_{eight} converted to a base ten number is
A 11   B 13   C 67   D 103

13 A car travels 100 km in $2\frac{1}{2}$ hours. Its average speed in km h^{-1} is
A 40   B 50   C 70   D 75

14 $3(a - 2b) - 2(3a - b) =$
A $-3a - b$ B $-3a - 3b$
C $-3a - 4b$ D $-3a - 8b$

15 The ratio of 2 h 20 min to 5 h 15 min is
A $44 : 103$ B $4 : 9$
C $9 : 4$ D $103 : 44$

16 In standard form, $0.00342 =$
A 3.42×10^{-2}
B 3.42×10^{-3}
C 3.42×10^{-4}
D 3.42×10^{-5}

17 Given P = {2, 3, 4}, the number of subsets in P is
A 3   B 6   C 8   D 9

18 Given that n represents an odd number, which of the following must be odd?
A $n + 1$ B $n + 2$
C $2n + 2$ D $3n + 1$

19 A boy is x years old. His father is four times as old. In y years, the father's age will be
A $4(x + y)$ B $4x - y$
C $4x + y$ D $4(y - x)$

20 Which of the following sets of numbers could *not* represent the angles of a triangle?
A 90, 38, 52
B 24, 76, 80
C 15, 60, 105
D 43, 65, 82

21 $1.504 \div 4.7 = 0.32$, then $320 \times 0.47 =$
A 1.504   B 15.04
C 150.4   D 1504

22 A shelf can hold 84 books each 3 cm wide. The number of books of width 4 cm it can hold is

   A 56   **B** 63   **C** 84   **D** 105

23 A rectangle measures 12 cm by 10 cm. How many 2 cm by 2 cm squares are needed to cover the rectangle completely?

   A 20   **B** 30   **C** 40   **D** 50

24 The nearest approximation to the cost of 37 books at $12.03 each is

   A $450   **B** $445   **C** $440   **D** $435

25 If $\frac{5}{6}$ of a certain number is $-6\frac{2}{3}$, the number is

   A $-5\frac{1}{3}$ **B** $-5\frac{5}{9}$

   C $-7\frac{1}{2}$ **D** -8

26 The result of taking 2 away from n and multiplying the result by 3 is equal to 9. Which of the following expresses this as an algebraic equation?

   A $3(2 - n) = 9$ **B** $2 - 3n = 9$
   C $3(n - 2) = 9$ **D** $3n - 2 = 9$

27 Which of the following statements *best* describes the diagonals of a rhombus?

   A diagonals are not equal
   B diagonals bisect each other
   C diagonals meet at right angles
   D diagonals bisect at right angles

28 The area of a rectangle is 144 cm². If the length of the rectangle is 8 cm, its perimeter is

   A 16   **B** 26   **C** 36   **D** 52

29 A man paid $1210 for 11 goats. He sold them at a profit of 32%. The selling price of each goat is

   A $125.00   **B** $145.20
   C $160.20   **D** $193.33

30 The table gives the average monthly rainfall in Bartago over a one-year period.

Month	Rainfall (mm)
January	86
February	52
March	45
April	65
May	85
June	152
July	159
August	180
September	191
October	205
November	180
December	94

Which of the following statements may *not* be true for the information?

   A The least amount of rainfall was recorded in March.
   B October was the rainiest month.
   C It rained every day in October.
   D The rainy season lasted from June to November.

Paper 2

Answer ALL the questions.

① (a) Simplify

$$\frac{3\frac{3}{5} \times 1\frac{5}{9}}{2\frac{1}{10}}$$

(b) A ball of string contains $13\frac{1}{2}$ metres. Lengths of 2.3 m, 1.8 m, 95 cm and 2.37 m are cut off. The remainder is divided into equal pieces each 32 cm long. Calculate the number of equal pieces.

② (a) A = {3, 6, 9, 12, 15}   B = {1, 3, 4}
   (i) Draw an arrow diagram from A to B to show the relation 'is three times'.
   (ii) State which set is the domain, and which is the range.
(b) Three clocks are programmed to ring every 20 minutes, 30 minutes and 45 minutes, respectively.
   (i) If they ring together at 7:00 p.m., calculate when they will next ring together.
   (ii) Write the answer for (i) using the 24-hour clock.

③ (a) Simplify the following.
   (i) $8x + 3y + 6 - 5x - 2y + 4$
   (ii) $3(x + 5) - 2(y - 1)$
   (iii) $\frac{x}{4} + \frac{2x}{3} - \frac{3x}{2}$
(b) Factorise the following.
   (i) $7a - 35$
   (ii) $3x - 15y + 21$
   (iii) $4x^2 - 28xy$

④ A lamp can be bought either in cash for $247.00 or by paying 52 weekly payments of $5.70.

(a) Calculate the instalment cost.
(b) By what percentage is the instalment cost greater than the cash price?

⑤ (a) Which of the following shapes has rotational symmetry? State the order for those which have rotational symmetry.

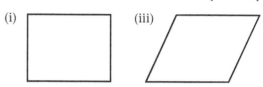

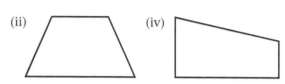

Fig. P1

(b) The perimeter of a school compound is 1805 m. If 201 fence posts are equally spaced around the perimeter, estimate the distance between any two posts.

⑥ In Fig. P2 △ABC is a right-angled isosceles triangle in which AB = BC. Calculate the size of the lettered angles. Give reasons.

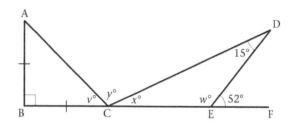

Fig. P2

⑦ (a) Solve the following equations.
   (i) $5x - 2 = 6 + x$
   (ii) $\frac{4x}{9} = -5\frac{1}{3}$
   (iii) $\frac{y + 2}{3} = \frac{3 - y}{2}$
(b) If you add one third of a number, n, to the number itself, the result is 24.
   (i) Write this information in an algebraic equation.
   (ii) Solve the equation for n.

8 Water is pulled up from a well in a bucket on a rope. The rope winds on a cylindrical drum 15 cm in diameter. It takes 28 turns of the drum to pull the bucket up from the bottom of the well. Calculate the depth of the well. (Use $\frac{22}{7}$ for π.)

9 Fig. P3 shows a cuboid cut in half to make a wedge.
Make a table to show
(a) the number of faces of the wedge,
(b) the shape of the faces,
(c) the number of edges.

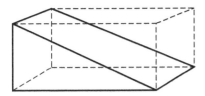

Fig. P3

10 Fig. P4 is a bar chart showing how the students in a class travel to school.

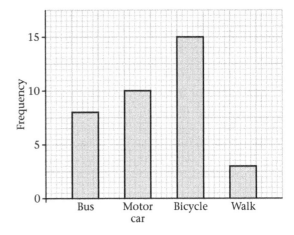

Fig. P4

(a) Use Fig. P4 to answer the following questions.
   (i) How many students walked to school?
   (ii) How many students rode bicycles to school?
   (iii) How many students are in the class?
(b) Calculate the size of the angles in a pie chart which represents the information in Fig. P4.
(c) Draw the pie chart.

SI units

Length

The **metre** is the basic unit of length.

unit	abbreviation	basic unit
1 kilometre	1 km	1000 m
1 hectometre	1 hm	100 m
1 decametre	1 dam	10 m
1 metre	1 m	1 m
1 decimetre	1 dm	0.1 m
1 centimetre	1 cm	0.01 m
1 millimetre	1 mm	0.001 m

The most common measures are the millimetre, the metre and the kilometre.
$1 \text{ m} = 1000 \text{ mm}$
$1 \text{ km} = 1000 \text{ m} = 1\,000\,000 \text{ mm}$

Area

The **square metre** is the basic unit of area. Units of area are derived from units of length.

unit	abbreviation	relation to other units of area
square millimetre	mm²	
square centimetre	cm²	$1 \text{ cm}^2 = 100 \text{ mm}^2$
square metre	m²	$1 \text{ m}^2 = 10\,000 \text{ cm}^2$
square kilometre	km²	$1 \text{ km}^2 = 1\,000\,000 \text{ m}^2$
hectare (for land measure)	ha	$1 \text{ ha} = 10\,000 \text{ m}^2$

Mass

The **gram** is the basic unit of mass.

unit	abbreviation	basic unit
1 kilogram	1 kg	1000 g
1 hectogram	1 hg	100 g
1 decagram	1 dag	10 g
1 gram	1 g	1 g
1 decigram	1 dg	0.1 g
1 centigram	1 cg	0.01 g
1 milligram	1 mg	0.001 g

The **tonne** (t) is used for large masses. The most common measures of mass are the milligram, the gram, the kilogram and the tonne.
$1 \text{ g} = 1000 \text{ mg}$
$1 \text{ kg} = 1000 \text{ g} = 1\,000\,000 \text{ mg}$
$1 \text{ t} = 1000 \text{ kg} = 1\,000\,000 \text{ g}$

Time

The **second** is the basic unit of time.

unit	abbreviation	basic unit
1 second	1 s	1 s
1 minute	1 min	60 s
1 hour	1 h	3600 s

Mensuration formulae

	perimeter	area
square side s	$4s$	s^2
rectangle length l, breadth b	$2(l+b)$	lb
circle radius r	$2\pi r$	πr^2
triangle base b, height h		$\frac{1}{2}bh$
parallelogram base b, height h		bh

Multiplication table

×	1	2	3	4	5	6	7	8	9	10
1	1	2	3	4	5	6	7	8	9	10
2	2	4	6	8	10	12	14	16	18	20
3	3	6	9	12	15	18	21	24	27	30
4	4	8	12	16	20	24	28	32	35	40
5	5	10	15	20	25	30	35	40	45	50
6	6	12	18	24	30	36	42	48	54	60
7	7	14	21	28	35	42	49	56	63	70
8	8	16	24	32	40	48	56	64	72	80
9	9	18	27	36	45	54	63	72	81	90
10	10	20	30	40	50	60	70	80	90	100

Divisibility tests

Any whole number is exactly divisible by

2	if its last digit is even
3	if the sum of its digits is divisible by 3
4	if its last two digits form a number divisible by 4
5	if its last digit is 5 or 0
6	if its last digit is even and the sum of its digits is divisible by 3
8	if its last three digits form a number divisible by 8
9	if the sum of its digits is divisible by 9
10	if its last digit is 0

Symbols

symbol	meaning
=	is equal to
≠	is not equal to
≃	is approximately equal to
>	is greater than
<	is less than
⩾	is greater than or equal to
⩽	is less than or equal to
°	degree (angle)
°C	degrees Celsius (temperature)
A, B, C ...	points
AB	the line joining the point A and the point B, *or the* distance between points A and B
△ABC	triangle ABC
$\widehat{ABC}$	the angle ABC
⌐	lines meeting at right angles
π	pi (3.14 ...)
%	per cent
A = {p, q, r}	A is the set p, q, r
B = {1, 2, 3, ...}	B is the infinite set 1, 2, 3 and so on
n (A)	number of elements in set A
∈	is an element of
∉	is not an element of
{ } or ∅	the empty set
U or ℰ	the universal set
A ⊂ B	A is a subset of B
A ⊃ B	A contains B
⊄, ⊅	negations of ⊂ and ⊃

Squares $x \to x^2$

x	0	1	2	3	4	5	6	7	8	9
5.5	30.25	30.36	30.47	30.58	30.69	30.80	30.91	31.02	31.14	31.25
5.6	31.36	31.47	31.58	31.70	31.81	31.92	32.04	32.15	32.26	32.38
5.7	32.49	32.60	32.72	32.83	32.95	33.06	33.18	33.29	33.41	33.52
5.8	33.64	33.76	33.87	33.99	34.11	34.22	34.34	34.46	34.57	34.69
5.9	34.81	34.93	35.05	35.16	35.28	35.40	35.52	35.64	35.76	35.88
6.0	36.00	36.12	36.24	36.36	36.48	36.60	36.72	36.84	36.97	37.09
6.1	37.21	37.33	37.45	37.58	37.70	37.82	37.95	38.07	38.19	38.32
6.2	38.44	38.56	38.69	38.81	38.94	39.06	39.19	39.31	39.44	39.56
6.3	39.69	39.82	39.94	40.07	40.20	40.32	40.45	40.58	40.70	40.83
6.4	40.96	41.09	41.22	41.34	41.47	41.60	41.73	41.86	41.99	42.12
6.5	42.25	42.38	42.51	42.64	42.77	42.90	43.03	43.16	43.30	43.43
6.6	43.56	43.69	43.82	43.96	44.09	44.22	44.36	44.49	44.62	44.76
6.7	44.89	45.02	45.16	45.29	45.43	45.56	45.70	45.83	45.97	46.10
6.8	46.24	46.38	46.51	46.65	46.79	46.92	47.06	47.20	47.33	47.47
6.9	47.61	47.75	47.89	48.02	48.16	48.30	48.44	48.58	48.72	48.86
7.0	49.00	49.14	49.28	49.42	49.56	49.70	49.84	49.98	50.13	50.27
7.1	50.41	50.55	50.69	50.84	50.98	51.12	51.27	51.41	51.55	51.70
7.2	51.84	51.98	52.13	52.27	52.42	52.56	52.71	52.85	53.00	53.14
7.3	53.29	53.44	53.58	53.73	53.88	54.02	54.17	54.32	54.46	54.61
7.4	54.76	54.91	55.06	55.20	55.35	55.50	55.65	55.80	55.95	56.10
7.5	56.25	56.40	56.55	56.70	56.85	57.00	57.15	57.30	57.46	57.61
7.6	57.76	57.91	58.06	58.22	58.37	58.52	58.68	58.83	58.98	59.14
7.7	59.29	59.44	59.60	59.75	59.91	60.06	60.22	60.37	60.53	60.68
7.8	60.84	61.00	61.15	61.31	61.47	61.62	61.78	61.94	62.09	62.25
7.9	62.41	62.57	62.73	62.88	63.04	63.20	63.36	63.52	63.68	63.84
8.0	64.00	64.16	64.32	64.48	64.64	64.80	64.96	65.12	65.29	65.45
8.1	65.61	65.77	65.93	66.10	66.26	66.42	66.59	66.75	66.91	67.08
8.2	67.24	67.40	67.57	67.73	67.90	68.06	68.23	68.39	68.56	68.72
8.3	68.89	69.06	69.22	69.39	69.56	69.72	69.89	70.06	70.22	70.39
8.4	70.56	70.73	70.90	71.06	71.23	71.40	71.57	71.74	71.91	72.08
8.5	72.25	72.42	72.59	72.76	72.93	73.10	73.27	73.44	73.62	73.79
8.6	73.96	74.13	74.30	74.48	74.65	74.82	75.00	75.17	75.34	75.52
8.7	75.69	75.86	76.04	76.21	76.39	76.56	76.74	76.91	77.09	77.26
8.8	77.44	77.62	77.79	77.97	78.15	78.32	78.50	78.68	78.85	79.03
8.9	79.21	79.39	79.57	79.74	79.92	80.10	80.28	80.46	80.64	80.82
9.0	81.00	81.18	81.36	81.54	81.72	81.90	82.08	82.26	82.45	82.63
9.1	82.81	82.99	83.17	83.36	83.54	83.72	83.91	84.09	84.27	84.46
9.2	84.64	84.82	85.01	85.19	85.38	85.56	85.75	85.93	86.12	86.30
9.3	86.49	86.68	86.86	87.05	87.24	87.42	87.61	87.80	87.98	88.17
9.4	88.36	88.55	88.74	88.92	89.11	89.30	89.49	89.68	89.87	90.06
9.5	90.25	90.44	90.63	90.82	91.01	91.20	91.39	91.58	91.78	91.97
9.6	92.16	92.35	92.54	92.74	92.93	93.12	93.32	93.51	93.70	93.90
9.7	94.09	94.28	94.48	94.67	94.87	95.06	95.26	95.45	95.65	95.84
9.8	96.04	96.24	96.43	96.63	96.83	97.02	97.22	97.42	97.61	97.81
9.9	98.01	98.21	98.41	98.60	98.80	99.00	99.20	99.40	99.60	99.80

Squares

x	0	1	2	3	4	5	6	7	8	9
1.0	1.00	1.02	1.04	1.06	1.08	1.10	1.12	1.14	1.17	1.19
1.1	1.21	1.23	1.25	1.28	1.30	1.32	1.35	1.37	1.39	1.42
1.2	1.44	1.46	1.49	1.51	1.54	1.56	1.59	1.61	1.64	1.66
1.3	1.69	1.72	1.74	1.77	1.80	1.82	1.85	1.88	1.90	1.93
1.4	1.96	1.99	2.02	2.04	2.07	2.10	2.13	2.16	2.19	2.22
1.5	2.25	2.28	2.31	2.34	2.37	2.40	2.43	2.46	2.50	2.53
1.6	2.56	2.59	2.62	2.66	2.69	2.72	2.76	2.79	2.82	2.86
1.7	2.89	2.92	2.96	2.99	3.03	3.06	3.10	3.13	3.17	3.20
1.8	3.24	3.28	3.31	3.35	3.39	3.42	3.46	3.50	3.53	3.57
1.9	3.61	3.65	3.69	3.72	3.76	3.80	3.84	3.88	3.92	3.96
2.0	4.00	4.04	4.08	4.12	4.16	4.20	4.24	4.28	4.33	4.37
2.1	4.41	4.45	4.49	4.54	4.58	4.62	4.67	4.71	4.75	4.80
2.2	4.84	4.88	4.93	4.97	5.02	5.06	5.11	5.15	5.20	5.24
2.3	5.29	5.34	5.38	5.43	5.48	5.52	5.57	5.62	5.66	5.71
2.4	5.76	5.81	5.86	5.90	5.95	6.00	6.05	6.10	6.15	6.20
2.5	6.25	6.30	6.35	6.40	6.45	6.50	6.55	6.60	6.66	6.71
2.6	6.76	6.81	6.86	6.92	6.97	7.02	7.08	7.13	7.18	7.24
2.7	7.29	7.34	7.40	7.45	7.51	7.56	7.62	7.67	7.73	7.78
2.8	7.84	7.90	7.95	8.01	8.07	8.12	8.18	8.24	8.29	8.35
2.9	8.41	8.47	8.53	8.58	8.64	8.70	8.76	8.82	8.88	8.94
3.0	9.00	9.06	9.12	9.18	9.24	9.30	9.36	9.42	9.49	9.55
3.1	9.61	9.67	9.73	9.80	9.86	9.92	9.99	10.05	10.11	10.18
3.2	10.24	10.30	10.37	10.43	10.50	10.56	10.63	10.69	10.76	10.82
3.3	10.89	10.96	11.02	11.09	11.16	11.22	11.29	11.36	11.42	11.49
3.4	11.56	11.63	11.70	11.76	11.83	11.90	11.97	12.04	12.11	12.18
3.5	12.25	12.32	12.39	12.46	12.53	12.60	12.67	12.74	12.82	12.89
3.6	12.96	13.03	13.10	13.18	13.25	13.32	13.40	13.47	13.54	13.62
3.7	13.69	13.76	13.84	13.91	13.99	14.06	14.14	14.21	14.29	14.36
3.8	14.44	14.52	14.59	14.67	14.75	14.82	14.90	14.98	15.05	15.13
3.9	15.21	15.29	15.37	15.44	15.52	15.60	15.68	15.76	15.84	15.92
4.0	16.00	16.08	16.16	16.24	16.32	16.40	16.48	16.56	16.65	16.73
4.1	16.81	16.89	16.97	17.06	17.14	17.22	17.31	17.39	17.47	17.56
4.2	17.64	17.72	17.81	17.89	17.98	18.06	18.15	18.23	18.32	18.40
4.3	18.49	18.58	18.66	18.75	18.84	18.92	19.01	19.10	19.18	19.27
4.4	19.36	19.45	19.54	19.62	19.71	19.80	19.89	19.98	20.07	20.16
4.5	20.25	20.34	20.43	20.52	20.61	20.70	20.79	20.88	20.98	21.07
4.6	21.16	21.25	21.34	21.44	21.53	21.62	21.72	21.81	21.90	22.00
4.7	22.09	22.18	22.28	22.37	22.47	22.56	22.66	22.75	22.85	22.94
4.8	23.04	23.14	23.23	23.33	23.43	23.52	23.62	23.72	23.81	23.91
4.9	24.01	24.11	24.21	24.30	24.40	24.50	24.60	24.70	24.80	24.90
5.0	25.00	25.10	25.20	25.30	25.40	25.50	25.60	25.70	25.81	25.91
5.1	26.01	26.11	26.21	26.32	26.42	26.52	26.63	26.73	26.83	26.94
5.2	27.04	27.14	27.25	27.35	27.46	27.56	27.67	27.77	27.88	27.98
5.3	28.09	28.20	28.30	28.41	28.52	28.62	28.73	28.84	28.94	29.05
5.4	29.16	29.27	29.38	29.48	29.59	29.70	29.81	29.92	30.03	30.14

Square roots from 1 to 9.99 $x \to \sqrt{x}$

x	0	1	2	3	4	5	6	7	8	9
1.0	1.00	1.00	1.01	1.01	1.02	1.02	1.03	1.03	1.04	1.04
1.1	1.05	1.05	1.06	1.06	1.07	1.07	1.08	1.08	1.09	1.09
1.2	1.10	1.10	1.10	1.11	1.11	1.12	1.12	1.13	1.13	1.14
1.3	1.14	1.14	1.15	1.15	1.16	1.16	1.17	1.17	1.17	1.18
1.4	1.18	1.19	1.19	1.20	1.20	1.20	1.21	1.21	1.22	1.22
1.5	1.22	1.23	1.23	1.24	1.24	1.24	1.25	1.25	1.26	1.26
1.6	1.26	1.27	1.27	1.28	1.28	1.28	1.29	1.29	1.30	1.30
1.7	1.30	1.31	1.31	1.32	1.32	1.32	1.33	1.33	1.33	1.34
1.8	1.34	1.35	1.35	1.35	1.36	1.36	1.36	1.37	1.37	1.37
1.9	1.38	1.38	1.39	1.39	1.39	1.40	1.40	1.40	1.41	1.41
2.0	1.41	1.42	1.42	1.42	1.43	1.43	1.44	1.44	1.44	1.45
2.1	1.45	1.45	1.46	1.46	1.46	1.47	1.47	1.47	1.48	1.48
2.2	1.48	1.49	1.49	1.49	1.50	1.50	1.50	1.51	1.51	1.51
2.3	1.52	1.52	1.52	1.53	1.53	1.53	1.54	1.54	1.54	1.55
2.4	1.55	1.55	1.56	1.56	1.56	1.57	1.57	1.57	1.57	1.58
2.5	1.58	1.58	1.59	1.59	1.59	1.60	1.60	1.60	1.61	1.61
2.6	1.61	1.62	1.62	1.62	1.62	1.63	1.63	1.63	1.64	1.64
2.7	1.64	1.65	1.65	1.65	1.66	1.66	1.66	1.66	1.67	1.67
2.8	1.67	1.68	1.68	1.68	1.69	1.69	1.69	1.69	1.70	1.70
2.9	1.70	1.71	1.71	1.71	1.71	1.72	1.72	1.72	1.73	1.73
3.0	1.73	1.73	1.74	1.74	1.74	1.75	1.75	1.75	1.75	1.76
3.1	1.76	1.76	1.77	1.77	1.77	1.77	1.78	1.78	1.78	1.79
3.2	1.79	1.79	1.79	1.80	1.80	1.80	1.81	1.81	1.81	1.81
3.3	1.82	1.82	1.82	1.82	1.83	1.83	1.83	1.84	1.84	1.84
3.4	1.84	1.85	1.85	1.85	1.85	1.86	1.86	1.86	1.87	1.87
3.5	1.87	1.87	1.88	1.88	1.88	1.88	1.89	1.89	1.89	1.89
3.6	1.90	1.90	1.90	1.91	1.91	1.91	1.91	1.92	1.92	1.92
3.7	1.92	1.93	1.93	1.93	1.93	1.94	1.94	1.94	1.94	1.95
3.8	1.95	1.95	1.95	1.96	1.96	1.96	1.96	1.97	1.97	1.97
3.9	1.97	1.98	1.98	1.98	1.98	1.99	1.99	1.99	1.99	2.00
4.0	2.00	2.00	2.00	2.01	2.01	2.01	2.01	2.02	2.02	2.02
4.1	2.02	2.03	2.03	2.03	2.03	2.04	2.04	2.04	2.04	2.05
4.2	2.05	2.05	2.05	2.06	2.06	2.06	2.06	2.07	2.07	2.07
4.3	2.07	2.08	2.08	2.08	2.08	2.09	2.09	2.09	2.09	2.10
4.4	2.10	2.10	2.10	2.10	2.11	2.11	2.11	2.11	2.12	2.12
4.5	2.12	2.12	2.13	2.13	2.13	2.13	2.14	2.14	2.14	2.14
4.6	2.14	2.15	2.15	2.15	2.15	2.16	2.16	2.16	2.16	2.17
4.7	2.17	2.17	2.17	2.17	2.18	2.18	2.18	2.18	2.19	2.19
4.8	2.19	2.19	2.20	2.20	2.20	2.20	2.20	2.21	2.21	2.21
4.9	2.21	2.22	2.22	2.22	2.22	2.22	2.23	2.23	2.23	2.23
5.0	2.24	2.24	2.24	2.24	2.24	2.25	2.25	2.25	2.25	2.26
5.1	2.26	2.26	2.26	2.26	2.27	2.27	2.27	2.27	2.28	2.28
5.2	2.28	2.28	2.28	2.29	2.29	2.29	2.29	2.30	2.30	2.30
5.3	2.30	2.30	2.31	2.31	2.31	2.31	2.32	2.32	2.32	2.32
5.4	2.32	2.33	2.33	2.33	2.33	2.33	2.34	2.34	2.34	2.34

x	0	1	2	3	4	5	6	7	8	9
5.5	2.35	2.35	2.35	2.35	2.35	2.36	2.36	2.36	2.36	2.36
5.6	2.37	2.37	2.37	2.37	2.37	2.38	2.38	2.38	2.38	2.39
5.7	2.39	2.39	2.39	2.39	2.40	2.40	2.40	2.40	2.40	2.41
5.8	2.41	2.41	2.41	2.41	2.42	2.42	2.42	2.42	2.42	2.43
5.9	2.43	2.43	2.43	2.44	2.44	2.44	2.44	2.44	2.45	2.45
6.0	2.45	2.45	2.45	2.46	2.46	2.46	2.46	2.46	2.47	2.47
6.1	2.47	2.47	2.47	2.48	2.48	2.48	2.48	2.48	2.49	2.49
6.2	2.49	2.49	2.49	2.50	2.50	2.50	2.50	2.50	2.51	2.51
6.3	2.51	2.51	2.51	2.52	2.52	2.52	2.52	2.52	2.53	2.53
6.4	2.53	2.53	2.53	2.54	2.54	2.54	2.54	2.54	2.55	2.55
6.5	2.55	2.55	2.55	2.56	2.56	2.56	2.56	2.56	2.57	2.57
6.6	2.57	2.57	2.57	2.57	2.58	2.58	2.58	2.58	2.58	2.59
6.7	2.59	2.59	2.59	2.59	2.60	2.60	2.60	2.60	2.60	2.61
6.8	2.61	2.61	2.61	2.61	2.62	2.62	2.62	2.62	2.62	2.62
6.9	2.63	2.63	2.63	2.63	2.63	2.64	2.64	2.64	2.64	2.64
7.0	2.65	2.65	2.65	2.65	2.65	2.66	2.66	2.66	2.66	2.66
7.1	2.66	2.67	2.67	2.67	2.67	2.67	2.68	2.68	2.68	2.68
7.2	2.68	2.69	2.69	2.69	2.69	2.69	2.69	2.70	2.70	2.70
7.3	2.70	2.70	2.71	2.71	2.71	2.71	2.71	2.71	2.72	2.72
7.4	2.72	2.72	2.72	2.73	2.73	2.73	2.73	2.73	2.73	2.74
7.5	2.74	2.74	2.74	2.74	2.75	2.75	2.75	2.75	2.75	2.75
7.6	2.76	2.76	2.76	2.76	2.76	2.77	2.77	2.77	2.77	2.77
7.7	2.77	2.78	2.78	2.78	2.78	2.78	2.79	2.79	2.79	2.79
7.8	2.79	2.79	2.80	2.80	2.80	2.80	2.80	2.81	2.81	2.81
7.9	2.81	2.81	2.81	2.82	2.82	2.82	2.82	2.82	2.82	2.83
8.0	2.83	2.83	2.83	2.83	2.84	2.84	2.84	2.84	2.84	2.84
8.1	2.85	2.85	2.85	2.85	2.85	2.85	2.86	2.86	2.86	2.86
8.2	2.86	2.87	2.87	2.87	2.87	2.87	2.87	2.88	2.88	2.88
8.3	2.88	2.88	2.88	2.89	2.89	2.89	2.89	2.89	2.89	2.90
8.4	2.90	2.90	2.90	2.90	2.91	2.91	2.91	2.91	2.91	2.91
8.5	2.92	2.92	2.92	2.92	2.92	2.92	2.93	2.93	2.93	2.93
8.6	2.93	2.93	2.94	2.94	2.94	2.94	2.94	2.94	2.95	2.95
8.7	2.95	2.95	2.95	2.95	2.96	2.96	2.96	2.96	2.96	2.96
8.8	2.97	2.97	2.97	2.97	2.97	2.97	2.98	2.98	2.98	2.98
8.9	2.98	2.98	2.99	2.99	2.99	2.99	2.99	2.99	3.00	3.00
9.0	3.00	3.00	3.00	3.00	3.01	3.01	3.01	3.01	3.01	3.01
9.1	3.02	3.02	3.02	3.02	3.02	3.02	3.03	3.03	3.03	3.03
9.2	3.03	3.03	3.04	3.04	3.04	3.04	3.04	3.04	3.05	3.05
9.3	3.05	3.05	3.05	3.05	3.06	3.06	3.06	3.06	3.06	3.06
9.4	3.07	3.07	3.07	3.07	3.07	3.07	3.08	3.08	3.08	3.08
9.5	3.08	3.08	3.09	3.09	3.09	3.09	3.09	3.09	3.10	3.10
9.6	3.10	3.10	3.10	3.10	3.10	3.11	3.11	3.11	3.11	3.11
9.7	3.11	3.12	3.12	3.12	3.12	3.12	3.12	3.13	3.13	3.13
9.8	3.13	3.13	3.13	3.14	3.14	3.14	3.14	3.14	3.14	3.14
9.9	3.15	3.15	3.15	3.15	3.15	3.15	3.16	3.16	3.16	3.16

$x \rightarrow \sqrt{x}$

Square roots from 10 to 99.9

x	.0	.1	.2	.3	.4	.5	.6	.7	.8	.9
10	3.16	3.18	3.19	3.21	3.22	3.24	3.26	3.27	3.29	3.30
11	3.32	3.33	3.35	3.36	3.38	3.39	3.41	3.42	3.44	3.45
12	3.46	3.48	3.49	3.51	3.52	3.54	3.55	3.56	3.58	3.59
13	3.61	3.62	3.63	3.65	3.66	3.67	3.69	3.70	3.71	3.73
14	3.74	3.75	3.77	3.78	3.79	3.81	3.82	3.83	3.85	3.86
15	3.87	3.89	3.90	3.91	3.92	3.94	3.95	3.96	3.97	3.99
16	4.00	4.01	4.02	4.04	4.05	4.06	4.07	4.09	4.10	4.11
17	4.12	4.14	4.15	4.16	4.17	4.18	4.20	4.21	4.22	4.23
18	4.24	4.25	4.27	4.28	4.29	4.30	4.31	4.32	4.34	4.35
19	4.36	4.37	4.38	4.39	4.40	4.42	4.43	4.44	4.45	4.46
20	4.47	4.48	4.49	4.51	4.52	4.53	4.54	4.55	4.56	4.57
21	4.58	4.59	4.60	4.62	4.63	4.64	4.65	4.66	4.67	4.68
22	4.69	4.70	4.71	4.72	4.73	4.74	4.75	4.76	4.77	4.79
23	4.80	4.81	4.82	4.83	4.84	4.85	4.86	4.87	4.88	4.89
24	4.90	4.91	4.92	4.93	4.94	4.95	4.96	4.97	4.98	4.99
25	5.00	5.01	5.02	5.03	5.04	5.05	5.06	5.07	5.08	5.09
26	5.10	5.11	5.12	5.13	5.14	5.15	5.16	5.17	5.18	5.19
27	5.20	5.21	5.22	5.22	5.23	5.24	5.25	5.26	5.27	5.28
28	5.29	5.30	5.31	5.32	5.33	5.34	5.35	5.36	5.37	5.38
29	5.39	5.39	5.40	5.41	5.42	5.43	5.44	5.45	5.46	5.47
30	5.48	5.49	5.50	5.50	5.51	5.52	5.53	5.54	5.55	5.56
31	5.57	5.58	5.59	5.59	5.60	5.61	5.62	5.63	5.64	5.65
32	5.66	5.67	5.67	5.68	5.69	5.70	5.71	5.72	5.73	5.74
33	5.74	5.75	5.76	5.77	5.78	5.79	5.80	5.81	5.81	5.82
34	5.83	5.84	5.85	5.86	5.87	5.87	5.88	5.89	5.90	5.91
35	5.92	5.92	5.93	5.94	5.95	5.96	5.97	5.97	5.98	5.99
36	6.00	6.01	6.02	6.02	6.03	6.04	6.05	6.06	6.07	6.07
37	6.08	6.09	6.10	6.11	6.12	6.12	6.13	6.14	6.15	6.16
38	6.16	6.17	6.18	6.19	6.20	6.20	6.21	6.22	6.23	6.24
39	6.24	6.25	6.26	6.27	6.28	6.28	6.29	6.30	6.31	6.32
40	6.32	6.33	6.34	6.35	6.36	6.36	6.37	6.38	6.39	6.40
41	6.40	6.41	6.42	6.43	6.43	6.44	6.45	6.46	6.47	6.47
42	6.48	6.49	6.50	6.50	6.51	6.52	6.53	6.53	6.54	6.55
43	6.56	6.57	6.57	6.58	6.59	6.60	6.60	6.61	6.62	6.63
44	6.63	6.64	6.65	6.66	6.66	6.67	6.68	6.69	6.69	6.70
45	6.71	6.72	6.72	6.73	6.74	6.75	6.75	6.76	6.77	6.77
46	6.78	6.79	6.80	6.80	6.81	6.82	6.83	6.83	6.84	6.85
47	6.86	6.86	6.87	6.88	6.88	6.89	6.90	6.91	6.91	6.92
48	6.93	6.94	6.94	6.95	6.96	6.96	6.97	6.98	6.99	6.99
49	7.00	7.01	7.01	7.02	7.03	7.04	7.04	7.05	7.06	7.06
50	7.07	7.08	7.09	7.09	7.10	7.11	7.11	7.12	7.13	7.13
51	7.14	7.15	7.16	7.16	7.17	7.18	7.18	7.19	7.20	7.20
52	7.21	7.22	7.22	7.23	7.24	7.25	7.25	7.26	7.27	7.27
53	7.28	7.29	7.29	7.30	7.31	7.31	7.32	7.33	7.33	7.34
54	7.35	7.36	7.36	7.37	7.38	7.38	7.39	7.40	7.40	7.41

x	.0	.1	.2	.3	.4	.5	.6	.7	.8	.9
55	7.42	7.42	7.43	7.44	7.44	7.45	7.46	7.46	7.47	7.48
56	7.48	7.49	7.50	7.50	7.51	7.52	7.52	7.53	7.54	7.54
57	7.55	7.56	7.56	7.57	7.58	7.58	7.59	7.60	7.60	7.61
58	7.62	7.62	7.63	7.64	7.64	7.65	7.66	7.66	7.67	7.67
59	7.68	7.69	7.69	7.70	7.71	7.71	7.72	7.73	7.73	7.74
60	7.75	7.75	7.76	7.77	7.77	7.78	7.78	7.79	7.80	7.80
61	7.81	7.82	7.82	7.83	7.84	7.84	7.85	7.85	7.86	7.87
62	7.87	7.88	7.89	7.89	7.90	7.91	7.91	7.92	7.92	7.93
63	7.94	7.94	7.95	7.96	7.96	7.97	7.97	7.98	7.99	7.99
64	8.00	8.01	8.01	8.02	8.02	8.03	8.04	8.04	8.05	8.06
65	8.06	8.07	8.07	8.08	8.09	8.09	8.10	8.11	8.11	8.12
66	8.12	8.13	8.14	8.14	8.15	8.16	8.16	8.17	8.17	8.18
67	8.19	8.19	8.20	8.20	8.21	8.22	8.22	8.23	8.23	8.24
68	8.25	8.25	8.26	8.26	8.27	8.28	8.28	8.29	8.29	8.30
69	8.31	8.31	8.32	8.32	8.33	8.34	8.34	8.35	8.35	8.36
70	8.37	8.37	8.38	8.38	8.39	8.40	8.40	8.41	8.41	8.42
71	8.43	8.43	8.44	8.44	8.45	8.46	8.46	8.47	8.47	8.48
72	8.49	8.49	8.50	8.50	8.51	8.51	8.52	8.53	8.53	8.54
73	8.54	8.55	8.56	8.56	8.57	8.57	8.58	8.59	8.59	8.60
74	8.60	8.61	8.61	8.62	8.63	8.63	8.64	8.64	8.65	8.65
75	8.66	8.67	8.67	8.68	8.68	8.69	8.69	8.70	8.71	8.71
76	8.72	8.72	8.73	8.73	8.74	8.75	8.75	8.76	8.76	8.77
77	8.77	8.78	8.79	8.79	8.80	8.80	8.81	8.81	8.82	8.83
78	8.83	8.84	8.84	8.85	8.85	8.86	8.87	8.87	8.88	8.88
79	8.89	8.89	8.90	8.91	8.91	8.92	8.92	8.93	8.93	8.94
80	8.94	8.95	8.96	8.96	8.97	8.97	8.98	8.98	8.99	8.99
81	9.00	9.01	9.01	9.02	9.02	9.03	9.03	9.04	9.04	9.05
82	9.06	9.06	9.07	9.07	9.08	9.08	9.09	9.09	9.10	9.10
83	9.11	9.12	9.12	9.13	9.13	9.14	9.14	9.15	9.15	9.16
84	9.17	9.17	9.18	9.18	9.19	9.19	9.20	9.20	9.21	9.21
85	9.22	9.22	9.23	9.24	9.24	9.25	9.25	9.26	9.26	9.27
86	9.27	9.28	9.28	9.29	9.30	9.30	9.31	9.31	9.32	9.32
87	9.33	9.33	9.34	9.34	9.35	9.35	9.36	9.37	9.37	9.38
88	9.38	9.39	9.39	9.40	9.40	9.41	9.41	9.42	9.42	9.43
89	9.43	9.44	9.44	9.45	9.46	9.46	9.47	9.47	9.48	9.48
90	9.49	9.49	9.50	9.50	9.51	9.51	9.52	9.52	9.53	9.53
91	9.54	9.54	9.55	9.56	9.56	9.57	9.57	9.58	9.58	9.59
92	9.59	9.60	9.60	9.61	9.61	9.62	9.62	9.63	9.63	9.64
93	9.64	9.65	9.65	9.66	9.66	9.67	9.67	9.68	9.69	9.69
94	9.70	9.70	9.71	9.71	9.72	9.72	9.73	9.73	9.74	9.74
95	9.75	9.75	9.76	9.76	9.77	9.77	9.78	9.78	9.79	9.79
96	9.80	9.80	9.81	9.81	9.82	9.82	9.83	9.83	9.84	9.84
97	9.85	9.85	9.86	9.86	9.87	9.87	9.88	9.88	9.89	9.89
98	9.90	9.90	9.91	9.91	9.92	9.92	9.93	9.93	9.94	9.94
99	9.95	9.95	9.96	9.96	9.97	9.97	9.98.	9.98	9.99	9.99

Tables

$x \rightarrow \dfrac{1}{x}$

x	0	1	2	3	4	5	6	7	8	9
5.5	0.182	181	181	181	181	180	180	180	179	179
5.6	.179	178	178	178	177	177	177	176	176	176
5.7	.175	175	175	175	174	174	174	173	173	173
5.8	.172	172	172	172	171	171	171	170	170	170
5.9	.169	169	169	169	168	168	168	168	167	167
6.0	0.167	166	166	166	166	165	165	165	164	164
6.1	.164	164	163	163	163	163	162	162	162	162
6.2	.161	161	161	161	160	160	160	159	159	159
6.3	.159	158	158	158	158	157	157	157	157	156
6.4	.156	156	156	156	155	155	155	155	154	154
6.5	0.154	154	153	153	153	153	152	152	152	152
6.6	.152	151	151	151	151	150	150	150	150	149
6.7	.149	149	149	149	148	148	148	148	147	147
6.8	.147	147	147	146	146	146	146	146	145	145
6.9	.145	145	145	144	144	144	144	143	143	143
7.0	0.143	143	142	142	142	142	142	141	141	141
7.1	.141	141	140	140	140	140	140	139	139	139
7.2	.139	139	139	138	138	138	138	138	137	137
7.3	.137	137	137	136	136	136	136	136	136	135
7.4	.135	135	135	135	134	134	134	134	134	134
7.5	0.133	133	133	133	133	132	132	132	132	132
7.6	.132	131	131	131	131	131	131	130	130	130
7.7	.130	130	130	129	129	129	129	129	129	128
7.8	.128	128	128	128	128	127	127	127	127	127
7.9	.127	126	126	126	126	126	126	125	125	125
8.0	0.125	125	125	125	124	124	124	124	124	124
8.1	.123	123	123	123	123	123	123	122	122	122
8.2	.122	122	122	122	121	121	121	121	121	121
8.3	.120	120	120	120	120	120	120	119	119	119
8.4	.119	119	119	119	118	118	118	118	118	118
8.5	0.118	118	117	117	117	117	117	117	117	116
8.6	.116	116	116	116	116	116	115	115	115	115
8.7	.115	115	115	115	114	114	114	114	114	114
8.8	.114	114	113	113	113	113	113	113	113	112
8.9	.112	112	112	112	112	112	112	111	111	111
9.0	0.111	111	111	111	111	110	110	110	110	110
9.1	.110	110	110	110	109	109	109	109	109	109
9.2	.109	109	108	108	108	108	108	108	108	108
9.3	.108	107	107	107	107	107	107	107	107	106
9.4	.106	106	106	106	106	106	106	106	105	105
9.5	0.105	105	105	105	105	105	105	104	104	104
9.6	.104	104	104	104	104	104	104	103	103	103
9.7	.103	103	103	103	103	103	102	102	102	102
9.8	.102	102	102	102	102	102	101	101	101	101
9.9	.101	101	101	101	101	101	100	100	100	100

Reciprocals

x	0	1	2	3	4	5	6	7	8	9
1.0	1.000	0.990	980	971	962	952	943	935	926	917
1.1	0.909	901	893	885	877	870	862	855	847	840
1.2	.833	826	820	813	806	800	794	787	781	775
1.3	.769	763	758	752	746	741	735	730	725	719
1.4	.714	709	704	699	694	690	685	680	676	671
1.5	0.667	662	658	654	649	645	641	637	633	629
1.6	.625	621	617	613	610	606	602	599	595	592
1.7	.588	585	581	578	575	571	568	565	562	559
1.8	.556	552	549	546	543	541	538	535	532	529
1.9	.526	524	521	518	515	513	510	508	505	503
2.0	0.500	498	495	493	490	488	485	483	481	478
2.1	.476	474	472	469	467	465	463	461	459	457
2.2	.455	452	450	448	446	444	442	441	439	437
2.3	.435	433	431	429	427	426	424	422	420	418
2.4	.417	415	413	412	410	408	407	405	403	402
2.5	0.400	398	397	395	394	392	391	389	388	386
2.6	.385	383	382	380	379	377	376	375	373	372
2.7	.370	369	368	366	365	364	362	361	360	358
2.8	.357	356	355	353	352	351	350	348	347	346
2.9	.345	344	342	341	340	339	338	337	336	334
3.0	0.333	332	331	330	329	328	327	326	325	324
3.1	.323	322	321	319	318	317	316	315	314	313
3.2	.313	312	311	310	309	308	307	306	305	304
3.3	.303	302	301	300	299	299	298	297	296	295
3.4	.294	293	292	292	291	290	289	288	287	287
3.5	0.286	285	284	283	282	282	281	280	279	279
3.6	.278	277	276	275	275	274	273	272	272	271
3.7	.270	270	269	268	267	267	266	265	265	264
3.8	.263	262	262	261	260	260	259	258	258	257
3.9	.256	256	255	254	254	253	253	252	251	251
4.0	0.250	249	249	248	248	247	246	246	245	244
4.1	.244	243	243	242	242	241	240	240	239	239
4.2	.238	238	237	236	236	235	235	234	234	233
4.3	.233	232	231	231	230	230	229	229	228	228
4.4	.227	227	226	226	225	225	224	224	223	223
4.5	0.222	222	221	221	220	220	219	219	218	218
4.6	.217	217	216	216	216	215	215	214	214	213
4.7	.213	212	212	211	211	211	210	210	209	209
4.8	.208	208	207	207	207	206	206	205	205	205
4.9	.204	204	203	203	202	202	202	201	201	200
5.0	0.200	200	199	199	198	198	198	197	197	196
5.1	.196	196	195	195	195	194	194	193	193	193
5.2	.192	192	192	191	191	190	190	190	189	189
5.3	.189	188	188	188	187	187	187	186	186	186
5.4	.185	185	185	184	184	183	183	183	182	182

Answers

Exercise 1a (page 1)

1. 14
2. 8
3. 16
4. 20
5. 37
6. 22
7. 48
8. 35
9. 59
10. 26
11. 33
12. 21

Exercise 1b (page 2)

1. (a) four hundreds (b) four units
   (c) two units (d) two thousands
   (e) two hundreds (f) nine hundreds
   (g) nine units (h) six thousands
   (i) six tens (j) zero tens
   (k) zero thousands (l) one ten-thousand

2. (a) eight tenths (b) eight hundredths
   (c) five hundredths (d) five units
   (e) six tenths (f) zero tenths
   (g) one unit (h) one thousandth
   (i) one ten (j) seven hundredths

3. The units and tens are under the tens and hundreds.

$$\begin{array}{r} 352 \\ +\ 79 \\ \hline \end{array}$$

4. (a) $\begin{array}{r} 3107 \\ 26 \\ +\ 147 \\ \hline \end{array}$ (b) $\begin{array}{r} 6203 \\ -\ 97 \\ \hline \end{array}$

   (c) $\begin{array}{r} 1429 \\ +\ 6580 \\ \hline \end{array}$ (d) $\begin{array}{r} 6700 \\ -\ 34 \\ \hline \end{array}$

5. (a) (i) 632 (ii) 236
   (b) (i) 872 (ii) 278
   (c) (i) 650 (ii) 056
   (d) (i) 8521 (ii) 1258
   (e) (i) 7420 (ii) 0247

6. (a) $\begin{array}{r} 60.91 \\ +\ 3.2 \\ \hline \end{array}$ (b) $\begin{array}{r} 26.3 \\ -\ 1.7 \\ \hline \end{array}$

   (c) $\begin{array}{r} \$4.49 \\ +\ \$56.20 \\ \hline \end{array}$ (d) $\begin{array}{r} 42.5 \\ -\ 9.65 \\ \hline \end{array}$

Exercise 1c (page 3)

1. $9 \times 1 + 8 \times 10 + 3 \times 100 + 2 \times 1000$
2. $7 \times 1 + 4 \times 8 + 6 \times 64$
3. $4 \times 1 + 5 \times 8 + 1 \times 64 + 5 \times 512 + 3 \times 4096$
4. $2 \times 1 + 0 \times 5 + 1 \times 25 + 4 \times 125$
5. $1 \times 1 + 1 \times 2 + 0 \times 4 + 1 \times 8$
6. $0 \times 1 + 1 \times 3 + 0 \times 9 + 2 \times 27 + 2 \times 81$
7. $3 \times 1 + 2 \times 8 + 5 \times 64 + 6 \times 512 + 2 \times 4096$
8. $0 \times 1 + 0 \times 2 + 1 \times 4 + 1 \times 8$
9. $2 \times 1 + 0 \times 3 + 1 \times 9 + 2 \times 27$
10. $1 \times 1 + 0 \times 2 + 0 \times 4 + 1 \times 8 + 1 \times 16$

Exercise 1d (page 4)

1. 25 days
2. 18 wk 2 d
3. (a) 5 wk 6 d (b) 4 wk 4 d
   (c) 3 wk 1 d (d) 6 wk
4. (a) Wednesday (b) Friday
   (c) Thursday (d) Saturday
5. (a) 120 s (b) 654 s
   (c) 202 s (d) 3600 s
6. (a) 4 min (b) 300 min
   (c) 154 min (d) 1440 min
7. (a) 7 h (b) 48 h
   (c) 83 h (d) 168 h
8. (a) 2 min 19 s (b) 30 min
   (c) 4 h 1 min (d) 2 h 20 min 18 s
   (e) 6 d 12 h (f) 6 d 0 h 20 min
9. (a) 6 min 45 s (b) 4 wk 4 d
   (c) 1 d 18 h (d) 1 h 33 min 40 s

Exercise 1e (page 6)

① 17   ② 40
③ 1011   ④ 110
⑤ 2101   ⑥ 11 011
⑦ 301   ⑧ 22
⑨ 1022   ⑩ 37
⑪ 110 001   ⑫ 144
⑬ 344   ⑭ 2002
⑮ 130   ⑯ 1 100 010
⑰ 200   ⑱ 430
⑲ 222 212   ⑳ 10 000 001
㉑ 400   ㉒ 1335
㉓ 100 000 000 ㉔ 202 010

Exercise 1f (page 6)

① 15   ② 25   ③ 153   ④ 314
⑤ 409   ⑥ 23   ⑦ 14   ⑧ 27
⑨ 89   ⑩ 23   ⑪ 15   ⑫ 13
⑬ 65   ⑭ 47   ⑮ 7   ⑯ 10
⑰ 30   ⑱ 20   ⑲ 52   ⑳ 19
㉑ 325   ㉒ 273   ㉓ 248   ㉔ 61

Practice exercise P1.1 (page 7)

①

Base number	Greatest digit
10	9
8	7
7	6
5	4
2	1

② (a) $8^4 \; 8^3 \; 8^2 \; 8^1 \; 8^0$
   (b) $5^4 \; 5^3 \; 5^2 \; 5^1 \; 5^0$
③ (a) 4000   (b) 8
   (c) 16   (d) $\frac{500}{100}$
   (e) $\frac{9}{10\,000}$ (f) $\frac{9}{10}$
④ (a) $1 \times 10^2 + 0 \times 10^1 + 1 \times 10^0$
   (b) $1 \times 8^2 + 0 \times 8^1 + 1 \times 8^0$
   (c) $1 \times 2^2 + 0 \times 2^1 + 1 \times 2^0$
⑤ 65, 5
⑥ (a) (i) 25_8, 352_8, 4451_8
       (ii) 77_8, 255_8, 1040_8
   (b) (i) 1001_2, 1101_2, $11\,010_2$, $100\,111_2$,
           $10\,001\,101_2$
       (ii) 1011_2, $10\,001_2$, $10\,111_2$, $1\,011\,001_2$,
           $10\,000\,100_2$.

⑦ (a) 215   (b) 37   (c) 74
   (d) not valid   (e) 61

Exercise 2a (page 9)

1 (a) Barbados: the only country
  (b) aeroplane: the only one made by humans
  (c) 7: the only odd number
  (d) chicken: the only one with 2 legs
  (e) (2×5): the others equal 6
2 (a) {driver, passenger, soldier, conductor}
  (b) {cow, driver, passenger, soldier, conductor}
  (c) {petrol}
  (d) {engine, driver, petrol, passenger, conductor, wheel}
3 (a) {things found on a farm}
  (b) {words beginning with p}
  (c) {things containing metal}
  (d) {things which cannot be touched}
4 (c) {2, 4, 6, ...}
5 (a) T = {cup, saucer, spoon} = {tea set}
  (b)

Exercise 2b (page 10)

① (a) {a, b, c, ..., z}   (b) {1, 3, 5, ..., 29}
   (c) {2, 4, 6, ...}
   (d) {January, February, ..., December}
② (a) 19, 23, 27   (b) 32, 64, 128
   (c) 35, 42, 49   (d) 54, 65, 76
   (e) 25, 36, 49
③ (a) ∈ (b) ∉ (c) ∉ (d) ∈ (e) ∈ (f) ∈
④ (a) pencil ∈ {writing instruments}
   (b) pencil ∉ {animals}
   (c) bread ∉ {vehicles}
   (d) 3 ∉ {a, b, c, ..., z}
   (e) 3 ∈ {digits}
   (f) 3 ∈ {digits}
   (g) eagle ∉ {mammals}
   (h) eagle ∈ {birds}
⑤ (a) 8   (b) 10   (c) 7   (d) 9
   (e) 5   (f) 1   (g) 15   (h) 8
⑥ (a) F   (b) T   (c) F   (d) T
   (e) F   (f) F   (g) T   (h) T
   (i) F   (j) T

Exercise 2c (page 11)

1. ∅
2. infinite
3. ∅
4. ∅
5. ∅
6. infinite
7. infinite
8. ∅
9. infinite
10. ∅

Exercise 2d (page 12)

1. (a) { }, {knife}, {fork}, {spoon},
      {knife, fork}, {knife, spoon},
      {fork, spoon}, {knife, fork, spoon}
   (b) ∅, {2}, {3}, {4}, {2, 3}, {2, 4}, {3 ,4},
       {2, 3, 4}
2. (a) ∅, {a}, {a, b}
   (b) 4
3. ∅ {e}, {a}, {f}, {w}, {e, a}, {e, f}, {e, w}, {a, f},
   {a, w}, {f, w}, {e, a, f}, {e, a, w}, {c, f, w},
   {a, f, w}, {e, a, f, w}
4. (a)

Number of elements in a set	Number of subsets in that set
0	1
1	2
2	4
3	8
4	16

   (b) 32
5. (a) and (d) only
6. (a) {dogs} ⊂ {mammals}
   (b) {cars} ⊄ {cities}
   (c) A ⊂ B   (d) Y ⊃ X
7. (a) F is a subset of G
   (b) A is not a subset of B
   (c) M does not contain N
   (d) D contains E
8. (a)

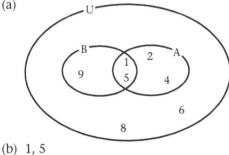

   (b) 1, 5

9.

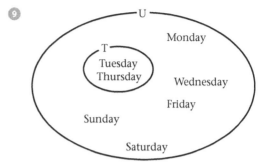

10. (a) F ⊂ G   (b) 10   (c) 2
    (d)

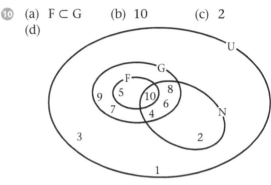

Practice exercise P2.1 (page 13)

1. (a) boys ∈ {males}
   (b) {girls} ⊂ {females}
   (c) keyboard ∈ {computer}
   (d) 7 ∈ {factors of 35}
   (e) {30, 45} ⊂ {multiples of 15}
   (f) metre ∉ {units of weight}
   (g) {solids} ⊃ {cube, prism, cone}
   (h) 5 ⊄ {factors of 9}
   (i) {1.6, $\frac{1}{3}$} ⊄ {counting numbers}
   (j) {animals} ⊅ {cat, dog, bird, tree}
2. (a) false (b) true (c) true (d) true (e) false

Practice exercise P2.2 (page 13)

1. Answers will vary.
2. (a) {1, 2, 4, 8, 16, 32}
   (b) {7, 14, 21, 28, ...}
   (c) {11, 13, 15, 17, 19}
   (d) {7, 11, 13, 17, 19}
3. (a) U, T, S   (b) $\frac{5}{6}, \frac{6}{7}, \frac{7}{8}$ (c) 1, $\frac{1}{2}, \frac{1}{4}$
4. (a) {2, 4, 6, 8, 10}
   (b) {2, 3, 5, 7, 11}
   (c) {10, 20, 30, 40, 50}
   (d) {0.1, 0.2, 0.3, 0.4, 0.5}

5 (a) T = {furniture in a house}
  (b) L = {languages}
  (c) M = {multiple of 5 less than 30}
  (d) F = {five three-letter words beginning
    with *d* and ending with *n*}

Practice exercise P2.3 (page 14)

1 (a) F = {f, e, d}; H = {h, i, d}
  (b)

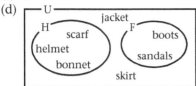

  (c) 3 and 3
  (d) No. They do not have the same
    members.

2 (i) (a) F = {1, 2, 4, 5, 10, 20}
    (b) n(F) = 6
  (ii) (a) O = {1, 3, 5, 7, 9}
    (b) n(O) = 5
  (iii) Q = {square, rhombus}
    (b) n(Q) = 2

3 (a) P is s   (b) n(P) = 3
  (a) W is I
  (a) E is s   (b) n(E) = 1
  (a) C is s   (b) n(C) = 21
  (a) N is i

4 Answers will vary.

Practice exercise P2.4 (page 14)

1 (a) Answers will vary.
  (b) H = {scarf, helmet, bonnet}
    F = {boots, sandals}
  (c) J
  (d)

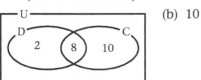

2 (a) D = {college degrees}
    C = {school certificates}

U

D   C

2   8   10

(b) 10

---

3 Answers will vary.

Exercise 3a (page 15)

1 {1, 2, 3, 4, 6, 12}
2 {1, 2, 3, 6, 9, 18}
3 {1, 2, 4, 5, 10, 20}
4 {1, 2, 3, 4, 6, 8, 12, 24}
5 {1, 2, 4, 7, 14, 28}
6 {1, 2, 3, 5, 6, 10, 15, 30}
7 {1, 2, 4, 8, 16, 32}
8 {1, 2, 3, 4, 6, 8, 12, 16, 24, 48}
9 {1, 3, 7, 9, 21, 63}
10 {1, 2, 3, 4, 6, 8, 9, 12, 18, 24, 36, 72}
11 3, 9
12 2, 3, 4, 6, 9
13 2, 3, 6, 9
14 2, 4, 7, 8
15 2, 3, 4, 5, 6   **16** 2, 3, 5, 6, 9
17 2, 3, 4, 5, 6, 8   **18** 2, 3, 4, 6, 8, 9
19 2, 3, 4, 5, 6, 9   **20** 2, 3, 5, 6, 7

Exercise 3b (page 15)

1 (a) 3, 9, 15   (b) 6, 12, 18
2 {2, 4, 6, 8, 10}
3 {1, 3}
4 {1}
5 {25, 27, 29, 31, 33}
6 {124, 126, 128, 130, 132, 134}

Exercise 3c (page 16)

1 {2, 3, 5, 7, 11, 13, 17, 19, 23, 29}
2 {14, 15, 16, 18, 20, 21, 22, 24, 25, 26, 27, 28,
    30, 32, 33, 34, 35, 36, 38, 39}

Exercise 3d (page 17)

1 (a) $2 \times 2 \times 3$ (b) {2, 3}
2 (a) $2 \times 3 \times 3$ (b) {2, 3}
3 (a) $2 \times 2 \times 7$ (b) {2, 7}
4 (a) $2 \times 3 \times 5$ (b) {2, 3, 5}
5 (a) $2 \times 2 \times 2 \times 3 \times 3$ (b) {2, 3}
6 (a) $2 \times 2 \times 3 \times 7$ (b) {2, 3, 7}
7 (a) $2 \times 2 \times 3 \times 3 \times 3$ (b) {2, 3}
8 (a) $3 \times 5 \times 7$ (b) {3, 5, 7}
9 (a) $2 \times 2 \times 3 \times 3 \times 5$ (b) {2, 3, 5}
10 (a) $2 \times 2 \times 2 \times 3 \times 3 \times 3$ (b) {2, 3}

⑪ (a) $2 \times 2 \times 2 \times 2 \times 2 \times 3 \times 3$ (b) {2, 3}
⑫ (a) $5 \times 5 \times 5 \times 7$ (b) {5, 7}
⑬ (a) $2 \times 2 \times 3 \times 3 \times 5 \times 5$ (b) {2, 3, 5}
⑭ (a) $2 \times 2 \times 2 \times 2 \times 5 \times 11$ (b) {2, 5, 11}
⑮ (a) $2 \times 2 \times 2 \times 3 \times 3 \times 3 \times 7$ (b) {2, 3, 7}
⑯ (a) $3 \times 3 \times 5 \times 5 \times 7 \times 11$
   (b) {3, 5, 7, 11}

Exercise 3e (page 17)

① (a) 7^3 (b) 3^5 (c) 2^2
   (d) 10^4 (e) 6^7 (f) 8^{10}
   (g) $4^3 \times 5^7$ (h) $11^2 \times 17^6$
② (1) $2^2 \times 3$ (2) 2×3^2
   (3) $2^2 \times 7$ (5) $2^3 \times 3^2$
   (6) $2^2 \times 3 \times 7$ (7) $2^2 \times 3^3$
   (9) $2^2 \times 3^2 \times 5$ (10) $2^3 \times 3^3$
   (11) $2^5 \times 3^2$ (12) $5^3 \times 7$
   (13) $2^2 \times 3^2 \times 5^2$ (14) $2^4 \times 5 \times 11$
   (15) $2^3 \times 3^3 \times 7$ (16) $3^2 \times 5^2 \times 7 \times 11$
③ (a) $2^3 \times 3$ (b) $2^4 \times 3$
   (c) $3^2 \times 7$ (d) $2^2 \times 23$
   (e) $2^3 \times 17$ (f) $3^2 \times 7^2$
   (g) $2^3 \times 3^2 \times 5$ (h) 5^4
   (i) 2^9 (j) $2^4 \times 3^2 \times 5$
   (k) 3^6 (l) $2^3 \times 5^3$
   (m) $2^4 \times 7 \times 11$ (n) $2^2 \times 3^3 \times 13$
   (o) $5^2 \times 7^2$ (p) $2 \times 3^6 \times 5$
④ (a) {2, 3} (b) {2, 3} (c) {3, 7}
   (d) {2, 23} (e) {2, 17} (f) {3, 7}
   (g) {2, 3, 5} (h) {5} (i) {2}
   (j) {2, 3, 5} (k) {3} (l) {2, 5}
   (m) {2, 7, 11} (n) {2, 3, 13}
   (o) {5, 7} (p) {2, 3, 5}

Exercise 3f (page 18)

① 15 ② 14 ③ 18 ④ 21
⑤ 22 ⑥ 20 ⑦ 30 ⑧ 40
⑨ 50 ⑩ 70 ⑪ 24 ⑫ 28
⑬ 27 ⑭ 25 ⑮ 35 ⑯ 44
⑰ 42 ⑱ 45 ⑲ 48 ⑳ 54
㉑ 55 ㉒ 60 ㉓ 63 ㉔ 90
㉕ 56 ㉖ 66 ㉗ 75 ㉘ 81
㉙ 84 ㉚ 88

Exercise 3g (page 18)

① 6 ② 6 ③ 5 ④ 11
⑤ 21 ⑥ 2 ⑦ 5 ⑧ 15
⑨ 3 ⑩ 3 ⑩ 7 ⑩ 14

Exercise 3h (page 19)

① 2 ② 3 ③ 7
④ 3 ⑤ 2 ⑥ 3
⑦ 2, 3, 6 ⑧ 2, 5, 10 ⑨ 2, 3, 6
⑩ 3, 9 ⑪ 2, 4 ⑫ 2, 4, 8
⑬ 2 ⑭ 7 ⑮ 3
⑯ 2, 11, 22 ⑰ 3, 5, 15 ⑱ 2, 3, 4, 6, 12

Exercise 3i (page 19)

① (a) 3×3 (b) 2×3 (c) 2×5
   (d) $2 \times 3 \times 3$ (e) $3 \times 5 \times 7$
② (a) $2^2 \times 3$ (b) 2×3^3 (c) $3^2 \times 5^2$
   (d) $2^3 \times 3 \times 5$ (e) 5
③ (a) 3 (b) 2 (c) 6 (d) 8
   (e) 5 (f) 15 (g) 14 (h) 9
   (i) 21 (j) 72 (k) 35 (l) 18
   (m) 32 (n) 27 (o) 36 (p) 108

Exercise 3j (page 20)

① 2, 4, 6, 8, 10, …
② 3, 6, 9, 12, 15, …
③ 5, 10, 15, 20, 25, …
④ 7, 14, 21, 28, 35, …
⑤ 8, 16, 24, 32, 40, …
⑥ 9, 18, 27, 36, 45, …
⑦ 10, 20, 30, 40, 50, …
⑧ 11, 22, 33, 44, 55, …
⑨ 12, 24, 36, 48, 60, …
⑩ 20, 40, 60, 80, 100, …

Exercise 3k (page 20)

① 12, 24, 36, … ② 10, 20, 30, …
③ 21, 42, 63, … ④ 30, 60, 90, …
⑤ 60, 120, 180, … ⑥ 42, 84, 126, …

Exercise 3l (page 21)

① (a) the cross-shaded numbers are the
       common multiples of 2 and 3.
   (b) (i) 15 (ii) 60
② (a) $2 \times 2 \times 3 \times 3 \times 5$
   (b) $2 \times 2 \times 2 \times 3 \times 3 \times 5$
   (c) $2 \times 2 \times 2 \times 3 \times 3 \times 5 \times 5$
   (d) $2 \times 3 \times 3 \times 5 \times 5 \times 7$
   (e) $2 \times 3 \times 3 \times 3 \times 5 \times 5 \times 7$
③ (a) $2^2 \times 3 \times 5^2$
   (b) $2^3 \times 3^2 \times 5^2 \times 7$
   (c) $2^2 \times 3^3 \times 5 \times 7^2$
   (d) $2^3 \times 3^2 \times 5^3 \times 7^2$
   (e) $2^4 \times 3^2 \times 5^3 \times 7^3$

④ (a) 12   (b) 24   (c) 18   (d) 56
    (e) 24   (f) 36   (g) 12   (h) 12
    (i) 36   (j) 24   (k) 120   (l) 120
    (m) 60   (n) 90   (o) 720   (p) 180
    (q) 120   (r) 180

Exercise 3m (page 22)

① 92, 2756, 9428, 56 016
② 267, 534, 1287, 28 104
③ 2872, 19 176, 172 744
④ 2331, 35 406, 52 866
⑤ (a) ✗ ✓ ✗ ✓ ✗ ✗ ✓ ✗
   (b) ✓ ✓ ✗ ✓ ✓ ✗ ✗ ✓
   (c) ✗ ✓ ✗ ✗ ✗ ✗ ✓ ✗
   (d) ✗ ✗ ✗ ✓ ✗ ✗ ✗ ✗
   (e) ✓ ✗ ✓ ✗ ✗ ✓ ✗ ✗
   (f) ✓ ✓ ✓ ✓ ✓ ✓ ✓ ✓
   (g) ✓ ✓ ✗ ✗ ✓ ✗ ✗ ✗
   (h) ✓ ✓ ✗ ✓ ✓ ✗ ✓ ✓
   (i) ✗ ✓ ✗ ✓ ✗ ✗ ✗ ✗
   (j) ✓ ✓ ✓ ✓ ✓ ✓ ✓ ✓
   (k) ✓ ✗ ✗ ✗ ✗ ✗ ✗ ✗
   (l) ✓ ✓ ✓ ✗ ✓ ✓ ✗ ✗
⑥ 3960, 51 840
⑦ (a) 8, 9   (b) 36, 72
⑧ (a) 76 356, 16 869, 22 374
   (b) 76 356, 36 116
   (c) 76 356, 22 374   (d) 76 356
⑨ (a) 45 295 is not even *or* the sum of its
      digits is not divisible by 9.
   (b) 141 is divisible by 3 but 2699 is not
      divisible by 3.
⑩ (a) 2   (b) 4   (c) 5
⑪ 96
⑫ $192 = 4^2 \times 4^2 \times 3$

Exercise 3n (page 23)

① 24 g   ② 72 g
③ 6 cm × 6 cm, 117   ④ 120 cm
⑤ $13.44   ⑥ 11
⑦ 540 g   ⑧ 360
⑨ 39 min 36 s   ⑩ 36

Practice exercise P3.1 (page 24)

① Answers vary. Examples are given.
   (i) (a) numbers that are multiples of 2
       (b) {2, 4, 6, 8, 10, 12, 14, 16, 18, 20}

   (ii) (a) numbers that cannot be exactly
         divided by 2
       (b) {1, 3, 5, 7, 9, 11, 13, 15, 17, 19}
   (iii) (a) the only factors of a prime number
         are 1 and the number itself
       (b) {2, 3, 5, 7, 11, 13, 17, 19, 23, 29}
   (iv) (a) a number that has other factors as
         well as 1 and itself
       (b) {4, 6, 8, 9, 10, 12, 14, 15, 16, 18}
   (v) (a) a number that results from
         multiplying a number by itself
       (b) {4, 9, 16, 25, 36, 49, 64, 81, 100,
         121}
② (a) 2, 12, 16, 24, 30, 36, 48, 64, 88, 100,
      102, 144, 256
   (b) 9, 11, 17, 23, 37, 49, 67, 81, 101, 221,
      239
   (c) 2, 11, 17, 23, 37, 67, 101
   (d) 9, 12, 16, 24, 30, 36, 48, 49, 64, 81, 88,
      100, 102, 144, 256
   (e) 9, 16, 36, 49, 64, 81, 100, 144, 256

Practice exercise P3.2 (page 24)

① (a) true   (b) false   (c) false
   (d) true   (e) true   (f) false
   (g) false   (h) true   (i) true
   (j) false   (k) true   (l) false
   (m) true   (n) false   (o) false
② (a) (i) 1, 2, 3, 4, 6, 8, 12, 24   (ii) 2, 3
   (b) (i) 1, 5, 25   (ii) 5
   (c) (i) 1, 2, 3, 6, 7, 14, 21, 42
      (ii) 2, 3, 7
   (d) (i) 1, 5, 19, 95   (ii) 5, 19
   (e) (i) 1, 2, 4, 5, 10, 20, 25, 50, 100
      (ii) 2, 5
   (f) (i) 1, 2, 3, 4, 6, 11, 12, 22, 33, 44, 66, 132
      (ii) 2, 3, 11
   (g) (i) 1, 5, 25, 125, 625   (ii) 5
   (h) (i) 1, 2, 4, 8, 16, 32, 64, 128, 256, 512,
      1024
      (ii) 2
   (i) (i) 1, 2, 3, 4, 6, 8, 9, 12, 16, 18, 24, 36,
      54, 72, 81, 108, 144, 162, 216, 324,
      432, 648, 1296
      (ii) 2, 3
   (j) (i) 1, 2, 4, 7, 8, 14, 28, 49, 56, 98, 196,
      343, 392, 686, 1372, 2744
      (ii) 2, 7

Practice exercise P3.3 (page 25)

1. (a) Answers will vary.
   (b) Answers will vary.
2. (a) 18, 728, 220, 1950, 70
   (b) 225, 635, 70, 220, 1950, 4145
   (c) 441, 12, 21, 3000, 225, 111
   (d) 98, 21, 8113, 126, 441, 35
3. (a)–(c) Answers will vary.

Practice exercise P3.4 (page 25)

1. 32
2. 64
3. 900
4. 1 000 000

Practice exercise P3.5 (page 25)

1. (a) HCF = 6   (b) LCM = 7560
2. (a) 105   (b) 720
3. (a) 22   (b) 21
4. (a) HCF, 40 m   (b) LCM, 60 beats
   (c) (i) LCM, 56 cm
       (ii) 7 rows of 8 cm, 4 rows of 14 cm
   (d) LCM, every 60 s
   (e) (i) HCF, 4 pupils   (ii) 19 teams
   (f) (i) HCF, 60 m   (ii) 17 drums
   (g) (i) HCF, 15
       (ii) 3 rows of roses, 8 rows of gerberas,
            20 rows of anthuriums
5. 56
6. any product of $3^2 \times 5$

Exercise 4a (page 28)

1. (a) 3 cm; 30 mm   (b) 3.5 cm; 35 mm
   (c) 7.9 cm; 79 mm   (d) 0.7 cm; 7 mm
   (e) 6.2 cm; 62 mm
2. (a) 40 mm; 4 cm   (b) 46 mm; 4.6 cm
   (c) 33 mm; 3.3 cm   (d) 15 mm; 1.5 cm
   (e) 49 mm; 4.9 cm

Exercise 4b (page 29)

1. (a) 0.364 km
       0.025 7 km
       8.092 km
       0.006 408 km
       0.205 km
   (b) 364 m
       25.7 m
       8092 m
       6.408 m
       205 m
   (c) 364 000 mm
       25 700 mm
       8 092 000 mm
       6408 mm
       205 000 mm
2. (a) 3000 m   (b) 5000 m   (c) 8000 m
   (d) 2000 m   (e) 6000 m   (f) 10 000 m
   (g) 4000 m   (h) 3500 m   (i) 4200 m
   (j) 6800 m   (k) 8100 m   (l) 5900 m
3. (a) 3850 m   (b) 8390 m   (c) 9140 m
   (d) 9400 m   (e) 3315 m   (f) 5050 m
4. (a) 5 km   (b) 10 km   (c) 2.4 km
   (d) 6.52 km   (e) 7.33 km
5. (a) 1.73 m   (b) 4.58 m   (c) 1.5 m
   (d) 1.05 m   (e) 1.01 m   (f) 1 m
   (g) 0.53 m   (h) 0.4 m   (i) 0.19 m
   (j) 0.05 m
6. (a) 1 m   (b) 7 m   (c) 4.1 m
   (d) 3.726 m   (e) 8.119 m   (f) 0.3 m
   (g) 0.051 m   (h) 0.003 m
7. (a) 3.65 m   (b) 210 m   (c) 1.922 4 km
   (d) 192.5 cm
8. 1.6 km
9. 2.66 m
10. 34.8 cm
11. 9 mm
12. 75.7 cm

Exercise 4c (page 30)

1. (a) 10.2 kg   (b) 1.487 kg
   (c) 2.885 kg   (d) 4620 kg
2. (a) 235.5 g   (b) 3.03 kg   (c) 0.267 48 kg
3. 6.89 t
4. 2.19 kg
5. 284 g
6. 855 g
7. 36 g

Exercise 4d (page 31)

1. (a) 150   (b) 75   (c) 100
   (d) 3   (e) $2\frac{1}{2}$
2. (a) 300   (b) 80   (c) 900
   (d) 1800   (e) 10 800
3. 1 h 17 min
4. 6 h 42 min
5. 1800 m
6. $1\frac{1}{2}$ h

7

Ordinary Time	24-hour clock	
	Digits	Words
2:45 p.m.	14:45 hrs	fourteen-forty-five
7:30 a.m.	07:30 hrs	zero-seven-thirty
7:30 p.m.	19:30 hrs	nineteen-thirty
6 a.m.	06:00 hrs	zero-six-hundred
10 p.m.	22:00 hrs	twenty-two-hundred
12:40 a.m.	00:40 hrs	zero-hundred-forty

8 (a) 14:00 hrs (b) 20:30 hrs
(c) 09:00 hrs (d) 16:00 hrs
(e) 17:30 hrs (f) 05:45 hrs
(g) 05:25 hrs (h) 23:55 hrs
(i) 20:55 hrs (j) 06:10 hrs
(k) 12:00 hrs (l) 00:03 hrs

8 (a) 5 a.m. (b) 11 p.m.
(c) 7:30 p.m. (d) 9:30 a.m.
(e) 9:45 p.m. (f) 2:15 p.m.
(g) 6:19 p.m. (h) 7:40 a.m.

10 (a) departs, arrives (b) $9\frac{1}{2}$ h
(c) $4\frac{3}{4}$ h (d) flight number
(e) JM017 (f) 18:55

Exercise 4e (page 33)

1 (a) (i) 27 °F (ii) 36 °F (iii) 81 °F
(b) (i) 5 °C (ii) 15 °C (iii) 25 °C
2 (a) (i) 15 °C (ii) 60 °C
(b) (i) 77 °F (ii) 104 °F
3 (a) 45 °F (b) 25 °C (c) 235 °C, 210 °C

Practice exercise P4.1 (page 34)

1 (a) 477 cm (b) 4.77 m
2 37.5 km
3 21.9 kg
4 8.5 km
5 7.8 tonnes
6 24.8 tonnes

7

12-hour clock	24-hour clock
7:40 a.m.	07:40
1:00 p.m.	**13:00**
2:30 p.m.	14:30
9:45 a.m.	09:45
5:45 p.m.	**17:45**
12:00 mid day	**12:00**

8 (a) 2 h 35 min (b) 19:15
(c) 19:00; 7:00 p.m.

7

Celcius scale	Fahrenheit scale
0 °C	**32 °F**
180 °C	**356 °F**
25 °C	**77 °F**
30 °C	86 °F
45 °C	**113 °F**

10 (a) 55 min (b) 8:15 a.m.
11 (a) 14 April (b) a.m. (c) 3
(d) 97 min (e) 1 call

Exercise 5a (page 37)

1 $\frac{6}{18} = \frac{8}{24} = \frac{50}{150} = \frac{300}{900} = \frac{100}{300}$
2 $\frac{3}{6} = \frac{4}{8} = \frac{5}{10} = \frac{50}{100} = \frac{25}{50}$
3 $\frac{220}{30} = \frac{4}{6} = \frac{600}{900} = \frac{16}{24} = \frac{10}{15}$
4 $\frac{2}{8} = \frac{3}{12} = \frac{15}{60} = \frac{25}{100} = \frac{7}{28}$
5 $\frac{15}{20} = \frac{24}{32} = \frac{21}{28} = \frac{75}{100} = \frac{18}{24}$
6 $\frac{2}{10} = \frac{4}{20} = \frac{20}{100} = \frac{24}{120} = \frac{5}{25}$

Exercise 5b (page 37)

1 (a) 12 (b) 24 (c) 9 (d) 18
(e) 21 (f) 22 (g) 3 (h) 12
(i) 40 (j) 49 (k) 18 (l) 20

2 (a) $\frac{7}{12}, \frac{2}{3}, \frac{3}{4}, \frac{5}{6}$ $\left(\frac{7}{12}, \frac{8}{12}, \frac{9}{12}, \frac{10}{12}\right)$
(b) $\frac{3}{4}, \frac{4}{5}, \frac{17}{20}, \frac{9}{10}$ $\left(\frac{15}{20}, \frac{16}{20}, \frac{17}{20}, \frac{18}{20}\right)$
(c) $\frac{5}{9}, \frac{7}{12}, \frac{11}{18}, \frac{2}{3}$ $\left(\frac{20}{36}, \frac{21}{36}, \frac{22}{36}, \frac{24}{36}\right)$

3 (a) $\frac{2}{9}, \frac{5}{18}, \frac{1}{3}$ $\left(\frac{4}{18}, \frac{5}{18}, \frac{6}{18}\right)$
(b) $\frac{1}{2}, \frac{8}{15}, \frac{17}{30}, \frac{3}{5}$ $\left(\frac{15}{30}, \frac{16}{30}, \frac{17}{30}, \frac{18}{30}\right)$
(c) $\frac{3}{5}, \frac{5}{8}, \frac{13}{20}, \frac{7}{10}$ $\left(\frac{24}{40}, \frac{25}{40}, \frac{26}{40}, \frac{28}{40}\right)$

Exercise 5c (page 38)

1 (a) 3 (b) 3; 7 (c) 4 (d) 5
(e) 5; 1 (f) 28; 4 (g) 3 (h) 5
(i) 9 (j) 25 (k) 5 (l) 1

2 (a) $\frac{1}{5}$ (b) $\frac{1}{6}$ (c) $\frac{1}{20}$ (d) $\frac{3}{8}$
(e) $\frac{2}{3}$ (f) $\frac{3}{4}$ (g) $\frac{5}{6}$ (h) $\frac{4}{5}$
(i) $\frac{4}{9}$ (j) $\frac{3}{7}$ (k) $\frac{5}{7}$ (l) $\frac{8}{11}$

Exercise 5d (page 39)

1 (a) $1\frac{1}{3}$ (b) $2\frac{1}{2}$ (c) $2\frac{3}{4}$ (d) $1\frac{2}{5}$
(e) $2\frac{1}{5}$ (f) $3\frac{4}{7}$ (g) $4\frac{1}{6}$ (h) $2\frac{3}{8}$
(i) $2\frac{8}{9}$ (j) $8\frac{1}{5}$ (k) $8\frac{4}{7}$ (l) $9\frac{1}{6}$

② (a) $\frac{7}{4}$ (b) $\frac{10}{3}$ (c) $\frac{13}{2}$ (d) $\frac{23}{5}$

    (e) $\frac{37}{4}$ (f) $\frac{23}{6}$ (g) $\frac{30}{7}$ (h) $\frac{38}{5}$

    (i) $\frac{43}{8}$ (j) $\frac{47}{7}$ (k) $\frac{38}{3}$ (l) $\frac{40}{3}$

Exercise 5e (page 40)

① $1\frac{5}{12}$ **②** $\frac{1}{12}$ **③** $1\frac{1}{18}$ **④** $\frac{11}{18}$

⑤ $\frac{5}{24}$ **⑥** $\frac{23}{24}$ **⑦** $\frac{1}{3}$ **⑧** $\frac{14}{15}$

⑨ $2\frac{1}{4}$ **⑩** $1\frac{1}{4}$ **⑪** $3\frac{5}{6}$ **⑫** $\frac{7}{12}$

⑬ $6\frac{5}{8}$ **⑭** $8\frac{11}{18}$ **⑮** $2\frac{19}{20}$ **⑯** $2\frac{13}{24}$

⑰ $7\frac{1}{12}$ **⑱** $4\frac{3}{8}$ **⑲** $5\frac{1}{8}$ **⑳** $8\frac{1}{12}$

Exercise 5f (page 40)

① $5\frac{5}{24}$ **②** $\frac{11}{24}$ **③** $5\frac{11}{20}, \frac{9}{20}$ **④** $\frac{3}{8}$

⑤ $\frac{1}{8}$ **⑥** $\frac{11}{20}$ **⑦** $\frac{1}{3}$ **⑧** $\frac{1}{30}$

⑨ $3\frac{3}{4}$ h   **⑩** $\frac{4}{5}$

Exercise 5g (page 41)

① $5\frac{1}{3}$ **②** 9   **③** $1\frac{1}{2}$ **④** $\frac{4}{5}$

⑤ $4\frac{1}{2}$ **⑥** 10   **⑦** $6\frac{2}{3}$ **⑧** $7\frac{1}{2}$

⑨ $\frac{2}{3}$ **⑩** $7\frac{1}{2}$

Exercise 5h (page 42)

① $\frac{3}{8}$ **②** $\frac{1}{2}$ **③** $\frac{1}{4}$ **④** $\frac{2}{5}$

⑤ 1   **⑥** $2\frac{1}{2}$ **⑦** $2\frac{1}{5}$ **⑧** $1\frac{1}{3}$

⑨ $4\frac{1}{2}$ **⑩** 12   **⑪** $\frac{1}{6}$ **⑫** $\frac{1}{2}$

⑬ $1\frac{2}{3}$ **⑭** $1\frac{5}{27}$ **⑮** 5   **⑯** 6

⑰ $6\frac{1}{4}$ **⑱** 2   **⑲** $2\frac{1}{2}$ **⑳** $\frac{3}{4}$

Exercise 5i (page 43)

① $\frac{1}{3}$ **②** $\frac{3}{4}$ **③** $1\frac{2}{5}$ **④** 35

⑤ $\frac{3}{5}$ **⑥** $1\frac{1}{4}$ **⑦** $1\frac{1}{2}$ **⑧** $\frac{8}{15}$

⑨ $\frac{2}{3}$ **⑩** 6   **⑪** $1\frac{4}{5}$ **⑫** $\frac{2}{5}$

⑬ $\frac{2}{7}$ **⑭** $1\frac{1}{2}$ **⑮** $\frac{5}{16}$ **⑯** $\frac{3}{4}$

⑰ $3\frac{1}{5}$ **⑱** $2\frac{2}{5}$ **⑲** $\frac{1}{2}$ **⑳** $2\frac{1}{3}$

Exercise 5j (page 43)

① $7\frac{4}{5}$ **②** $3\frac{3}{4}$ **③** $2\frac{4}{7}$

④ $2\frac{1}{2}, \frac{2}{5}; 1\frac{4}{5} \div 4\frac{1}{2}$ **⑤** 35 kg   **⑥** 50

⑦ 800   **⑧** $1.14   **⑨** $\frac{3}{5}$

⑩ 6

Practice exercise P5.1 (page 44)

① (a) $\frac{5}{6}$, five-sixths

    (b) $\frac{2}{3}$, two-thirds

    (c) $\frac{6}{7}$, six-sevenths

    (d) $\frac{1}{6}$, one-sixth

    (e) $\frac{7}{10}$, seven-tenths

    (f) $\frac{3}{14}$, three-fourteenths

    (g) $\frac{1}{4}$, one-quarter

    (h) $\frac{3}{5}$, three-fifths

② (a) $\frac{1}{2}$ (b) $\frac{1}{4}$ (c) $\frac{1}{3}$ (d) $\frac{3}{4}$

    (e) $\frac{6}{7}$ (f) $\frac{3}{5}$ (g) $\frac{4}{7}$ (h) $\frac{5}{8}$

③ (a) $\frac{1}{4} = \frac{3}{12} = \frac{5}{20} = \frac{6}{24} = \frac{8}{32} = \frac{10}{40}$

    (b) $\frac{2}{5} = \frac{4}{10} = \frac{8}{20} = \frac{10}{25} = \frac{16}{40} = \frac{24}{60}$

    (c) $\frac{5}{8} = \frac{10}{16} = \frac{15}{24} = \frac{25}{40} = \frac{75}{120}$

④ (a) $\frac{2}{8}$ (b) $\frac{3}{5}$ (c) $\frac{5}{8}$

    (d) $\frac{14}{42}$ (e) $\frac{7}{8}$

Practice exercise P5.2 (page 44)

① (a) $45 (b) $30   (c) 20 m (d) $18

    (e) 24 h (f) 7.5 kg (g) $18 (h) $106.66

② (a) $21   (b) $10.50

    (c) 16 cm   (d) 32 litres

③ (a) $\frac{1}{4}$ of 100 ml   (b) $\frac{3}{5}$ of 40 days

    (c) $\frac{5}{6}$ of 12 books (d) $\frac{2}{15}$ of 60 cm

    (e) $\frac{5}{6}$ of 54 Mb

Practice exercise P5.3 (page 45)

① (a) $\frac{5}{2}$ (b) $\frac{13}{5}$ (c) $\frac{7}{6}$ (d) $\frac{45}{7}$

    (e) $\frac{15}{4}$ (f) $\frac{13}{8}$ (g) $\frac{61}{10}$ (h) $\frac{53}{12}$

② (a) $3\frac{1}{2}$ (b) $2\frac{1}{7}$ (c) $1\frac{2}{3}$ (d) $3\frac{1}{8}$

    (e) $2\frac{1}{4}$ (f) $1\frac{11}{12}$ (g) $2\frac{4}{9}$ (h) $3\frac{3}{10}$

Practice exercise P5.4 (page 45)

① (a) $\frac{2}{5}$ (b) $\frac{3}{4}$ (c) $\frac{6}{9}$ (d) $1\frac{1}{5}$

    (e) $1\frac{1}{13}$ (f) $1\frac{3}{5}$ (g) $\frac{2}{3}$ (h) $\frac{1}{3}$

    (i) $\frac{1}{2}$ (j) $\frac{4}{5}$ (k) $\frac{1}{2}$ (l) $1\frac{1}{2}$

② (a) $\frac{2}{3}$ (b) $\frac{5}{12}$ (c) $\frac{3}{8}$ (d) $\frac{4}{9}$

    (e) $\frac{9}{10}$ (f) $1\frac{3}{7}$

③ (a) $\frac{9}{10}$ (b) $\frac{5}{8}$ (c) $\frac{3}{14}$ (d) $\frac{11}{15}$

    (e) $1\frac{1}{8}$ (f) $2\frac{11}{30}$ (g) 7   (h) $3\frac{1}{24}$

④ Answers will vary

Answers

⑤ (a) $\frac{2}{5}$ (b) $\frac{1}{6}$ (c) $\frac{7}{12}$

    (d) $\frac{5}{36}$ (e) $\frac{2}{3}$ (f) $3\frac{3}{4}$

⑥ (a) 2   (b) $\frac{2}{3}$ (c) 4

    (d) $1\frac{1}{5}$ (e) $1\frac{1}{2}$ (f) $\frac{7}{15}$

⑦ (a) $7\frac{1}{6}$ (b) $3\frac{7}{8}$ (c) $1\frac{5}{7}$ (d) $6\frac{1}{2}$

    (e) $\frac{7}{12}$ (f) $2\frac{1}{10}$ (g) $1\frac{1}{7}$ (h) $2\frac{7}{10}$

⑧ (a) $\frac{7}{9}$ (b) $1\frac{7}{24}$ (c) $4\frac{17}{36}$ (d) $19\frac{1}{3}$

    (e) $2\frac{37}{42}$ (f) $1\frac{1}{14}$ (g) $17\frac{1}{2}$ (h) $3\frac{5}{13}$

    (i) $1\frac{7}{20}$ (j) $6\frac{1}{12}$ (k) $\frac{6}{55}$ (l) $1\frac{1}{2}$

⑨ 72 balls

Exercise 6a (page 46)

① (a) 9   (b) 12   (c) 3   (d) 0

    (e) 13   (f) 4   (g) 6   (h) 15

② (a) 14   (b) 0   (c) 20   (d) 17

    (e) 19   (f) 12   (g) 23   (h) 16

    (i) 11   (j) 6   (k) 1   (l) 2

Exercise 6b (page 48)

① $x + 2$

② $(d - 10)$ m

③ (a) $(b \times 5)$ g   (b) $\frac{b}{200}$ kg

④ (a) $(x \times 12)$ c   (b) $\$\frac{x \times 12}{100}$

Exercise 6c (page 48)

① 21 years, $(13 + y)$ years

② $7, n + 4, x + 7$

③ $6 + z$

④ $(48 + m)$ kg

⑤ $k + 1$

⑥ $(6 + n)$th of September

⑦ $(d + d + d + d)$ cm

⑧ $x + 7, x + 7 + y$

Exercise 6d (page 49)

① $7, n - 2, r - 5$

② 20 years, $(30 - y)$ years

③ $x - 30$

④ $m - 5$

⑤ $(15 - x)$ cm

⑥ $\$(16\,000 - x)$

⑦ $(20 - x)$ km

⑧ $\$(s - t)$

Exercise 6e (page 49)

① 1000 m, 4000 m, $(x \times 1000)$ m

② 7 days, 35 days, $(w \times 7)$ days

③ $(x \times 12)$ months

④ $d \times 9$

⑤ $\$32, \$(x \times 4), \$(t \times 5)$

⑥ $(x \times 1000)$ g

⑦ $(x \times 3)$ mm

⑧ $(m \times 10)$ cents

Exercise 6f (page 49)

① 2 weeks, $\frac{d}{7}$ weeks   ② 3 m, $\frac{k}{1000}$ m

③ $\frac{b}{6}$ (or $\frac{1}{6}b$) kg   ④ 30c, $\frac{p}{5}$ cents

⑤ $\frac{h}{9}$ ⑥ $\frac{140}{x}$ cm

⑦ $\frac{n}{3}$ or $\frac{1}{3}n$ ⑧ $\frac{x}{5}$ cm

Exercise 6g (page 50)

① $3a$ ② $3x$ ③ $2p$ ④ $5r$

⑤ $8t$ ⑥ $4m$ ⑦ $10z$ ⑧ $2y$

⑨ $6c$ ⑩ $5k$

Exercise 6h (page 50)

① (a) $4 + 4 + 4$ (b) $30 + 4$

    (c) $6 + 6 + 6$ (d) $30 + 6$

    (e) $9 + 9$ (f) $20 + 9$

    (g) 5   (h) $50 + 1$

② (a) $a + a + a$

    (b) $x + x + x + x + x$

    (c) $y + y$

    (d) $n + n + n + n$

    (e) $m + m + m$

    (f) $d + d + d + d + d + d$

    (g) $f + f$

    (h) $e + e + e + e + e + e + e + e + e + e$

Exercise 6i (page 50)

① 3   ② 7   ③ 4   ④ 8

⑤ 15   ⑥ 9   ⑦ 18   ⑧ 2

⑨ 1   ⑩ 10   ⑪ $\frac{1}{3}$ ⑫ $\frac{1}{2}$

⑬ $\frac{1}{4}$ ⑭ $\frac{2}{3}$ ⑮ $\frac{3}{4}$ ⑯ $\frac{1}{5}$

⑰ $\frac{1}{10}$ ⑱ $\frac{2}{3}$ ⑲ $\frac{3}{4}$ ⑳ $\frac{3}{5}$

Exercise 6j (page 50)

1. $(x + 20)$ min
2. $6 - n$
3. $(2a + 10)$ m
4. $\dfrac{x}{1000}$ kg
5. (a) $50x$ c, (b) $\dfrac{x}{2}$
6. $\$\dfrac{3800}{d}$
7. $20 - x$
8. $(5x + 12)$ km
9. Team A, $\dfrac{n}{3}$ pts
10. $9x$ m^2

Practice exercise P6.1 (page 51)

1. 10
2. 0
3. 8
4. 7
5. -2
6. 4
7. 12
8. 7
9. 8

Practice exercise P6.2 (page 51)

1. add 3 years to Jenny's age
2. subtract 5 years from Mona's age
3. multiply Toby's age by 2
4. multiply Tony's age by 4
5. divide her mother's age by 2
6. add 3 kg to Ken's weight.
7. subtract 4 kg from Sarah's weight
8. multiply the number of chairs by 4
9. multiply 3 tons by the number of lorries
10. (a) subtract 23 from her mother's age
    (b) add 23 to Pam's age.

Practice exercise P6.3 (page 51)

1. multiply by 2
2. add 2
3. multiply by 2 and then subtract 1
4. divide by 2
5. square the number
6. divide the number by 3

Practice exercise P6.4 (page 52)

1. (a) $3x$ (b) 3
2. (a) x,   (b) 1
3. (a) 0   (b) 0
4. (a) $-2x$ (b) -2
5. (a) $2x$ (b) 2
6. (a) $2x$ (b) 2
7. (a) $4x$ (b) 4
8. (a) 0   (b) 0
9. (a) 0   (b) 0
10. (a) $2x$ (b) 2

Practice exercise P6.5 (page 52)

1. 2
2. -4
3. $\dfrac{1}{3}$
4. -1
5. 4
6. $\dfrac{3}{7}$
7. 7
8. $-\dfrac{5}{8}$
9. $\dfrac{1}{2}$
10. $\dfrac{1}{3}$
11. $\dfrac{1}{3}$
12. -3

Practice exercise P6.6 (page 52)

1. $6d$
2. $\dfrac{m}{5}$
3. $t + 3$
4. $w - 8$
5. $15n$

Exercise 7c (page 55)

1. most boxes, room, book, rubber, brick (there are many more)
2. (a) 4   (b) 4   (c) 5   (d) 4
3.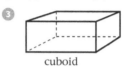

   cuboid   cube
4. rectangle
5. square
6. (a) 9   (b) 36
   (c) 8   (d) 16
   (e) 10   (f) 2
7. (a), (c), (d), (f)
8. Here are 9 more nets.

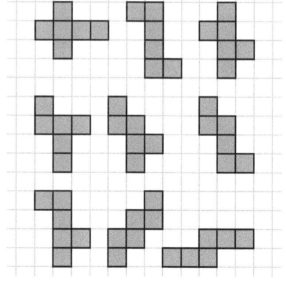

10. 140 cm
11. 5.5 cm
12. (a) 3   (b) 9   (c) 7

Answers

	Vertices	Faces	Edges
cuboid	8	6	12
cube	8	6	12

Exercise 7d (page 58)

① most tin cans, cotton reel, coin, new round pencil, torch battery (there are many more)
② new hexagonal pencil, any cuboid, roofs of some buildings, laboratory glass prism, nut from nut and bolt
③ (a) 4　　(b) 3　　(c) 3　　(d) 5
④

cylinder　　　triangular prism

⑤ circle
⑥ triangle, rectangle
⑦ (a) hexagonal prism
　 (b) triangular prism
　 (c) cylinder
⑧ 5 faces, 9 edges, 6 vertices
⑨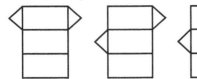

etc.

⑩ 14 cm
⑪

View	Vertices	Faces	Edges
(a)	5	2	6
(b)	6	3	8

⑫

Prism	Vertices	Faces	Edges
triangular	6	5	9
hexagonal	12	8	18
I-shaped	24	14	36

Exercise 7e (page 59)

① (a) number of faces equals number of sides of base
　 (b) triangular

② sharp end of a round pencil, ice-cream cone, some hats, end of a trumpet, funnel
③ sharp end of hexagonal pencil, top of some roofs, some hats, corner of a box, tower
④ (a) 2 cones
　 (b) 2 hexagonal pyramids
　 (c) 3 square-based pyramids
　 (d) 2 cones
⑤

cone　　　square-based pyramid

⑥ triangular-based (*or* tetrahedron)
⑦ 7, hexagon
⑧ (a) square-based pyramid
　 (b) cone　　　　(c) tetrahedron
⑨ (a) DC　　(b) CB　　(c) GF
　 (d) HA　　(e) C, E
⑩ 48 cm
⑪ 5 vertices, 3 faces, 7 edges
⑫

Solid	Vertices	Faces	Edges
triangular pyramid	4	4	6
square pyramid	5	5	8
hexagonal pyramid	7	7	12
triangular prism	6	5	9
solid in Fig 7.31	9	9	16

Exercise 7f (page 61)

① various balls, domes on buildings, some bowls, tip of ball-point pen
② (a) cone, hemisphere
　 (b) cone, cylinder
　 (c) hemisphere, cube, cylinder
　 (d) 2 cuboids, triangular prism, square-based pyramid
③ number of edges = $f + v - 2$
④ (b), (c), (f)
⑤ 5 cm
⑥ square-based pyramid, triangular prism
⑦ (a) BC　　(b) JK　　(c) IJ
　 (d) AN　　(e) C, K　　(f) F, J

8 cylinder, sphere, hemisphere

9

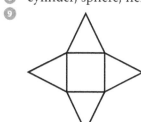
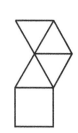

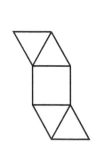
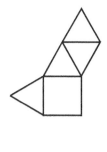

10 (a) triangular-based pyramid (tetrahedron)
(b) cuboid
(c) cube

Practice exercise P7.1 (page 62)

1 Answers will vary.

2 (a) true (b) false (c) true
(d) false (e) false (f) false
(g) true (h) true (i) true
(j) true

3 (a) 1 (b) 3 (c) 5 opposite 2 (d) yes

Exercise 8a (page 63)

2

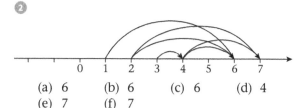

(a) 6 (b) 6 (c) 6 (d) 4
(e) 7 (f) 7

3

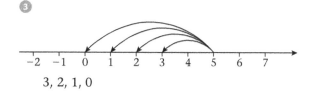

3, 2, 1, 0

Exercise 8b (page 65)

1 (a) (i) (+3) (ii) (−2)
(iii) (−1) (iv) (+5)
(b) (i) Tom (ii) Bob (iii) Ismith
(iv) Clive (v) Sam

2 (a) +2 m (b) −5 m (c) +4 m
(d) +3 m (e) −3 m (f) −4 m
(g) −1 m (h) +5 m (i) 0
(j) +1 m

Exercise 8c (page 65)

1 (a) +7 m, −2 m
(b) 9 m

2 −15 °C

3 (a) (i) −8 km (ii) +9 km
(iii) +15 km (iv) −11 km
(b) (i) 17 km (ii) 6 km
(iii) 26 km (iv) 3 km

4 (a) −5 h (b) +7 h
(c) +1½ h (d) −¾ h
(e) +3 h (f) −2 h
(g) +9 h (h) −3½ h
(i) +4¾ h

5 13 min

6 24 years

7 −210 m

8 8 mm

9 Gino: $58; Sita: $89 (19 + 64 + 6)

10 (a) $8
(b) $25

Exercise 8d (page 66)

1 (a) +12 (b) +5 (c) −3
(d) +15 (e) +4 (f) 0

2 (a) −8, −3, −2, 0, +1, +5
(b) −5, −4, −2, +5, +7, +9
(c) −20, −14, −8, 0, +5, +6
(d) −18, −11, −6, +6, +11, +18
(e) −14, −10, −5, −3, −1, 0
(f) −11, −10, −9, +1, +16, +17

3 (b) −15 < +8 (c) +4 > −13
(d) +18 > +1 (e) −18 < −1
(f) 0 > −2 (g) −1 < +1
(h) −7 > −17 (i) −13 < +13
(j) 0 > −4 (k) +5 > −5
(l) −9 < 0

④	start	move	finish
(a)	+2	5 units to the left	−3
(b)	+5	7 units to the left	−2
(c)	−1	7 units to the left	−8
(d)	+2	4 units to the right	+6
(e)	−3	5 units to the right	+2
(f)	−10	10 units to the right	0
(g)	+12	9 units to the left	+3
(h)	+14	30 units to the left	−16
(i)	−20	28 units to the right	+8
(j)	+1	2 units to the left	−1
(k)	+2	6 units to the left	−4
(l)	+5	2 units to the left	+3
(m)	+6	12 units to the left	−6
(n)	−2	2 units to the right	0
(o)	−15	30 units to the right	+15
(p)	+15	30 units to the left	−15
(q)	−11	24 units to the right	+13
(r)	+12	14 units to the left	−2
(s)	0	6 units to the right	+6
(t)	0	17 units to the left	−17
(u)	+3	2 units to the right	+5
(v)	+10	8 units to the left	+2
(w)	+6	10 units to the left	−4
(x)	−15	35 units to the right	+20
(y)	−4	11 units to the right	+7
(z)	+8	13 units to the left	−5

Exercise 8e (page 67)

① (a) +10 (b) +11 (c) +12
   (d) +13 (e) +14 (f) +15
② (a) −2 (b) −1 (c) 0
   (d) +1 (e) +2 (f) +3
③ +19 ④ +5
⑤ +2 ⑥ +18
⑦ 0 ⑧ 10

Exercise 8f (page 67)

① (a) 4 (b) 5 (c) 6
   (d) 7 (e) 8
② (a) −2 (b) −1 (c) 0
   (d) 1 (e) 2
③ 4 ④ 7
⑤ 3 ⑥ −1
⑦ 8 ⑧ −9
⑨ 11 ⑩ 12

Exercise 8g (page 68)

1 (a) +5 (b) 0 (c) +8
    +4   −1   +6
    +3   −2   +4
    +2   −3   +2
    +1   −4   0
    0   −5   −2

2 (a) −3 (b) +2 (c) −8
  (d) −14 (e) 0 (f) +3
  (g) −18 (h) −21 (i) −5
  (j) +20 (k) −20 (l) +20
  (m) −39 (n) −44 (o) −89
  (p) −3 (q) +6

Exercise 8h (page 69)

① (a) 2 (b) 1 (c) 0
   (d) −1 (e) −2
② (a) −4 (b) −5 (c) −6
   (d) −7 (e) −8
③ 4 ④ −4 ⑤ −8
⑥ −7 ⑦ 7 ⑧ −5
⑨ 7 ⑩ 14

Exercise 8i (page 69)

① 8 ② −5 ③ −13
④ −9 ⑤ −1 ⑥ −30
⑦ −34 ⑧ 15 ⑨ −8
⑩ −24

Exercise 8j (page 70)

① (a) +15, +10, +5, 0, −5, −10, −15
  (b) +6, +4, +2, 0, −2, −4, −6
  (c) +30, +20, +10, 0, −10, −20, −30
② (a) $(+5) \times (+3) = -15$
    $(+5) \times (+2) = +10$
    $(+5) \times (+1) = +5$
    $(+5) \times 0 = 0$
    $(+5) \times (-1) = -5$
    $(+5) \times (-2) = -10$
    $(+5) \times (-3) = -15$
  (b) $(+2) \times (+3) = +6$
    $(+2) \times (+2) = +4$
    $(+2) \times (+1) = +2$
    $(+2) \times 0 = 0$
    $(+2) \times (-1) = -2$
    $(+2) \times (-2) = -4$
    $(+2) \times (-3) = -6$

(c) $(+10) \times (+3) = +30$
$(+10) \times (+2) = +20$
$(+10) \times (+1) = +10$
$(+10) \times \ 0 \ = \ \ 0$
$(+10) \times (-1) = -10$
$(+10) \times (-2) = -20$

③ (a) $+18$ (b) $+50$ (c) -40
(d) -28 (e) $+34$ (f) $+120$
(g) -100 (h) -60 (i) $+7\frac{1}{2}$
(j) $+\frac{1}{4}$ (k) -7 (L) $-\frac{2}{3}$
(m) $+3$ (n) $+15.3$ (o) -4.8
(p) -16 (q) 56 (r) -56
(s) 56 (t) -56

Exercise 8k (page 71)

① (a) $+3, 0, -3, -6$
(b) $(+1) \times (+3), 0 \times (+3), (-1) \times (+3),$
$(-2) \times (+3)$
(c) $(+1) \times (+3) = +3$
$0 \ \times (+3) = \ \ 0$
$(-1) \times (+3) = -3$
$(-2) \times (+3) = -6$

② (a) $-3, 0, +3, +6$
(b) $(+1) \times (-3), 0 \times (-3), (-1) \times (-3),$
$(-2) \times (-3)$
(c) $(+1) \times (-3) = -3$
$0 \ \times (-3) = \ \ 0$
$(-1) \times (-3) = +3$
$(-2) \times (-3) = +6$

Exercise 8l (page 71)

① (a) $+15, +10, +5, 0, -5, -10, -15$
(b) $-6, -4, -2, 0, +2, +4, +6$
(c) $-30, -20, -10, 0, +10, +20, +30$

② (a) $(+3) \times (+5) = +15$
$(+2) \times (+5) = +10$
$(+1) \times (+5) = +5$
$0 \ \times (+5) = \ \ 0$
$(-1) \times (+5) = -5$
$(-2) \times (+5) = -10$
$(-3) \times (+5) = -15$
(b) $(+3) \times (-2) = -6$
$(+2) \times (-2) = -4$
$(+1) \times (-2) = -2$
$0 \ \times (-2) = \ \ 0$
$(-1) \times (-2) = +2$
$(-2) \times (-2) = +4$
$(-3) \times (-2) = +6$

(c) $(+3) \times (-10) = -30$
$(+2) \times (-10) = -20$
$(+1) \times (-10) = -10$
$0 \ \times (-10) = \ \ 0$
$(-1) \times (-10) = +10$
$(-2) \times (-10) = +20$

③ (a) -18 (b) $+50$ (c) $+40$
(d) $+28$ (e) -32 (f) -80
(g) $+64$ (h) $+65$ (i) -3
(j) $-1\frac{1}{3}$ (k) $+1$ (l) $+2$
(m) -6 (n) -7.2 (o) $+2.7$
(p) $+18.6$ (q) -48 (r) -48
(s) $+48$ (t) $+48$

Practice exercise P8.1 (page 72)

① (a) (i) $+4$ (ii) $+5$ (iii) $+5$
(b) (i) -6 (ii) -7 (iii) -2

② (a) $-5, -1, +1, +3$
(b) $-6, -2, 0, +1, +4$
(c) $-5, -4, -3, +3, +4, +5$

③ (a) $7 > -5$ (b) $2 < 3$
(c) $-5 > -9$ (d) $-23 < 13$
(e) $21 > -19$ (f) $11 > -13$
(g) $-7 < -5$ (h) $-7 < 3$
(i) $-5 > -9$ (j) $-23 < 13$
(k) $21 > -12$ (l) $-1 > -13$

Practice exercise P8.2 (page 72)

① (a) 5 (b) 3 (c) 2 (d) -2
(e) 2 (f) -2 (g) 0 (h) -14
(i) 12 (j) -21 (k) -16 (l) -5

② (a) 35 (b) -21 (c) $+45$
(d) $+12$ (e) $+12$ (f) $+65$
(g) -6 (h) -6 (i) -12
(j) -48 (k) -72 (l) -36
(m) -72 (n) -299 (o) -221

Practice exercise P8.3 (page 73)

① 1 ② -5 ③ -14 ④ 4
⑤ 0 ⑥ -18 ⑦ -35 ⑧ 18
⑨ -45 ⑩ 52 ⑪ 165 ⑫ 120

Practice exercise P8.4 (page 73)

① (a) $1, 2, 4$ (b) $-5, -3, -1$
(c) $1, -1, -3, -5$ (d) $-3, -1, 1, 2, 4$

② (a) $-6\,°C \ -5\,°C \ -3\,°C \ -1\,°C \ 2\,°C$
$4\,°C \ 7\,°C$
(b) $-6\,°C \ -5\,°C \ -3\,°C \ -1\,°C \ 2\,°C$

Answers

Revision exercise 1 (page 74)

1. (a) 5 tens   (b) 7 units
   (c) 2 thousands   (d) 9 hundreds
2. (a) 6 tenths   (b) 8 thousandths
   (c) 1 hundredth   (d) 5 hundredths
3. $2 \times 8^3 + 0 \times 8^2 + 6 \times 8 + 5 \times 1$
4. (a) 223_{five} (b) 2100_{three}
5. 27_{ten}, 33_{eight}
6. (a) $2 \times 3 \times 3$ (b) 2×13
   (c) $3 \times 3 \times 5$ (d) $3 \times 5 \times 5$
7. (a) 49   (b) 125   (c) 72   (d) 108
8. 9   9. 60   10. 28

Revision test 1 (page 74)

1. C   2. A   3. C   4. D   5. C
6. (a) $1\,110\,001_{\text{two}}$ (b) 161_{eight}
7. $2 \times 2 \times 3 \times 5$ 8. 8   9. 1089
10. (a) 2, 5, 9   (b) 45, 90

Revision exercise 2 (page 74)

1. (a), (b)
2. $\varnothing$, {x}, {y}, {z}, {x, y}, {x, z}, {y, z},
   {x, y, z}
3. (a) 7   (b) 4   (c) {q, r, s, t}
   (d) {s, t}, {s, u}, {t, u}
4. (a) 19   (b) 5   (c) 5   (d) 4
5. (a) $m + 5$ (b) $m - 5$
   (c) $5m$ (d) $\frac{m}{5}$ or $\frac{1}{5}m$
6. (a) 5   (b) -9 (c) 0
7. (a) -15 (b) 16   (c) -30 (d) 9
8. (a) -11 (b) -3 (c) 21   (d) -58
9. $2x - 6$ 10. $(15x + 20y)$ cents

Revision test 2 (page 75)

1. D   2. B   3. A   4. C   5. A
6.

7. $\frac{n}{2} - 8$ 8. $-1.5\,°C$
9. (a) 8   (b) -5 (c) -7 (d) -12
10. $\$30y$ $\$(30y - x)$

Revision exercise 3 (page 75)

1. 95
2. 125 min (2 h 5 min)
3. (a) FJ251   (b) 19:45   (c) 4 h
4. (a) $1\frac{1}{2}$ h   (b) 600 km/h
5. (a) $\frac{3}{5}$ (b) $\frac{2}{9}$ (c) $\frac{4}{5}$
6. (a) $\frac{7}{12}$ (b) $\frac{7}{20}$ (c) $\frac{1}{6}$ (d) $1\frac{1}{4}$
   (e) $4\frac{3}{8}$ (f) $\frac{1}{10}$
7. 125 cm
8. $\frac{3}{5}$ 9. 5 faces, 3 rectangles
10. (a) (i) 3.4 °F   (ii) 1.9 °C
    (b) (i) 37 °C   (ii) 38.9 °C

Revision test 3 (page 76)

1. C   2. D   3. D   4. B   5. A
6. 48 km
7. (a) $2\frac{3}{4}$ (b) 10   (c) $1\frac{1}{5}$
8. $\frac{7}{10}$
9. (a) BG   (b) CH, EH, GH   (c) B
10. (a) DC   (b) E

General revision test A (page 77)

1. B   2. D   3. C   4. B   5. A
6. D   7. B   8. A   9. C   10. D
11.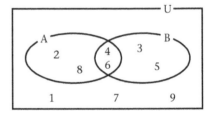
12. (a) $1\frac{4}{5}$ (b) $\frac{5}{8}$ (c) $5\frac{5}{8}$ (d) $1\frac{13}{20}$
13. $\frac{1}{6}$ 14. $\frac{7}{12}$
15. $\frac{17}{24}$
16. (a) $2\frac{1}{4}$ h   (b) 12:45 p.m.
17. (a) 29   (b) 49   (c) $\frac{7}{6}$ (d) $5\frac{2}{5}$
18. (a) 06:25 hrs   (b) 18:45 hrs
    (c) 19:55 hrs   (d) 23:50 hrs
19. 41 min, 12:06
20. 3 p.m.

Exercise 9a (page 78)

1. 0.38   2. 4.25   3. 0.52   4. 1.5
5. 1.7   6. 1.82   7. 0.76   8. 1.08

Exercise 9b (page 78)

1. (a) 0.8   (b) 0.1   (c) 2.1   (d) 1.5
    (e) 6.5   (f) 3.4   (g) 21.01 (h) 0.38
    (i) 12.53 (j) 10.08
2. (a) 0.8   (b) 0.5   (c) 2.7   (d) 1.4
    (e) 8.1   (f) 0.22   (g) 10.4 (h) 3.4
    (i) 47.8   (j) 10.95
3. (a) 0.9   (b) 0.3   (c) 2.3   (d) 0.3
    (e) 8.0   (f) 0.04   (g) 26.21 (h) 2.5
    (i) 12.04 (j) 1.78
4. (a) 0.5   (b) 0.2   (c) 1.4   (d) 2.7
    (e) 8.1   (f) 5.0   (g) 21.00 (h) 0.59
    (i) 12.72 (j) 10.04
5. (a) 0.8   (b) 0.1   (c) 2.0   (d) 0.2
    (e) 9.2   (f) 3.1   (g) 13.01 (h) 0.27
    (i) 76.923 (j) 36.45
6. (a) $7.56   (b) $2.72
7. (a) 11.37 m   (b) 2.207 kg *or* 2207 g
8. 5.62 m; 7.88 m
9. 0.59 kg (or 590 g)
10. 3.05

Exercise 9c (page 80)

1. 267
2. 45 500
3. 8030
4. 0.005 6
5. 0.811
6. 0.070 4
7. 270
8. 36
9. 0.052
10. 0.006 3
11. 4.23
12. 50
13. 65 000
14. 9 000 000
15. 4370
16. 20.7
17. 45.02
18. 0.000 000 36
19. 0.0514
20. 0.6
21. 0.03
22. 0.026
23. 0.7
24. 0.000 242
24. 40.007
26. 0.070 63
27. 620 cents
28. $3.44
29. 5.92 m
30. (a) 6240 kg   (b) 6 240 000 g
31. (a) 0.92 t   (b) 920 000 g
32. (a) 0.42 kg   (b) 0.000 42 t
33. (a) 0.504 g   (b) 0.000 504 kg
34. (a) 20 900 mm   (b) 0.020 9 km
35. (a) 7.05 km   (b) 7 050 000 mm

Exercise 9d (page 81)

1. 5.6
2. 0.8
3. 1.2
4. 2.7
5. 4
6. 3.6
7. 4.9
8. 4.2
9. 3
10. 4
11. 0.18
12. 0.28
13. 0.06
14. 0.12
15. 0.09
16. 0.018
17. 0.042
18. 0.003 2
19. 0.003
20. 0.000 15

Exercise 9e (page 81)

1. (a) 1.08   (b) 0.579   (c) 0.002 5
    (d) 1.6   (e) 4.234 6
2. (a) 1.2   (b) 13.95   (c) 0.04
    (d) 21.08   (e) 1.051 2
3. (a) 0.56   (b) 1.02   (c) 0.64
    (d) 0.280 8   (e) 0.677 5
4. (a) 3   (b) 0.51   (c) 0.000 016
    (d) 0.943   (e) 0.266 22
5. (a) 2.7   (b) 0.91   (c) 0.09
    (d) 0.115 5   (e) 33.858
6. 1.59   7. 0.111 65   8. 6.72 cm
9. $66.50
10. (a) 259.44 g   (b) 0.259 44 kg

Exercise 9f (page 82)

1. 3
2. 0.3
3. 0.03
4. 30
5. 3
6. 300
7. 3
8. 30 000
9. 6
10. 0.6
11. 60
12. 0.06
13. 600
14. 600
15. 600
16. 60

Exercise 9g (page 82)

1. (a) 0.03   (b) 9   (c) 3650
    (d) 0.7   (e) 3.4
2. (a) 0.2   (b) 0.3   (c) 46
    (d) 0.03   (e) 6.1
3. (a) 0.04   (b) 8   (c) 2.05
    (d) 40   (e) 35
4. (a) 0.2   (b) 9   (c) 53
    (d) 60   (e) 230
5. (a) 0.1   (b) 0.5   (c) 1.08
    (d) 400   (e) 780
6. 0.35
7. (a) 0.47 kg   (b) 470 g
8. 45   9. $1.80   10. 13.6 km

Exercise 9h (page 83)

1. 0.5
2. 0.25
3. 0.375
4. 0.4
5. 0.05
6. 0.35
7. 0.04
8. 0.12
9. 0.625
10. 0.26
11. 0.92
12. 0.85

Exercise 9i (page 83)

1. $0.\dot{3}$
2. $0.\dot{6}$
3. $0.\dot{2}$
4. $0.\dot{5}$
5. $0.\dot{8}$
6. $0.2\dot{7}$
7. $0.4\dot{5}$
8. $0.8\dot{3}$
9. $0.58\dot{3}$
10. $0.\dot{1}4\dot{2}8\dot{5}\dot{7}$

Answers

Exercise 9j (page 84)

1. $\frac{4}{5}$ 2. $\frac{3}{4}$ 3. $\frac{1}{5}$ 4. $\frac{9}{20}$

5. $\frac{11}{25}$ 6. $\frac{27}{50}$ 7. $\frac{14}{25}$ 8. $\frac{33}{50}$

9. $\frac{21}{25}$

Exercise 9k (page 84)

1. $\frac{4}{9}$ 2. $\frac{2}{11}$ 3. $\frac{83}{99}$

4. $\frac{1}{7}$ 5. $\frac{4}{13}$

Exercise 9l (page 84)

1. 50%   2. 25%   3. $33\frac{1}{3}$%   4. 75%

5. 20%   6. 10%   7. 5%   8. 4%

9. 30%   10. 40%   11. $13\frac{1}{3}$%   12. 85%

13. $62\frac{1}{2}$%   14. $16\frac{2}{3}$%   15. 36%

Exercise 9m (page 85)

1. $\frac{1}{20}$ 2. $\frac{1}{2}$ 3. $\frac{1}{4}$ 4. $\frac{2}{5}$

5. $\frac{7}{10}$ 6. $\frac{1}{25}$ 7. $\frac{7}{20}$ 8. $\frac{3}{25}$

9. $\frac{4}{5}$ 10. $\frac{3}{4}$ 11. $\frac{6}{25}$ 12. $\frac{13}{20}$

13. $\frac{16}{25}$ 14. $\frac{3}{5}$ 15. $\frac{9}{20}$

Exercise 9n (page 85)

1.
(a) 0.5   (b) 0.5   (c) 0.25
(d) 0.4   (e) 0.7   (f) 0.04
(g) 0.15   (h) 0.12   (i) 0.8
(j) 0.75   (k) 0.24   (l) 0.35
(m) 0.64   (n) 0.6   (o) 0.45

2.
(a) 25%   (b) 85%   (c) 36%
(d) 81%   (e) 8%   (f) $12\frac{1}{2}$%
(g) 30.8%   (h) 70%   (i) 7%
(j) $66\frac{2}{3}$%   (k) 50%   (l) 2.2%
(m) $\frac{1}{2}$%   (n) 80%   (o) $13\frac{1}{3}$%

Exercise 9o (page 85)

1. $\frac{1}{4}$ 2. $\frac{1}{12}$ 3. $\frac{1}{4}$ 4. $\frac{1}{7}$

5. $\frac{2}{5}$ 6. $\frac{2}{3}$ 7. $\frac{13}{20}$ 8. $\frac{3}{8}$

9. $\frac{1}{4}$ 10. $\frac{1}{25}$

Exercise 9p (page 86)

1.
(a) 10%   (b) 25%
(c) 40%   (d) 80%
(e) 30%   (f) $12\frac{1}{2}$%
(g) 60%   (h) $2\frac{1}{2}$%
(i) 20%   (j) 40%

2. 75%
3. 12%
4. 20%
5. 9%, 91%
6. 12 cm, 25%
7. 16%, 84%
8. 6%

Practice exercise P9.1 (page 87)

(a) 4.85 5.18 5.24 5.26 6.69
(b) 0.08 0.68 0.86 6.08 6.8 8.06 8.6
(c) 0.09 0.9 0.999 999.09 9.99
(d) 0.1 0.13 0.18 0.31 0.38 0.81 0.83
(e) 9.41 9.43 9.44 9.45 9.47 9.48 9.49
(f) 0.069 14 0.069 41 0.080 63 0.086 03
0.086 32 0.088 03 0.600 9

Practice exercise P9.2 (page 87)

1. $25.35
2.
(a) (i) 5   (ii) $1.55
(b) (i) 2   (ii) $3.34
(c) (i) 1   (ii) $1.55

Practice exercise P9.3 (page 87)

1.
(e) 20.8   (c) 25.4   (a) 31
(b) 33.2   (d) 46.2
2.
(a) 1.85   (c) 1.86   (e) 1.87
(b) 1.95   (d) 2.85
3.
(e) 97.5   (d) 112.96   (c) 124.8
(b) 126.14   (a) 129.26

Practice exercise P9.4 (page 87)

1. (a) 15.48 ÷ 18
2. (c) 349.2 ÷ 36
3. (b) 403.2 ÷ 45

Practice exercise P9.5 (page 87)

1.
(a) 0.259   (b) 6.3   (c) 0.002 9
(d) 7 000 000 000   (e) 0.092
(f) 0.057
2.
(a) 7.2   (b) 0.24   (c) 0.027
(d) 0.056   (e) 0.06   (f) 6.3
(g) 0.000 28   (h) 1500
3.
(a) 300   (b) 0.3   (c) 50
(d) 400   (e) 0.005   (f) 0.000 03
(g) 0.001 6   (h) 0.000 000 36

Practice exercise P9.6 (page 88)

1 (a) $0.2 \times 10 = 2$
$0.2 \times 1 = 0.2$
$0.2 \times 0.1 = \mathbf{0.02}$
$0.2 \times \mathbf{0.01} = 0.002$
$0.2 \times 0.001 = \mathbf{0.000\,2}$

(b) $0.3 \times 20 = \mathbf{6}$
$0.3 \times \mathbf{2} = 0.6$
$0.3 \times \mathbf{0.2} = 0.06$
$0.3 \times 0.02 = \mathbf{0.006}$
$0.3 \times \mathbf{0.002} = 0.000\,6$

2 (a) 0.08 (b) 0.15 (c) 0.42
(d) 0.002 4 (e) 0.001 (f) 0.000 3

3 (a) 8.481 (b) 84.81 (c) 8.481
(d) 8.481 (e) 0.848 1 (f) 0.084 81

4 (a) 5.328 7 (b) 1.052 7 (c) 0. 021 084
(d) 625.1

Practice exercise P9.7 (page 88)

1 56c
2 3 400 g
3 0.004 5 cm
4 600 000 m
5 20 000 kb
6 (a) 12, 17, 28 (b) $0.1\,l$, $0.05\,l$, $0.1\,l$
(c) 16

Practice exercise P9.8 (page 88)

1 (a) (i) 0.73 kg (ii) 0.16 kg (iii) 14.4 kg
(b) (i) 0.77 kg (ii) 0.57 kg (iii) 24.64 kg
2 Copper A: 2.8 kg B: 2.76 kg
Zinc A: 1.68 kg B: 1.2 kg
Tin A: 0.63 kg B: 0.66 kg
3 the base, 5 blocks more

Practice exercise P9.9 (page 89)

1 Packers : 6000; Warriors : 4000
Lifers : 2125, Readers : 375

2

Percentages						
Amount (100%)	50%	25%	10%	30%	63%	8%
100 litres (l)	50 l	25 l	10 l	30 l	63 l	8 l
$20	$10	$5	$2	$6	$12.60	$1.60
1300 dogs	650	325	130	390	819	104
435 km	217.5	108.75	43.5	130.5	274.05	34.8

3

Type of use	Number of litres used
cooking	32 400
toilet	48 600
laundry	40 500
bathing	59 400
gardening	89 100

4 £7
5 (a) $2125 (b) $138
6 500 ml

Practice exercise P9.10 (page 89)

(a)

Fractions	$\frac{3}{5}$	$\frac{5}{8}$	$\frac{2}{3}$	$\frac{7}{10}$	$\frac{3}{4}$	$\frac{4}{5}$	$\frac{7}{8}$	$\frac{9}{10}$	$\frac{1}{1}$
Decimals	0.6	0.625	0.667	0.7	0.75	0.8	0.875	0.9	1.0
Percentages	60	62.5	$66\frac{2}{3}$	70	75	80	87.5	90	100

Practice exercise P9.11 (page 89)

1 (a) (i) 9% (ii) 63% (iii) 70%
(iv) 40% (v) 15% (vi) 68%
(vii) 40% (viii) 111% (ix) 130%

(b) (i) 0.9 (ii) 0.63 (iii) 0.7
(iv) 0.4 (v) 0.15 (vi) 0.68
(vii) 0.4 (viii) 1.11 (ix) 1.3

2 (a) (i) $\frac{31}{100}$ (ii) $\frac{93}{100}$ (iii) $\frac{3}{10}$
(iv) $\frac{7}{20}$ (v) $\frac{16}{25}$ (vi) $1\frac{2}{5}$

(b) (i) 0.31 (ii) 0.93 (iii) 0.3
(iv) 0.35 (v) 0.64 (vi) 1.4

3 (a) (i) $\frac{1}{2}$ (ii) $2\frac{3}{5}$ (iii) $19\frac{3}{5}$
(iv) $\frac{9}{200}$ (v) $15\frac{1}{20}$

(b) (i) 50% (ii) 260% (iii) 1960%
(iv) 4.5 % (v) 1505%

4 60% $\frac{5}{8}$ $\frac{2}{3}$ $\frac{3}{4}$ $\frac{7}{9}$ 0.875 $\frac{13}{10}$ 2.5

Answers

Exercise 10a (page 91)

1

Revs	Degrees	Revs	Degrees
1	360°	$\frac{1}{3}$	120°
2	720°	$\frac{1}{4}$	90°
3	1080°	$\frac{1}{10}$	36°
10	3600°	$\frac{1}{8}$	45°
$\frac{1}{2}$	180°	$1\frac{1}{2}$	540°

2 (a) $\frac{1}{6}$ revolution; 60°

(b) $\frac{1}{4}$ 90°

(c) $\frac{1}{3}$, 120°

(d) $\frac{5}{12}$, 150°

(e) $\frac{1}{2}$, 180°

(f) $\frac{5}{12}$, 150° (or $\frac{7}{12}$, 210°)

(g) $\frac{1}{4}$, 90° (or $\frac{3}{4}$, 270°)

(h) $\frac{1}{6}$, 60° (or $\frac{5}{6}$, 300°)

(i) $\frac{1}{12}$, 30° (or $\frac{11}{12}$, 330°)

3 mid-way between 2 and 3, 105°

4 (a) 75° (b) 135°

(c) 75° (d) 165°

5 quarter-way between 2 and 3 (nearer 2), $22\frac{1}{2}°$

6 (a) $112\frac{1}{2}°$ (b) $172\frac{1}{2}°$

(c) $7\frac{1}{2}°$ (d) $7\frac{1}{2}°$

Exercise 10b (page 93)

1 (a) HÂB, AB̂C, BĈD, DÊF, FĜH

(b) CD̂E, EF̂G, GĤA

(c) HÂB, AB̂C, BĈD, CD̂E, DÊF, EF̂G, FĜH, GĤA

2 *a* right   *b* obtuse   *c* acute

d acute   *e* reflex   *f* obtuse

g acute   *h* straight   *k* obtuse

l reflex   *m* right   *n* reflex

3 (a) 30° (b) 43° (c) 62° (d) 79°

(e) 140° (f) 156° (g) 117° (h) 161°

Exercise 10c (page 95)

1 40°   **2** 65°   **3** 58°   **4** 80°

5 28°   **6** 110°   **7** 135°   **8** 147°

9 102°   **10** 167°

Exercise 10d (page 95)

3 BĈD = 112°; BĈD + AĈD = 180°

Exercise 10e (page 96)

1 (a) 120' (b) 210' (c) 300'

(d) 495' (e) 1350' (f) 5400'

2 (a) 3° (b) $1\frac{1}{2}°$ (c) 4°

(d) $\frac{1}{3}°$ (e) 10° (f) $7\frac{1}{2}°$

3 (a) 70° 53' (b) 55° 35' (c) 23° 25'

(d) 6° 39' (e) 56° 12' (f) 6° 28'

Exercise 10f (page 96)

1 (a) obtuse (b) obtuse (c) acute

(d) reflex (e) reflex (f) acute

(g) acute (h) obtuse (i) acute

(j) obtuse (k) reflex (l) obtuse

(m) reflex (n) acute (o) obtuse

2 (a) 20° (b) 130° (c) 270°

(d) 220° (really 218°) (e) 60° (really 57°)

(f) 330° (really 331°)

3 see answers to question 2 above

6 6', 12', 18', 24', 30', 36', 42', 48', 54'

7 (a) 0.1° (b) 0.8° (c) 0.25°

(d) 0.35° (e) 0.55° (f) 0.95°

8 (a) 18° 12' (b) 77° 42' (c) 45° 45'

(d) 67° 30' (e) 28° 40' (f) 32° 10'

9 (a) 51.6° (b) 13.9°

(c) 32.25° (d) 80.65°

10 the sum of the three angles should be 180°

Practice exercise P10.1 (page 97)

1 (a) 60° (b) 212° (c) 25°

(d) 110° (e) 145° (f) 242°

2 (a) 150° (b) 30° (c) 338°

(d) 120° (e) 76° (f) 90°

(g) 278° (h) 68° (i) 143°

(j) 16° (k) 159°

Exercise 11a (page 98)

1 5*a*   **2** 2*x*   **3** 8*b*

4 4*y*   **5** 5*c*   **6** 6*z*

7 3*p*   **8** *k*   **9** 10*q*

10 2*r*   **11** 0   **12** 12*x*

13 20*y*   **14** 11*z*   **15** 10*a*

Exercise 11b (page 98)

1. 8b
2. 0
3. 3d
4. 15e
5. 7f
6. 7g
7. 11h
8. 2j
9. 4k
10. 3m
11. 11t
12. 6u
13. 6v
14. 5w
15. 8x
16. 4y
17. 10z
18. 5n
19. 4p
20. 6q

Exercise 11c (page 99)

1. $5x + 7$
2. $3x + 2$
3. $11x + 8y$
4. $6x + 1$
5. $5x - 2$
6. $5x - 2y$
7. $10x + 2$
8. $x - 8$
9. $13y + 12x$
10. $y - x$
11. $14x + 17$
12. $3x + 9$
13. $11y - 3x$
14. $8y - 9x$
15. $8a + 13$
16. $3a + 8$
17. $11x + 5y$
18. $9x + 8y$
19. $3a + b$
20. $9m + 4$

Exercise 11d (page 99)

1. $9x$
2. $2x + 6y$
3. $3n$
4. $8k$
5. $6e$
6. $6g$
7. $15a - 2$
8. $31p - 9q$
9. $28f - 11$
10. $3c$
11. $7m + 11n$
12. $4a + 6b$
13. $3h + 6$
14. $7x + 5y - 4$
15. $4x + 9y$
16. $4a + 9b$
17. $f + 2g$
18. $2m + 5n + 9$
19. $a + 3b + 3c$
20. $4p + 3q + 4r + 3s$

Exercise 11e (page 100)

1. $4x$ cm
2. $x + y + z$, $\$(2x + 3y + 5z)$
3. $180 - 2n$
4. $\$(85 - 5x)$
5. $\frac{9}{10}k$ kg
6. $3x + 3y$
7. $\frac{1}{4}x + 5$, $\frac{3}{4}x - 5$
8. $(5n - 79)$ dollars
9. $\frac{2}{3}x - 2$
10. for: $3x + 4y$, against: $x + 4y$

Exercise 11f (page 101)

1. $5a$
2. $4x$
3. xy
4. xy
5. a^2
6. x^2
7. $6a$
8. $6a$
9. $6a$
10. $6a$
11. $28x$
12. $40n$
13. $2x^2$
14. $12y^2$
15. $32x^2$
16. $20ab$

17. $35pq$
18. $27ab$
19. $6x^2$
20. $8y^2$
21. $3x^2$
22. $2pq$
23. $6ab$
24. $7mn$
25. $12a^2$
26. $35n^2$
27. $30x^2$
28. $36n^2$
29. $50pq$
30. $20ab$
31. $28a^2b$
32. $33ab^2$
33. $18xy^2$
34. $30xy^2$
35. $14p^2q$
36. $24a^2b$

Exercise 11g (page 101)

3. $2a$
3. $6a$
3. $6x$
3. $3y$
3. $3x$
3. $4c$
3. $4x$
3. $7y$
3. $5y$
10. $\frac{x}{9}$
11. $\frac{d}{4}$
12. $\frac{x}{2}$
13. $7ab$
14. $16y$
15. $17m$
16. $2kl$
17. d
18. $\frac{1}{a}$
19. z
20. $3x$
21. $6x$
22. $7x^2$
23. $5x$
24. $4x$
25. $6a^2$
26. $11n$
27. $6x$
28. $9ab$
29. $4x$
30. $5q$

Exercise 11h (page 102)

1. (a) 5   (b) 27   (c) 4   (d) 1
   (e) 2   (f) 1   (g) 0   (h) 6
2. (a) 8   (b) 30   (c) 20   (d) 12
   (e) 9   (f) 60   (g) 7   (h) 17
   (i) 7   (j) 2   (k) $\frac{1}{3}$ (l) $\frac{1}{2}$

Exercise 11i (page 102)

1. 6
2. 5
3. 0
4. 22
5. 2
6. 24
7. 23
8. 35
9. 20
10. 10

Exercise 11j (page 102)

1. $11x$
2. $17p$
3. $70n$
4. $14b - 3$
5. $6 + 10m$
6. $32a$
7. $24y$
8. $7a$
9. $7x$
10. $4n$
11. $3 + 6y$
12. 12
13. $8a$
14. $4x - 2$
15. 1
16. $84a^2$
17. $17x$
18. $53x$
19. $10x$
20. $-28x$

Practice exercise P11.1 (page 103)

1. (a) $8x$ (b) 8
2. (a) $6x$ (b) 6
3. (a) $-4x$ (b) -4
4. (a) $-7x$ (b) -7
5. (a) $3x$ (b) 3
6. (a) $4\frac{1}{3}x$ (b) $4\frac{1}{3}$
7. (a) $3\frac{1}{2}x$ (b) $3\frac{1}{2}$
8. (a) $8\frac{1}{4}x$ (b) $8\frac{1}{4}$

Practice exercise P11.2 (page 103)

1. $2\frac{1}{2}a$
2. $-6\frac{3}{4}x$
3. $3\frac{1}{3}d$
4. $-2\frac{7}{8}c$
5. $2\frac{3}{8}b$
6. $2\frac{1}{5}x$
7. $8\frac{1}{3}x$
8. $3\frac{3}{8}y$
9. $2\frac{3}{10}t$
10. p
11. $-\frac{16}{3}u$
12. $1\frac{2}{3}x$

Practice exercise P11.3 (page 103)

1. $4a + 3$
2. $5a - b$
3. $-4c + 7 - 4b$
4. $-3k - 2 - 2n$
5. $\frac{3}{4}f - \frac{1}{3}g + 2$
6. $4x + y$
7. $3h - 3k + 2$

Practice exercise P11.4 (page 103)

1. $w = 7h$
2. $s = c + 45$
3. $l = c - 5$
4. $m = 10\,l$
5. $n = l \div 5$
6. $t = 90c$
7. $l = 20 - p$
8. $c = 5n$
9. $p = u - 35$
10. $a + c = 120$
11. $l = c + t$
12. $d = 4s$

Practice exercise P11.5 (page 104)

1. Total cost = $\$c$, number of books = b
   $c = 8b$
2. Total weight = w g, number of books = b
   $w = 250b$
3. Amount for each person = $\$a$, prize = $\$p$
   $a = p \div 10$
4. Length of train = t m, length of carriage = c m
   $t = 8c$
5. Sale price = $\$p$, usual price = $\$u$
   $p = u - 20$

Practice exercise P11.6 (page 104)

1. 8
2. 2
3. 16
4. 26
5. $1\frac{1}{2}$
6. $\frac{5}{6}$
7. $\frac{3}{7}$
8. $4\frac{1}{3}$

Practice exercise P11.7 (page 104)

1. -3
2. 6
3. -1
4. -24
5. $\frac{1}{16}$
6. -3
7. undefined
8. $3\frac{2}{3}$

Practice exercise P11.8 (page 104)

1. $3uv$
2. $5u^2$
3. $3y^2$
4. xy
5. $5uv$
6. $4uw - 3v$
7. ab
8. $-2bc$
9. $3a^2 + 3a$
10. $-4ac + 3a$
11. $yx^2 - 5x^2y$
12. $7xy^2 - xy$

Practice exercise P11.9 (page 104)

1. (a) $[22 + (31 \times 5)]$
   (b) 177
2. (a) $[44 + (95 \times 4) - 15]$
   (b) 409
3. (a) $[(4 \times 18) + (12 \times 2) \div 3]$
   (b) 32
4. (a) $[27 + (15 \times 12) + (8 \times 24) - 20]$
   (b) 377

Exercise 12a (page 105)

1. (a) 2.7 cm; 3.9 cm; 4.1 cm
   (b) 39°, 67°, 74°
   (c) 180°   (d) scalene acute-angled
2. (a) 2.8 cm; 2.8 cm; 2.8 cm
   (b) 60°, 60°, 60°
   (c) 180°   (d) equilateral
3. (a) 2.5 cm, 4.2 cm; 4.9 cm
   (b) 30°, 60°, 90°
   (c) 180°   (d) right-angled
4. (a) 2.4 cm; 3.8 cm; 5.0 cm
   (b) 28°, 49°, 103°   (c) 180°
   (d) scalene obtuse-angled
5. (a) 1.6 cm; 3.7 cm; 3.7 cm
   (b) 26°, 77°, 77°
   (c) 180°   (d) isosceles
6. (a) 3.1 cm; 3.1 cm; 4.5 cm
   (b) 45°, 45°, 90°
   (c) 180°   (d) isosceles right-angled

Exercise 12b (page 106)

1. AB = 16 mm
   BC = 28 mm
   AC = 32 mm
2. PQ = 49 mm
   QR = 18 mm
   PR = 53 mm
3. SU = 32 mm
   UT = 37 mm
   ST = 49 mm
   The longest side is always opposite the right angle.

Exercise 12c (page 106)

1. normally only one line of symmetry
2. two of the angles are the same
3. $A\widehat{C}M = B\widehat{C}M$, $A\widehat{M}C = B\widehat{M}C$, AM = BM
4. your sketch should show that $B\widehat{C}M = 20°$ and AM = 4 cm; you may also have found that $C\widehat{M}A = C\widehat{M}B = 90°$ and that $C\widehat{A}M = C\widehat{B}M = 70°$

⑤ in isosceles △ACB, the equal angles are $\widehat{A}$ and $\widehat{B}$; the order of the letters helps to show the symmetry of the triangle

⑥ △PQR is a right-angled isosceles triangle.

Exercise 12d (page 107)

② each angle is 60°

③ an equilateral triangle has 3 equal sides, an isosceles triangle has only 2 equal sides; similarly for angles; an equilateral triangle has 3 lines of symmetry, an isosceles triangle has only 1 line of symmetry; all equilateral triangles have the same shape but isosceles triangles can have different shapes

④ there are 3 lines of symmetry all passing through the same point.

⑤
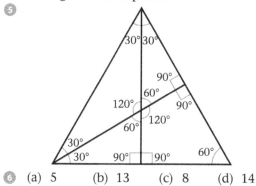

⑥ (a) 5   (b) 13   (c) 8   (d) 14

Exercise 12e (page 108)

① here are some rectangular shapes you may find in your classroom: door; door frame; window pane; window frame; ceiling; floor; wall; black-board; desk top; book cover; pencil box top; there will be many more

② 4   ③ 4   ④ 360°

⑤ (a) the diagonals are the same length
(b) the centre is the same distance from each corner

⑥ 2 acute angles and 2 obtuse angles

⑦ yes, opposite sides of a rectangle are parallel to each other

⑧ a rectangle generally has 2 axes of symmetry

Exercise 12f (page 109)

③ both diagonals of a square are the same length

④ the angles at the centre are all right angles

⑤ each diagonal makes an angle of 45° with the sides of the square

⑥ a square has 4 lines of symmetry

⑦ opposite sides equal; opposite sides parallel; diagonals of equal length; 4 angles of 90°; 4 sides

⑧ in a square, all 4 sides are the same length; in a square, the diagonals cross at right angles; in a square, there are 4 lines of symmetry; in a square, the diagonals meet the sides at an angle of 45°; none of these is true for a rectangle

Exercise 12g (page 109)

② 4

③ 4

⑤ (a) opposite sides are equal in length
(b) opposite angles are equal in size
(c) the angles should total 360°
(e) the diagonals are of different lengths
(f) each diagonal is bisected by the other
(g) 2 acute, 2 obtuse
(j) a parallelogram has no lines of symmetry

Exercise 12h (page 110)

② 4

③ 4

⑤ (a) all 4 sides are of equal length
(b) opposite angles are equal in size
(c) the angles should total 360°
(c) the diagonals are of different lengths
(f) each diagonal is bisected by the other
(g) right angles (90°)
(h) 2   (i) $a = b$ (j) $x = y$

⑥ opposite sides are parallel; opposite angles are equal; diagonals bisect each other

⑦ in a rhombus all 4 sides are equal in length, the diagonals cross at right angles and there are 2 lines of symmetry; none of these is true for a parallelogram

⑧ each angle of a square is a right angle; a square is a special kind of rhombus

Exercise 12i (page 111)

② 4 of each

④ (a) a kite has 2 pairs of sides of the same length

(b) one pair of opposite angles are of equal size

(c) 360°

(d) the diagonals are of different lengths

(e) right angles (90°)

(f) 1

⑤ their diagonals cross at right angles; they have bilateral symmetry

⑥ in a rhombus all 4 sides are equal in length, there are 2 pairs of opposite angles equal and there are 2 lines of symmetry; none of these is true for a kite; a rhombus is a special kind of kite

Exercise 12j (page 112)

① ACQP, ADRP, AEQP, BCQP, BDQP, BDRP are all trapeziums

② (a) rectangle   (b) trapezium
(c) parallelogram   (d) trapezium
(e) trapezium   (f) parallelogram

③ Yes, Fig. 12.28(d)

④ (a) $AD = BC$, $PQ = SR$
(b) $\widehat{A} = \widehat{B}$ and $\widehat{D} = \widehat{C}$, $\widehat{P} = \widehat{S}$ and $\widehat{Q} = \widehat{R}$

Exercise 12k (page 114)

② a circle can have as many lines of symmetry as you like; every diameter is a line of symmetry.

④ heart-shaped

⑤ (a) 12   (b) 1

Practice exercise P12.1 (page 144)

① (a) false   (b) false
(c) false   (d) true
(e) true   (f) true
(g) true   (h) true
(i) false   (j) true
(k) true   (l) false
(m) false   (n) true

② (a) parallelogram, trapezium, scalene triangle
(b) isosceles triangle, kite
(c) rectangle, rhombus
(d) equilateral triangle
(e) square
(f) circle

④ (a) rectangle, regular polygon
(b) square, equilateral triangle
(c) isosceles triangle
(d) equilateral triangle
(e) square
(f) trapezium
(g) isosceles trapezium
(h) parallelogram
(i) rhombus, square
(j) kite
(k) kite
(l) octagon
(m) decagon
(n) scalene triangle

⑤

Shape	Equal sides	Equal angles	Parallel sides	Lines of symmetry
Scalene triangle	none	none	none	none
Isosceles triangle	2	2	none	1
Equilateral triangle	3	3	none	3
Square	4	4	2 pairs	4
Rhombus	4	2 pairs	2 pairs	2
Kite	2 pairs	2	none	1

Exercise 13a (page 116)

① (a) (i) 19 000   (ii) 18 600   (iii) 18 620
(b) (i) 25 000   (ii) 25 200   (iii) 25 250
(c) (i) 33 000   (ii) 32 800   (iii) 32 780
(d) (i) 7000   (ii) 7200   (iii) 7160
(e) (i) 3000   (ii) 3000   (iii) 2970
(f) (i) 9000   (ii) 9500   (iii) 9480
(g) (i) 15 000   (ii) 14 900   (iii) 14 940
(h) (i) 27 000   (ii) 26 900   (iii) 26 890
(i) (i) 45 000   (ii) 45 100   (iii) 45 070
(j) (i) 10 000   (ii) 9900   (iii) 9900
(k) (i) 30 000   (ii) 30 100   (iii) 30 100
(l) (i) 8000   (ii) 8400   (iii) 8350

② (a) 350   (b) 380   (c) 700
(d) 710   (e) 720   (f) 1000

Exercise 13b (page 116)

1. (a) 7   (b) 12   (c) 79
   (d) 30   (e) 10   (f) 99
2. (a) 0.7   (b) 0.1   (c) 0.2
   (d) 0.5   (e) 1.0   (f) 0.0
3. (a) 0.16   (b) 0.17   (c) 0.71
   (d) 0.70   (e) 0.30   (f) 0.40
4. (a) (i) 1   (ii) 1.4
   (b) (i) 4   (ii) 4.1
   (c) (i) 10   (ii) 9.7
5. (a) (i) 0.4   (ii) 0.37
   (b) (i) 0.6   (ii) 0.63
   (c) (i) 0.2   (ii) 0.16
6. (a) (i) 30   (ii) 26   (iii) 26.5
   (b) (i) 10   (ii) 8   (iii) 8.4
   (c) (i) 10   (ii) 6   (iii) 5.8
7. (a) 1   (b) 0.8
   (c) 0.80
8. (a) 100   (b) 70
   (c) 70   (d) 69.6

Exercise 13c (page 117)

1. (a)   1500 to 2499
   (b)   6500 to 7499
   (c)   14 500 to 15 499
   (d)   22 500 to 23 499
2. (a)   250 to 349
   (b)   1150 to 1249
   (c)   750 to 849
   (d)   2650 to 2749
3. (a)   265 to 274
   (b)   385 to 394
   (c)   1535 to 1544
   (d)   715 to 724
4. (a)   642.5 to 643.4
   (b)   27.5 to 28.4
   (c)   1140.5 to 1141.4
   (d)   728.5 to 729.4
5. (a)   4.55 to 4.64
   (b)   14.75 to 14.84
   (c)   261.65 to 261.74
   (d)   325.45 to 325.54
6. (a)   2.305 to 2.314
   (b)   4.625 to 4.634
   (c)   0.285 to 0.294
   (d)   21.455 to 21.464

Exercise 13d (page 118)

1. (a) 12.9   (b) 12.93   (c) 12.935
2. (a) 24.1   (b) 24.12   (c) 24.118
3. (a) 5.1   (b) 5.07   (c) 5.073
4. (a) 1.9   (b) 1.94   (c) 1.938
5. (a) 2.0   (b) 1.99   (c) 1.988
6. (a) 0.6   (b) 0.58   (c) 0.585
7. (a) 0.1   (b) 0.06   (c) 0.063
8. (a) 0.0   (b) 0.04   (c) 0.038
9. (a) 0.9   (b) 0.90   (c) 0.900
10. (a) 0.0   (b) 0.01   (c) 0.009

Exercise 13e (page 118)

1. (a) (i) 7000   (ii) 7300   (iii) 7280
   (b) (i) 6000   (ii) 6000   (iii) 6040
   (c) (i) 10 000   (ii) 15 000   (iii) 14 600
   (d) (i) 4000   (ii) 3600   (iii) 3600
   (e) (i) 8000   (ii) 8000   (iii) 8010
   (f) (i) 5000   (ii) 5100   (iii) 5050
   (g) (i) 30 000   (ii) 28 000   (iii) 28 300
   (h) (i) 10 000   (ii) 9900   (iii) 9850
   (i) (i) 9000   (ii) 9400   (iii) 9400
   (j) (i) 30 000   (ii) 26 000   (iii) 26 000
2. (a) (i) 7   (ii) 7.0   (iii) 7.04   (iv) 7.038
   (b) (i) 20   (ii) 19   (iii) 18.5   (iv) 18.50
   (c) (i) 10   (ii) 13   (iii) 12.7   (iv) 12.68
   (d) (i) 4   (ii) 3.8   (iii) 3.80   (iv) 3.800
   (e) (i) 200   (ii) 230   (iii) 234   (iv) 234.1
3. (a) (i) 0.07   (ii) 0.068   (iii) 0.067 5
   (b) (i) 0.3   (ii) 0.31   (iii) 0.306
   (c) (i) 0.006   (ii) 0.006 3   (iii) 0.006 31
   (d) (i) 0.000 7   (ii) 0.000 67   (iii) 0.000 667
   (e) (i) 0.03   (ii) 0.034   (iii) 0.033 6

Exercise 13f (page 119)

1. (a) $600 000, $60
   (b) $10 000, $100
2. 39 years (there is a tendency to round down ages)
3. $184 million, $180 million, $200 million
4. between $21 750 000 and $21 840 000
5. (a) $11.4 million or $11 million
   (b) $9.2 million or $9 million
   (c) $25 million
   (d) $22 million
   (e) $40 million

Answers

Exercise 13g (page 119)

1 in order: degree, °C, gram, minute or second
2 (a) 185 mm   (b) 166 cm
   (c) 6380 km   (d) 985 m
   (e) 8850 m   (f) 14.3 km
   (g) 67.9 kg   (h) 6.63 g
   (i) 2.87 kg   (j) 7.64 t
   (k) 125 g   (l) 11.4 h
   (m) 40.1 litres   (n) $51 200
   (o) 26.2 °C

3

country	33 000 000
large city	1 300 000
large town	390 000
small town	8000
village	

Exercise 13h (page 120)

Rough answers from rounded numbers:
1 $20 + 20 = 40$
2 $70 - 50 = 20$
3 $20 \times 40 = 800$
4 $600 \div 20 = 30$
5 $40 + 50 = 90$
6 $700 - 300 = 400$
7 $100 \times 100 = 10\ 000$
8 $200 \div 20 = 10$
9 $50 + $60 = $110
10 $50 - $20 = $30
11 $10 \times 6 = $60
12 $40\ kg \div 8 = 5\ kg$
13 $100\ g + 500\ g = 600\ g$
14 $900\ m - 500\ m = 400\ m$
15 $50 \times 20 = $1000
16 $700 \div 20 = $35
17 $6 + 4 = 10$
18 $13 - $5 = $8
19 $3 \times 6 = 18$
20 $14 \div 7 = 2$
21 $3 + 13 = 16$
22 $6 - 3 = 3$
23 $17 \times 1 = 17$
24 $16 \div 4 = 4$
25 $7 + 5 = 12$
26 $8 - 3 = 5$
27 $8 \times 8 = 64$

28 $10 \div 4 = 2\frac{1}{2}$
29 $1 + 2 = 3$
30 $5 + $1 = $6
31 $18 - 7 = 11$
32 $3 \times 5 = $15
33 $0.4 \times 0.9 = 0.36$
34 $0.08 \times 0.03 = 0.002\ 4$
35 $0.3 \times 0.7 = 0.21$
36 $0.05 \times 0.02 = 0.001$
37 $0.6 \div 0.03 = 20$
38 $0.5 \div 0.1 = 5$
39 $0.07 \div 0.3 = 0.23$
40 $0.1 \div 0.03 = 3\frac{1}{3}$

Exercise 13i (page 121)

1 Compare the following accurate answers
with the estimates to the corresponding
questions in Exercise 13h.
   5 89   10 $34   15 $954   20 2.1
   25 $11\frac{7}{8}$ 30 $5.61   35 0.222   40 3.3
2 $262 000
3 33 (approx.)
4 $400 \times 9 = $3 600, $652
5 204 000
6 estimate: $40 \times 10 = 400$
   accurate: $36 \times 12 = 432$
7 (e)
8 (a) no
   (b) $4 \times 8 = $32
   (c) he forgot to change cents to dollars.
9 estimate $5 \times 300 = $1 500
10 estimate: $3000\ km \div 6 = 500\ km$
   accurate: 484 km

Exercise 13j (page 122)

7 (e) only

Practice exercises P13.1 (page 123)

1 (a) 23   (b) 113   (c) 29
   (d) 19   (e) 20   (f) 121
   (g) 95   (h) 37
2 (a) 30   (b) 50   (c) 100
   (d) 230
3 (a) 400   (b) 200   (c) 300
   (d) 900
4 (a) 6000   (b) 4000   (c) 8000
   (d) 9000

5 (a)

Number	Nearest 10	Nearest 100	Nearest 1000
2251	2250	2300	2000
16 235	16 240	16 200	16 000
841	840	800	1000
25	30	0	0

(b)

Country	Rounded to nearest	
	10	1000
Japan	56 650	57 000
Germany	2 104 490	2 104 000
Italy	1 531 580	1 532 000
Belgium	361 010	361 000
Denmark	67 500	67 000
U.S.A.	17 028 650	17 029 000

Country	Rounded to nearest	
	10 000	1 000 000
Japan	60 000	0
Germany	2 100 000	2 000 000
Italy	1 530 000	2 000 000
Belgium	360 000	0
Denmark	70 000	0
U.S.A.	17 030 000	17 000 000

Practice exercise P13.2 (page 124)

1 (a) 310 000 000   (b) 6600
(c) 1 200 000   (d) 260
2 (a) 13.45   (b) 2.92
(c) 3.16   (d) 3.76
3

	(a)			(b)		
	Powers			Roots		
x	x^2	x^3	x^6	$\sqrt{\ }$	$\sqrt[3]{\ }$	$\sqrt[5]{\ }$
12	144	1728	2 985 984	3.46	2.29	1.64
15	225	3375	11 390 625	3.87	2.47	1.72
20	400	8000	64 000 000	4.47	2.71	1.82

4 (a) 3.3   (b) 12.6   (c) 6.2
(d) 8.4   (e) 11.1   (f) 34.6
(g) 13.0   (h) 32.6

5 (a) 3.22   (b) 5.63   (c) 14.33
(d) 25.28   (e) 24.30   (f) 0.03
(g) 0.01   (h) 10.89

6

Element	Rounded to			
	Nearest whole number	1 d.p.	2 d.p	3 d.p.
Aluminium	27	27.0	26.98	26.982
Chromium	52	52.0	52.00	51.996
Helium	4	4.0	4.00	4.003
Lanthanum	139	138.9	138.91	138.906
Scandium	45	45.0	44.96	44.956

7 (a) 60   (b) 240 000
(c) 0.000 56   (d) 9.00

Practice exercise P13.3 (page 124)

1 (a) $124.33   (b) $1269.90
(c) $256.38   (d) $78.24
(e) $456.91   (f) $220
2 (a) 259.6   (b) 0.006
(c) 95.0   (d) 0.000 301
3 (a) more than 300 000 bricks, 1 s.f..
(b) more than 500 000 000 000 bytes, 1 s.f.
(c) less than 0.5 litres, 1 s.f..
(d) $0.015, 2 s.f.
4

	AGRIPRO	INSURCOM	CARITECH
Total annual sales ($) Rounded to 2 s.f.	78 000 000	35 000 000	600 000 000
% of sales to charity Rounded to 2 d.p.	0.11	0.04	0.12
Number of employees Rounded to 2 s.f.	77	940	1100
Annual profit ($millions) Rounded to 2 d.p.	6.10	11.29	743.60

Answers

⑤ answers will vary, for example,
(a) 9.5213, 9.57   (b) 3444, 3495
⑥ (a) 0.028 (b)   7700
(c) 41 420 000
(d) 0.01   (e)   32 100
⑦ (a) 20   (b) 20   (c) 200
(d) 4000   (e) 70   (f) 0.03
(g) 0.006 (h) 10

Practice exercise P13.4 (page 125)

① (a) wrong   (b) correct   (c) correct
(d) wrong   (e) correct   (f) wrong
(g) correct (h) wrong
② (a) (i) 436.38   (b) (ii) 361.43
(c) (i) 93.08   (d) (ii) 110.51
(e) (ii) 4.04   (f) (ii) 3.24
③ (a) $4.90   (b) 4.5 m
(c) 4 200 000 000 megabytes
(d) $5200   (e) $14 000
④ (i) wrong, estimate = 2350 × 200 = 470 000
(ii) correct, estimate = 100 000 ÷ 400 = 250
(iii) correct, estimate = 200^3 = 8 000 000
(iv) wrong, estimate = 0.09 ÷ 0.2 = 0.45
⑤ (i) 90% of $30, $27
(ii) 6% of 10 m, 0.6 m
(iii) 30% of 40 kg, 12 kg
(iv) 60% of 0.005 m², 0.003 m²
(v) 0.3% of $5 000 000, $15 000 000
⑥ (a) 6561   (b) 6.192
(c) 6.083   (d) 0.707

Exercise 14a (page 128)

① (a) 77 mm   (b) 124 mm   (c) 69 mm
(d) 88 mm
② (a) 5.4 cm   (b) 8.4 cm   (c) 8.5 cm
③ The following measurements are only
approximate:
(a) 75 mm   (b) 75 mm
(c) 63 mm   (d) 126 mm

Exercise 14b (page 129)

① (a) 10 cm   (b) 4 cm   (c) 52 mm
(d) 7 km   (e) 6 m   (f) $25\frac{1}{2}$ cm
(g) 16.6 cm (h) 10.2 cm (i) 7.15 m
(j) 9 km
② (a) 24 km   (b) 56 cm   (c) 9 mm
(d) 33.2 cm (e) 10 cm   (f) 90 m
(g) 4.5 mm   (h) 3.6 cm

② (a) 1900 m   (b) 1.9 km
③ 300 m
④ $16.80
⑤ length = 9 m; breadth = 4 m;
perimeter = 26 m; although each rectangle
is made from the same number of tiles, it
will be found that their perimeters are all
different
⑥ (a) 22 cm   (b) 28 m
(c) 34 m   (d) 42 m

Exercise 14c (page 130)

② (a) the circumference is greater than the
diameter
(b) just over three times as great
(c) it should be true for all three objects
(d) if everyone has worked carefully, you
should all get similar results

Exercise 14d (page 131)

① (a) 7 m   14 m   44 m
(b) $3\frac{1}{2}$ cm   7 cm   22 cm
(c) 14 mm   28 mm   88 mm
(d) 21 m   42 m   132 m
(e) 350 mm   700 mm   2200 mm
(f) 2.8 cm   5.6 cm   17.6 cm
② 440 m
③ (a) 36 cm   (b) 40 cm
(c) 32 cm   (d) 44 cm
④ 44 mm
⑤ 66 cm; 33 cm; 23 cm
⑥ 31.4 m
⑦ (a) 28 cm   (b) 22 cm   (c) 50 cm
⑧ 7850 cm (78.5 m)
⑨ 198 m
⑩ 31 m

Practice exercise P14.1 (page 133)

① (a) 113 m   (b) 1017 m
② 540 cm, 5.4 m
③ 25.1 m
④ (a) 420 cm   (b) 440 cm
⑤ 36 m
⑥ 352 m
⑦ (a) 600 cm   (b) 390 cm
(c) 1980 cm   (d) 200 tiles
⑧ (a) 5.9 cm   (b) 37.1 cm

Exercise 15a (page 134)

1. true   2. true   3. false   4. false
5. true   6. false   7. true   8. true
9. false   10. true   11. true   12. true
13. false   14. true   15. false   16. false
17. true   18. false   19. true   20. false

Exercise 15b (page 134)

1. $x = 9$ 2. $x = 3$ 3. $y = 14$
4. $y = 10$ 5. $p = 20$ 6. $p = 11$
7. $q = 16$ 8. $q = 9$ 9. $m = 3$
10. $m = 1$ 11. $n = 3$ 12. $n = 5$
13. $x = 23$ 14. $x = 14$ 15. $c = 19$
16. $c = 30$ 17. $b = 14$ 18. $b = 29$
19. $a = 1$ 20. $a = 1$ 21. $q = 8$
22. $q = 4$ 23. $e = 15$ 24. $e = 24$
25. $d = 7$ 26. $d = 27$ 27. $x = 4$
28. $y = 6$ 29. $b = 9$ 30. $c = 8$
31. $e = 1$ 32. $f = 4$ 33. $j = 8$
34. $m = 6$ 35. $p = 14$

Exercise 15c (page 135)

1. $x = 4$ 2. $x = 5$ 3. $x = 8$
4. $x = 6$ 5. $x = 3$ 6. $x = 3$
7. $x = 17$ 8. $x = 11$ 9. $x = 5$
10. $x = 25$ 11. $x = 4$ 12. $x = 8$
13. $x = 10$ 14. $x = 36$ 15. $x = 4$
16. $x = 8$ 17. $x = 6$ 18. $x = 1$
19. $x = 11$ 20. $x = 24$ 21. $x = 1$
22. $x = 6$ 23. $x = 0$ 24. $x = 3$

Exercise 15d (page 136)

1. $x = 7$ 2. $x = 4$ 3. $x = 4$
4. $x = 3$ 5. $x = 5$ 6. $x = 6$
7. $x = 10$ 8. $x = 6$ 9. $x = 10$
10. $x = 36$ 11. $x = 6$ 12. $x = 30$
13. $x = 21$ 14. $x = 4$ 15. $x = 20$
16. $x = 6$ 17. $x = 7$ 18. $x = 6$
19. $x = 2$ 20. $x = 6$

Exercise 15e (page 137)

1. $y = 3$ 2. $a = 3$ 3. $x = 4$
4. $p = 3$ 5. $n = 4$ 6. $m = 10$
7. $t = 4$ 8. $x = 2$ 9. $a = 6$
10. $y = 3$ 11. $d = 7$ 12. $q = 3$

13. $b = 2$ 14. $a = 10$ 15. $a = 1$
16. $x = 2$ 17. $a = 6$ 18. $x = 3$
19. $x = 3$ 20. $x = 0$

Exercise 15f (page 138)

1. $x = 4\frac{1}{3}$ 2. $x = 1\frac{5}{6}$ 3. $x = \frac{3}{5}$
4. $x = \frac{3}{4}$ 5. $x = 3\frac{1}{2}$ 6. $x = \frac{6}{7}$
7. $x = 1\frac{4}{9}$ 8. $x = \frac{1}{8}$ 9. $x = \frac{4}{5}$
10. $x = 1\frac{1}{2}$

Exercise 15g (page 139)

1. $n = 4$ 2. $y = 11$ 3. $c = 6$
4. $b = 3$ 5. $x = 12$ 6. $s = 4$
7. $u = 11$ 8. $m = 4$ 9. $p = 6$
10. $a = 1$ 11. $t = 2$ 12. $f = 5$
13. $h = 8$ 14. $y = 5$ 15. $d = 7$
16. $k = 2$ 17. $q = 3$ 18. $n = 8$
19. $t = 5\frac{1}{2}$ 20. $a = 1\frac{3}{5}$ 21. $z = 2\frac{1}{3}$
22. $x = 2\frac{1}{2}$ 23. $y = \frac{4}{9}$ 24. $b = 10\frac{1}{3}$
25. $h = \frac{2}{3}$ 26. $k = \frac{7}{10}$ 27. $m = \frac{1}{4}$
28. $z = 1\frac{5}{6}$ 29. $y = 1\frac{5}{11}$ 30. $y = 2\frac{1}{2}$

Practice exercise P15.1 (page 140)

1. 6   2. 10   3. 10   4. 15
5. 6   6. 24   7. 23   8. 6
9. 12   10. 16

Practice exercise P15.2 (page 140)

1. 4   2. 38   3. 6   4. 35
5. 6   6. 2   7. 4   8. 36

Practice exercise P15.3 (page 140)

1. 50   2. 252   3. 6   4. 4
5. 8   6. 8   7. 13   8. 13
9. 84   10. 144

Practice exercise P15.4 (page 140)

1. -1 2. $-3\frac{3}{4}$ 3. 6   4. -3
5. 2   6. $\frac{1}{4}$ 7. $-3\frac{3}{5}$ 8. $-\frac{1}{4}$
9. $\frac{1}{2}$ 10. 25

Revision exercise 4 (page 141)

1. (a) 3460   (b) 8000
   (c) 0.0515   (d) 0.247

② (a) 15.74   (b) 2.34
   (c) 4   (d) 0.024
   (e) 0.06   (f) 90 000
③ 30
④ $338 000
⑤ (a) $\frac{11}{40}$ (b) $27\frac{1}{2}\%$
⑥ 54
⑦ 15%
⑧ (a) $\frac{1}{4}$ (b) $\frac{7}{25}$ (c) $\frac{11}{20}$ (d) $\frac{5}{8}$
⑨ 20%
⑩ $558.00

Revision test 4 (page 141)

① A   ② C   ③ D
④ C   ⑤ D
⑥ (a) $0.8\dot{3}$ (b) $0.\dot{5}$ (c) $0.4\dot{5}$ (d) $0.41\dot{6}$
⑦ 30%
⑧ (a) $\frac{3}{10}$ (b) $\frac{7}{40}$ (c) $\frac{3}{5}$ (d) $\frac{3}{40}$
⑨ $4.50
⑩ (a) 40c   (b) 16 m   (c) 15 g   (d) $30

Revision exercise 5 (page 141)

① (a) 90   (b) 60   (c) 900   (d) 405
② (a) reflex   (b) acute
   (c) obtuse   (d) acute
   (e) obtuse   (f) reflex
③ $A\widehat{O}B = 34°$, $B\widehat{O}C = 56°$
⑤ $B\widehat{O}C = 114°$
⑥ 135°
⑦ (a) 2   (b) 4   (c) 3   (d) 1
⑧ 10 cm
⑨ 26 cm
⑩ (a) kite   (b) kite
   (c) square   (d) rhombus
   (e) trapezium   (f) kite
   (g) trapezium   (h) rectangle

Revision test 5 (page 142)

① D   ② C   ③ B
④ C   ⑤ C   ⑥ 125°
⑧ trapezium, parallelogram, rectangle, rhombus, square
⑨ (a) isosceles
   (b) $A\widehat{B}C = 58°$, $A\widehat{C}B = 64°$
⑩ any three of: trapezium, parallelogram, rhombus, kite

Revision exercise 6 (page 143)

① (a) $9a$ (b) $6x$ (c) $2n$ (d) $20x$
   (e) $8y$ (f) $4c$
② (a) $10n - 5$ (b) $5 - 6x$ (c) $2a + 6b$
③ (a) 6   (b) 5   (c) 7   (d) 4
④ (a) 5   (b) 4   (c) $5x$ (d) $-13x$
⑤ (a) false   (b) true
⑥ (a) $x = 7$ (b) $x = 4\frac{1}{2}$ (c) $c = 3$
   (d) $x = 100$ (e) $x = 4$ (f) $w = 20$
⑦ (a) $x = 6$ (b) $x = 16$ (c) $x = 12$
   (d) $a = 6$ (e) $b = 4$ (f) $c = 6$
⑧ (a) $x = 4\frac{1}{2}$ (b) $x = 1\frac{1}{3}$ (c) $x = 1\frac{2}{5}$
⑨ (a) $(l + 0.5)$ m   (b) $(4l + 2)$ m
⑩ (a) $4l + 2 = 10$ (b) $l = 2$

Revision test 6 (page 144)

① C   ② D   ③ A   ④ A   ⑤ B
⑥ (a) -2 (b) 1   (c) 2
⑦ (a) $2a$ (b) $3x - 7y$
⑧ (a) $27xy^2$ (b) $10an^3$
   (c) $3a$ (d) $5x$
⑨ (a) $x = 3$ (b) $a = 9$
   (c) $y = 12$ (d) $n = 5$
   (e) $d = 7$ (f) $x = 1$
⑩ (a) $+ 3$
   (b) $+ 3 = 14$, $n = 2$ (c) 7

Revision exercise 7 (page 144)

① (a) 30 000   (b) 29 800   (c) 29 840
② (a) 0.8   (b) 0.85   (c) 1
③ (a) $9 \times 5 = 45$
   (b) $10 \times 6 = 60$
   (c) $0.8 \div 0.4 = 2$
④ $500 \times 400 = $200 000
⑤ $100 \text{ kg} \div 8 = 12\frac{1}{2}$ kg
⑥ 2.94
⑦ (a) 1 cm   (b) 6 cm   (c) 9 cm
⑧ 3.14 m
⑨ 500 times
⑩ (a) 145 to 154   (b) 10.25 cm to 10.34 cm

Revision test 7 (page 144)

① A   ② C   ③ C   ④ B   ⑤ C
⑥ 5.48 m
⑦ $20 000 \times 20 = $400 000
⑧ $600 \div 10 = $60
⑨ 0.3 m × 10 = 3 m
⑩ (a) 10   (b) 150 lines

General revision test B (page 145)

① D　②　D　③　B　④　B　⑤　D
⑥　A　⑦　B　⑧　C　⑨　A　⑩　C

⑪　(a)　$6m^2n$　(b)　$4x^2y$　(c)　$\dfrac{2a}{9}$

　　(d)　$8p$　(e)　$\dfrac{2s}{3t}$

⑫　942 cm

⑬　(a)　70　　　　(b)　600

⑭　(a)　0.44　　　(b)　44%

⑮　(a)　$a = 4$　(b)　$a = 5$　(c)　$a = 16$

　　(d)　$x = 1\frac{2}{3}$　(e)　$x = \frac{2}{5}$　(d)　$x = \frac{1}{4}$

⑯　748 m

⑰　about 40

⑱　(a)　0.505　(b)　0.50　(c)　0.5

⑲　$585.75

⑳　$\widehat{A} + \widehat{B} = 114°$

Exercise 16a (page 147)

①　(a)　12 squares　　　(b)　$7\frac{1}{2}$ squares
　　(c)　15 squares　　　(d)　4 squares
　　(e)　approx. 28 squares (f)　10 squares
　　(g)　6 squares　　　(h)　10 squares
　　(i)　6 squares　　　(j)　approx. 3 squares

Exercise 16b (page 148)

①　(a)　3 cm　　2 cm　　　6 cm²
　　(b)　5 m　　4 m　　　20 m²
　　(c)　6 m　　$2\frac{1}{2}$ m　　15 m²
　　(d)　5.2 m　　3 m　　　15.6 m²
　　(e)　2.1 cm　　3.4 cm　　7.14 cm²
　　(f)　4 m　　4.5 m　　　18 m²
　　(g)　5 m　　3 m　　　15 m²
　　(h)　4 m　　3 m　　　12 m²
②　(a)　6 cm　　36 cm²　　(b)　9 m　　81 m²
　　(c)　7 m　　49 m²　　(d)　4 cm　　16 cm²
　　(e)　1.2 m　1.44 m²　　(f)　$2\frac{1}{2}$ cm　$6\frac{1}{4}$ cm²

Exercise 16c (page 149)

②　(a)　60 m²　　(b)　72 m²　　(c)　16 m²
　　(d)　28 m²　　(e)　26 m²　　(f)　13 m²
③　(a)　18 m²　　(b)　25 m²　　(c)　21 m²
　　(d)　34 m²　　(e)　63 m²　　(f)　19 m²
④　3.44 m²　⑤　236 m²　⑥　132 m²
⑦　752 cm²　⑧　1008 cm²; 256 cm²
⑨　45 m²　⑩　3 litres
⑪　20 m²; 8 boards
⑫　perimeter = 16 units, area = 12 units²; the boy can also make rectangles 1 × 7, 3 × 5 and 4 × 4; all the perimeters are 16 units;

the areas are 7, 15 and 16 units²; this shows that shapes can have the same perimeters but different areas

Exercise 16d (page 151)

①　35 m²　　②　48 cm²　　③　33 cm²
④　15 cm²　　⑤　63 cm²　　⑥　3 cm
⑦　5 cm　　⑧　4 m　　　⑨　5 cm
⑩　6 cm
⑪　1.6 cm
⑫　(a)　40 cm²　　　　(b)　height = 5 cm
⑬　(a)　30 cm²　　　　(b)　base = 5 cm
⑭　(a)　36 cm²　　　　(b)　height = 4 cm
⑮　(a)　12 cm²　　　　(b)　base = 6 cm

Exercise 16e (page 153)

①　(a)　16 cm²　　(b)　9 cm²　　(c)　35 cm²
　　(d)　1.8 cm²　　(e)　12.5 cm²　(f)　7 cm²
　　(g)　6 cm²
②　(a)　30 cm²　　(b)　16 cm²　　(c)　18 cm²
　　(d)　20 cm²　　(e)　15 cm²　　(f)　26 cm²
　　(g)　40 cm²　　(h)　37.5 cm²

Exercise 16f (page 154)

①　(a)　15 cm²　　(b)　26 cm²　　(c)　40 cm²
　　(d)　$37\frac{1}{2}$ cm²　(e)　5.31 cm²
②　(a)　5　(b)　4　(c)　$6\frac{1}{2}$　(d)　7　(e)　$5\frac{1}{2}$

Exercise 16g (page 155)

①　(a)　280　(b)　120　(c)　238　(d)　630
　　(e)　952　(f)　252
②　(a)　135　(b)　$87.75
③　$7.20　④　187　⑤　14
⑥　1132 tiles　⑦　(a)　19 cm　(b)　1395

Exercise 16h (page 157)

①　(a)　7 m　　　14 m　　　154 m²
　　(b)　$3\frac{1}{2}$ cm　　7 cm　　　$38\frac{1}{2}$ cm²
　　(c)　140 mm　　280 mm　　61 600 mm²
　　(d)　14 m　　　28 m　　　616 m²
　　(e)　$1\frac{3}{4}$ cm　　$3\frac{1}{2}$ cm　　$9\frac{5}{8}$ cm²
　　(f)　2.1 cm　　4.2 cm　　13.86 cm²
②　(a)　$19\frac{1}{4}$ cm²　　　　(b)　$38\frac{1}{2}$ cm²
　　(c)　$7\frac{7}{32}$ cm² (7.2 cm² approx.)
③　radius = 10 cm;　area = 314 cm²
④　27.9 m²　　⑤　39.25 cm²
⑥　$\frac{1}{9}$

7 (a) 15 cm   (b) 5 cm   (c) 157 cm²

8 (a) 10 150 m²   (b) 400 m

Practice exercise P16.1 (page 158)

1 18 000 cm²

2 (a) 158 cm²   (b) 169 cm²
(c) 63 cm²

3 (i) 82 cm²
(ii) 182.25 cm²
(iii) 72 cm²

4 616 cm²

5 (a) 1.4 cm   (b) 6.16 cm²
(c) 14 cm   (d) 5.6 cm
(e) 61.6 cm²

6 (a) 12 cm
(b) 31 cm² to the nearest whole number

7 (a) 3600 cm²   (b) 1424.5 cm²
(c) 2175.5 cm²

8 (a) 173.25 cm²   (b) 113 cm²
(c) 257 cm²   (d) 190.9 cm²

9 (a) 75.46 m²   (b) 10.8 m
(c) 91.65 m²   (d) 16.19 m²

Exercise 17a (page 160)

1 (b) $A\widehat{C}D + B\widehat{C}D = 180°$
(c) the sum of the two angles should be 180°

2 (a) $A\widehat{O}C = D\widehat{O}B$ (b) $B\widehat{O}C = A\widehat{O}D$

3 (b) the sum of the angles should be 360°
(c) the sum of the angles at a point is 360°

Exercise 17b (page 161)

1 (a) $a = 130°$ (angles on a straight line)
(b) $b = 40°$ (angles on a straight line)
(c) $c = 115°$ (angles on a straight line)
$d = 65°$ (vertically opposite angles)
$e = 115°$ (vertically opposite angles)
(d) $f = 120°$ (angles at a point)
(e) $g = 140°$ (angles at a point)
(f) $h = 150°$ (adjacent angles on a straight line)
(g) $i = 55°$ (angles on a straight line)
$j = 142°$ (angles on a straight line)
(h) $k = 45°$ (angles on a straight line)
$l = 135°$ (vertically opposite to 100° and 35° given)
$m = 45°$ (vertically opposite to k found)

(i) $n = 34°$ (vertically opposite angles)
$p = 56°$ (angles on a straight line)
$q = 56°$ (vertically opposite to p)
$r = 90°$ (vertically opposite to given right angle)
(j) $s = 25°$ (angles at a point; $2s = 50°$)
(k) $t = 122°$ (angles on a straight line)
$v = 23°$ ($v + 35°$ vertically opposite to 58°)
$w = 122°$ (vertically opposite to t)
(l) $x = 90°$ (angles on a straight line)
$y = 62°$ ($y + 28°$ vertically opposite to right angle)
$z = 49°$ ($z + 41°$ vertically opposite to x)

2 $S\widehat{O}Q = 37°$ (vertically opposite to $P\widehat{O}R$)
$P\widehat{O}S = 143°$ (adjacent to $P\widehat{O}R$ on straight line)
$R\widehat{O}Q = 143°$ (vertically opposite to $P\widehat{O}S$)

3 $R\widehat{X}Q = 85°$ **4** $R\widehat{X}S = 47°$

5 $P\widehat{X}Q = 22°$

6 $C\widehat{X}D = 64°$, $A\widehat{X}B = 144°$, $A\widehat{X}C = 170°$

7 $2x + 2y = 180°$, $x + y = M\widehat{O}N = 90°$

8 $x + 2x + 3x + 4x = 360$, $x = 36$;
$E\widehat{K}F = 36°$, $F\widehat{K}G = 72°$, $G\widehat{K}H = 108°$,
$H\widehat{K}E = 144°$

9 $x + 2x + 6x = 180$, $x = 20$; $U\widehat{X}V = 20°$

Exercise 17c (page 163)

1 (a) DC, EF, WX, ZY
(b) WZ, DE, CF, PS, QR, HG
(c) FG (this includes the segments FS, FR, SR, SG, RG)
(d) yes, AD   (e) no

2 (a) 8 angles
(b) corresponding angles are equal
(c)

(d) alternate angles are equal
(e) alternate angles are equal

(a) CŶX, DŶQ, YX̂B, AX̂Y
(b) XŶC, YŶA

 (a) ŵ (b) k̂ (c) û (d) n̂
(e) ŵ (f) k̂ (g) t̂ (h) n̂

5

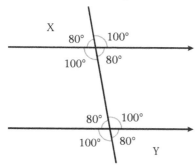

Exercise 17d (page 164)

(it is possible that you may have different but correct reasons)

1. (a) a = 110° (angles on a straight line)
   b = 70° (vertically opposite angles)
   c = 110° (vertically opposite to a)
   d = 110° (alternate to c)
   e = 70° (corresponding to b)
   f = 110° (corresponding to c)
   g = 70° (corresponding angles)
   (b) h = 40° (alternate angles)
   i = 140° (angles on a straight line)
   (c) j = 75° (angles on a straight line)
   k = 75° (corresponding angles)
   (d) l = 45° (alternate angles)
   m = 30° (alternate angles)
   (e) r = 113° (angles on a straight line)
   s = 67° (corresponding angles)
   t = 67° (alternate angles)
   u = 113° (alternate to r)
   v = 113° (corresponding to u)
   (f) w = 42° (alternate angles)
   x = 138° (angles on a straight line)
   y = 63° (angles on a straight line)
   z = 63° (alternate angles)

2

(a) 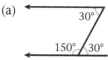 (b)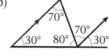

(c)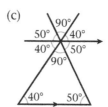

(d)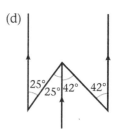

(e) 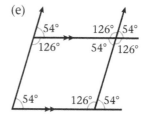 (f)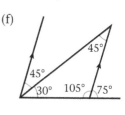

Exercise 17e (page 165)

1. (a) In △ABC, Â = 70°, B̂ = 70°, Ĉ = 40°
   In △PQR, P̂ = 42°, Q̂ = 63°, R̂ = 75°
   (b) 180°
   (c) 180°
   (d) in both cases the sum of the angles of the triangles was 180°
2. (c) the sum of the angles of both triangles was 180°
3. (d) the three angles should form a straight line along one edge; the sum of adjacent angles on a straight line is 180°

Exercise 17f (page 166)

1. (a) a = 90° (sum of angles of triangle is 180°)
   (b) b = 130° (sum of angles of triangle is 180°)
   (c) c = 60° (angles on a straight line)
   d = 80° (sum of angles of triangle is 180°)
   (d) e = 75° (base angles of isosceles triangle)
   f = 30° (sum of angles of triangle is 180°)
   (e) g = 40° (angles on a straight line)
   h = k = 70° (sum of angles of an isosceles triangle)
   (f) notice that the triangle is isosceles
   i = j = 25° (sum of angles of an isosceles triangle)
   (g) l = m = 65° (sum of angles of an isosceles triangle)
   p = n = 65° (alternate to l and m respectively)
   q = 50° (sum of angles of triangle is 180°)

(h) r = 25° (alternate angles)

s = 40° (s + 25° corresponding to 65°)

t = 115° (sum of angles of triangle is 180°)

u = 115° (angles on a straight line)

v = 40° (alternate to s)

(i) w = 25° (alternate angles)

x = 60° (sum of angles of triangle is 180°)

y = 110° (angles on a straight line)

z = 45° (sum of angles of triangle is 180°)

② (a) 68° (b) 79° (c) 106° (d) 120°
(e) 64° (f) 35° (g) 71° (h) 43°
(i) 60° (j) 90°

③

	A$\widehat{B}$C	B$\widehat{A}$C	A$\widehat{C}$B	A$\widehat{C}$D
(a)	58°	47°	75°	105°
(b)	65°	53°	62°	118°
(c)	67°	59°	54°	126°
(d)	46°	99°	35°	145°
(e)	43°	94°	43°	137°
(f)	60°	60°	60°	120°
(g)	35°	90°	55°	125°
(h)	58°	97°	25°	155°

④ $x + 2x + 3x = 180°$, $x = 30°$;
the angles are 30°, 60°, 90°

⑤ (a) 72°   (b) isosceles

⑥ (a)
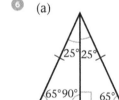

(b)
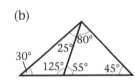

(c)
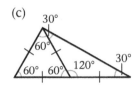

(d)
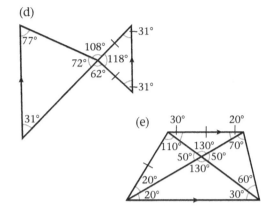

(e)

Exercise 17g (page 168)

① (a) $x = 140°$, $y = 100°$, $z = 120°$
(b) $p = 160°$, $q = 110°$, $r = 90°$
(c) $i = 125°$, $j = 135°$, $k = 100°$
(d) $l = 158°$, $m = 140°$, $n = 62°$

② (a) $x = 80°$, $y = 70°$, $z = 30°$
(b) $p = 40°$, $q = 78°$, $r = 62°$
(c) $i = 85°$, $j = 25°$, $k = 70°$
(d) $l = 16°$, $m = 43°$, $n = 121°$

③ in each case, exterior angle = sum of the opposite interior angles

④ in each case, sum of exterior angles = 360°

Exercise 17h (page 168)

① (a) m = 100° (exterior angle of triangle)
(b) n = 80° (exterior angle of triangle)
(c) p = 70° (exterior angle of triangle)
(d) q = 62° (base angle of isosceles triangle)
r = 124° (exterior angle of triangle)
(e) s = 65° (angles on a straight line)
t = 40° (angles on a straight line)
u = 105° (exterior angle of triangle)
or (sum of exterior angles of triangle)
(f) x = 51° (exterior angle of triangle)
y = 129° (angles on a straight line)
z = 123° (angles on a straight line)

② A$\widehat{B}$C = 40°

③ B$\widehat{A}$C = 65°, A$\widehat{C}$B = 72°
A$\widehat{C}$R = 108°, A$\widehat{B}$Q = 137°

④ $y = 42°$, $x = 48°$, $z = 52°$
$a = 57°$, $b = 75°$
$c = 20°$, $d = 70°$
$m = 47°$, $n = 133°$

⑤ $w = 75°$, $x = 65°$, $y = 26°$, $z = 38°$

Practice exercise P17.1 (page 170)

1. All except (i)
2. (a) 81° (b) 25° (c) 35° (d) 17°
   (e) 67.9°
3. (a) 168° (b) 151° (c) 117° (d) 139°
4. (a) false – the exterior angle of a triangle depends on the size of its adjacent interior angle.
   (c) false – adjacent angles on a straight line add up to 180°
   (g) false – the sum of the exterior angles of a triangle is 360°

Exercise 18a (page 171)

1. (a) Giantkillers (b) Locomotives
   (c) yes (d) 79
   (e) 58 (f) 13
   (g) 2 (h) $\frac{1}{4}$
   (i) nearly 2
   (j) 3 (to the nearest whole number)
   (k) 23 pts
   (l) 9 pts
2. (a) yes, from 901 students in 2001 to 1046 students in 2006
   (b) 145
   (c) 25 classes in 2001; 25 classes in 2002
   (d) 2003
   (e) 2007
   (f) about 1080
3. (a) bicycles
   (b) 63 (bicycles and motor bikes)
   (c) 4
   (d) motor bike
   (e) 0
   (f) no; all that can be said is that the girl did not see any buses among the 100 vehicles which passed her; some buses may travel on the road at other times.
   (g) possibly in a small village; we would expect more buses and taxis in a city
4. (a) rent (b) food (c) 3200
   (d) $939 (e) $6350 (f) month 3
   (g) month 2 (h) $\frac{1}{11}$
5. (a) 1st year: 3823, 2nd year: 3910
   (b) 1st year (c) 1725 (d) 800
6. (a) once a year
   (b) no
   (c) between year 13 and 14

Exercise 18c (page 175)

1. (a) 15 students (b) 87
   (c) 41 (d) 70
   (e) 4th (f) 9th equal
   (g) 12th (although he has the 10th best mark, there are 11 students in front of him)
   (h) 10 students (i) 14 students
   (j) 1 student (k) 64
2. (a) car: 5; lorry: 9; bus: 1; taxi: 2; others: 2
   (b) lorry
   (c) 19
   (d) motor bikes, bicycles, army vehicles
   (e) yes: 14 accidents to cars and lorries; only 5 accidents to all other vehicles.
3. (a) June
   (b) 300 mm
   (c) March
   (d) March, April
   (e) June, August, July
   (f) 2260 mm
4. (a) no
   (b) Universities
   (c) Primary Schools
   (d) there are far more students and teachers in Primary Schools than in Universities; there are many more Primary Schools than Universities
   (e) $\frac{1}{2}$ (f) $\frac{1}{4}$ (g) $\frac{1}{6}$

Exercise 18d (page 177)

1. (a) A, A, B, B, B, B, C, C, C, D
   (b) grade B
   (c) 6 students
   (d)

grade	A	B	C	D
frequency	2	4	3	1

   (e)

Grade A	🧍🧍
Grade B	🧍🧍🧍🧍
Grade C	🧍🧍🧍
Grade D	🧍

2 (a)

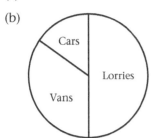

(b) sizes 12, 14 and 16

3 (a) 12 vehicles

(b)

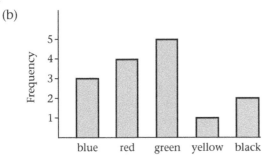

4 (a) green

(b)

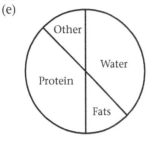

5 (a) $\frac{1}{8}$ (b) $\frac{3}{8}$
(c) 45°   (d) 135°
(e)

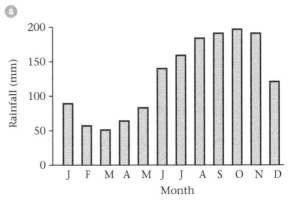

8

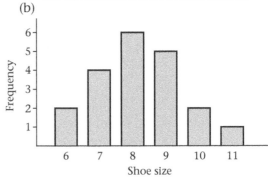

9 (a) industry   (b) $\frac{2}{9}$
(c) they are equal   (d) no
(e) advantage: gives a quick comparison between types of expenditure; disadvantage: does not show how much is being spent

10 (a) *Table A11*

size	6	7	8	9	10	11
frequency	2	4	6	5	2	1

(b)

(c) sizes 7, 8 and 9

Practice Exercise P18.1 (page 178)

1 (a) Brazil
2 (a) 40   (b) 27   (c) Grade C
4 (a) 35
(b) Highest mark 39, lowest mark 12
(c) 34   (d) 1
(e) 1   (f) 26
5 (a) 184   (b) $18.40
(c) Najier   (d) $1.60
6 (a) 2160
(b) August and September
(c) No

7 (a)

Collector	No. collected
Jan	14
Renee	11
Sydney	12
Kim	19

(b) 56   (c) Kim   (d) 14

Exercise 19a (page 182)

6

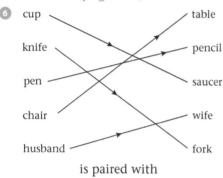

is paired with

Exercise 19b (page 182)

6 (a) (i)   (b) (iii)   (c) (ii)   (d) (i)

Practice exercise P19.1 (page 183)

1 many-to-one   **2** many-to-one
3 one-to-one   **4** many-to-many
5 one-to-many   **6** many-to-one

Practice exercise P19.2 (page 183)

1 one-to-one; is 1 more than
2 many-to-many; is a factor of

Practice exercise P19.3 (page 183)

1 $x \to 2x - 1$
$-1 \to -3$
$0 \to -1$
$2 \to 3$
$3 \to 5$

2 $x \to x^3$
$-3 \to -27$
$-1 \to -1$
$0 \to 0$
$1 \to 1$

3 $x \to 3x - 2$
$-3 \to -11$
$-1 \to -5$
$0 \to -2$
$1 \to 1$
$2 \to 4$

Practice exercise P19.4 (page 184)

1 (i) divided by 4
(ii) $x \to \dfrac{x}{4}$

2 (i) subtract 1
(ii) $x \to x - 1$

3 (i) square, then multiply by 2
(ii) $x \to 2x^2$

Practice exercise P19.5 (page 184)

Answers will vary in the naming of sets

(a) (i) U = {family};   P = {parents};
C = {children}; F = {females};
[or may name M = {males}, etc]

(ii) P = {Verna, Desmond}
C = {Richard, Andrew, Brian, Alanna, Genna}
F = {Verna, Alanna, Genna}

(iii)

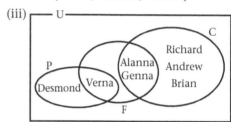

(b) (1) (i)

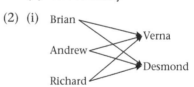

(ii) one to many

(2) (i)   Brian
Andrew
Richard
→ Verna
→ Desmond
(ii) many-to-many

(3) (i)   Brian
Andrew
Richard
→ Genna
→ Brian
→ Andrew
→ Richard
→ Alanna
(ii) many-to-many

(4) (i)

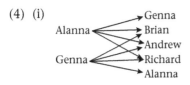

(ii) many-to-many

Practice exercise P19.6 (page 184)

① (a)

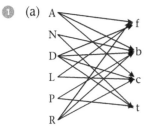

(b) many-to-many

② (a)

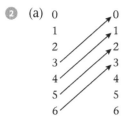

(b) many-to-many

③ (a)

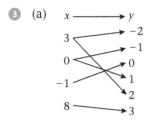

(b) many-to-many

④ (a)

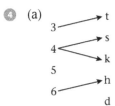

(b) many-to-many

Practice exercise P19.7 (page 184)

(a)

	Domain	Relation	Range
①	$\{1, 3, 5, 9\}$	$x \longrightarrow \dfrac{x}{3}$	$\{-4, -2, 0, 4\}$
②	$-3, -2, 0, 4\}$	$x \longrightarrow x - 5$	$\{4, -1, -5, 11\}$
③	$\{-3, -1, 1, 3\}$	$x \longrightarrow x^2 - 5$	$\{2, 18\}$
④	$\{0, 2, 4, 6\}$	$x \longrightarrow 2x^2$	$\{0, -2, -4, -6\}$
⑤	$\{-5, -1, 3, 2\}$	$x \longrightarrow -x$	$\{4\}$

(b)

①

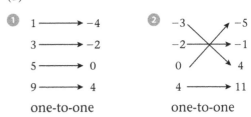

one-to-one

②

one-to-one

③

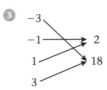

many-to-one

④

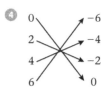

one-to-one

⑤

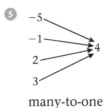

many-to-one

Exercise 20a (page 185)

①	1 : 2	②	2 : 3	③	5 : 4	④	4 : 3
⑤	2 : 3	⑥	4 : 1	⑦	1 : 10	⑧	2 : 5
⑨	2 : 3	⑩	3 : 2	⑪	4 : 3	⑫	5 : 3
⑬	2 : 7	⑭	1 : 7	⑮	3 : 4		

Exercise 20b (page 185)

①	8	②	8	③	24	④	24	⑤	10
⑥	30	⑦	6	⑧	6	⑨	80	⑩	27

Exercise 20c (page 186)

1. (a) $5, $10   (b) 10 kg, 16 kg
   (c) 16 cm, 4 cm   (d) $12, $18
   (e) 24 ml, 56 ml   (f) 18, 4

Exercise 20d (page 186)

1. (a) 35c, 63c,
   (b) 35 kg, 21 kg
   (c) 6.6 kg, 4.2 kg
   (d) 6.75 m, 16.2 m
2. 20 eggs, 40 eggs
3. Abel: 9 mangoes, Gino: 6 mangoes
4. $136.50, $91
5. $A:B = 11:9$

Exercise 20e (page 186)

1. 30
2. 6
3. $2.50
4. $1.60
5. 750 g
6. 140 m
7. 3.5 km
8. 4.5 litres
9. 3 h
10. $1.05

Exercise 20f (page 187)

1. $3:2$ (increase)
2. $3:4$ (decrease)
3. $5:8$ (decrease)
4. $8:3$ (increase)
5. $7:10$ (decrease)
6. $4:5$ (decrease)
7. $11:8$ (increase)
8. $3:4$ (decrease)
9. $4:3$ (increase)
10. $5:6$ (decrease)

Exercise 20g (page 187)

1. $5.40
2. 15
3. $45.90
4. 24
5. $172.80
6. 8
7. 20
8. 2 h 12 min
9. $20\frac{1}{4}$ m
10. $357

Exercise 20h (page 188)

1. $9.50
2. 68 km/h
3. 26.5 kg/m
4. 1.03 g/cm^3
5. 1300/km^3
6. $6.29
7. 18 litres
8. 950
9. (a) 120 km   (b) 2 h 24 min
10. 51 t/m
11. $10.40/h
12. 300/km^2
13. $393.60
14. 40 min
15. $76.95

Exercise 20i (page 189)

1. (i) pay for 5 days = $100
   (ii) direct   (iii) $20
2. (i) 6 notebooks cost 90c
   (ii) direct   (iii) 15c
3. (i) 5 men take 10 days
   (ii) inverse   (iii) 50 days
4. (i) a 2-yr-old boy has 6 sisters
   (ii) neither   (iii) impossible
5. (i) in 2 h a car travels 84 km
   (ii) direct   (iii) 42 km
6. (i) 15 cows have grass for 4 days
   (ii) inverse   (iii) 60 days
7. (i) temperature of 6 litres = 30 °C
   (ii) neither   (iii) 30 °C
8. (i) 9 bottles hold $4\frac{1}{2}$ litres
   (ii) direct   (iii) $\frac{1}{2}$ litre
9. (i) 9 people have water for 4 days
   (ii) inverse   (iii) 36 days
10. (i) 10 litres are used in 80 km
    (ii) direct   (iii) 8 km

Exercise 20j (page 190)

1.   1 (b) $40   (c) $440
     2 (b) 75c   (c) $6.60
     3 (b) 5 days   (c) 2 days
     5 (b) 126 km   (c) $42x$ km
     6 (b) 10 days   (c) $\frac{60}{y}$ days
     8 (b) $2\frac{1}{2}$ litres   (c) $\frac{1}{2}x$ litres
     9 (b) $4\frac{1}{2}$ days   (c) $\frac{36}{z}$ days
   10 (b) 40 km   (c) $8x$ km
2. $224
3. 4 days
4. $1.10
5. 6 journeys
6. impossible to say
7. 15 days
8. 27 buckets
9. 3 h 20 min
10. 1.9 cm

Practice exercise P20.1 (page 190)

1. (a) $3:8 = 9:24$ (b) $4:3 = 16:12$
2. (a) $3:5$ (b) $5:4$
3. (a) (i) 1.4 litres   (ii) 0.56 litres
   (b) (i) $8.61   (ii) $3.69
4. (a) $72:$18   (b) $300:$45:$15
   (c) $28.13:$39.38:$22.50
5. (a) $100:$150   (b) 10.5 m:4.5 m
   (c) $2\,l:2\,l:1\,l$ (d) 6 kg:9 kg:15 kg

Answers

Practice exercise P20.2 (page 190)

① $\frac{5}{3}:1$　　② $1:5$　　③ $1:20\,000$
④ $2.5:1$　　⑤ $200:1$　　⑥ $30:1$
⑦ $1.5:1$　　⑧ $2.5:1$　　⑨ $1:\frac{9}{7}$
⑩ $5:1$　　⑪ $1:\frac{11}{3}$　　⑫ $1:10$
⑬ $1:6$　　⑭ $2.8:1$

Practice exercise P20.3 (page 191)

① $300\,\text{m}$ to $0.375\,\text{km} = 4:5$
② $3:4:2 = 9:12:6$
③ $3:11:2 = \$6:\$22:\$4$
④ $15:6 = 2.5:1$

Practice exercise P20.4 (page 191)

① $32\,\text{cm}^2$　　② $\$87.50$　　③ $\$2015$
④ $16\,\text{h}$　　⑤ $6\,l$　　⑥ $360\,\text{cm}$

Practice exercise P20.5 (page 191)

① 16 pens
② 50 days
③ (a) 266 copies　　(b) 19 cartridges
④ (a) 32 km　　(b) 375 min
⑤ (a) 1.25 m　　(b) 0.96 m
⑥ (a) A: $\$1320$, B: $\$880$, C: $\$1760$
　(b) $5:2:3$
　(c) A: $\$1980$, B: $\$792$, C: $\$1188$
⑦ (a) $T - 10:6:9$, $A - 8:8:9$
　(b) $T - \frac{2}{5}:\frac{6}{25}:\frac{9}{25}$, $A - \frac{8}{25}:\frac{8}{25}:\frac{9}{25}$
　(c) Maths $-$ T: A $= 5:4$,
　Eng $-$ T: A $= 3:4$, Sc $-$ T: A $= 1:1$
⑧ (a) 11 kg, 16.5 kg, 5.5 kg　　(b) $4:4:3$
⑨ (a) $\$1200$　　(b) $\$3200$
　(c) $7:3$　　(d) J's $\$2240$, K's $\$960$

Exercise 21a (page 192)

② 2 lines of symmetry　④ (a), (c), (e), (h), (i)
⑤ protractor, 45° set square
⑥ the numbers of lines of symmetry are as follows
　(a) 2　　(b) 1　　(c) 2　　(d) 3
　(e) 1　　(f) 2　　(g) 1　　(h) 1
　(i) 3　　(j) 4　　(k) 5　　(l) 1
⑦ 4 lines of symmetry
⑧ (a)　　　(b)　　　(c)

Exercise 21b (page 195)

① (a), (b), (d), (f), (h), (i), (j), (k), (l)

②

Shape	Lines of symmetry	Order of symmetry
(a)	6	6
(b)	0	2
(c)	1	1
(d)	0	2
(e)	0	1
(f)	0	4
(g)	1	1
(h)	0	2
(i)	3	3
(j)	2	2
(k)	2	2
(l)	0	2

③

Shape	Has point symmetry only	Has line symmetry only	Both	Neither
Isosceles △		✓		
Equilateral △			✓	
Scalene △				✓
Rectangle			✓	
Square			✓	
Parallelogram	✓			
Rhombus			✓	
Kite		✓		
Trapezium				✓
Circle			✓	

④ no

⑤

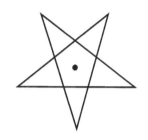

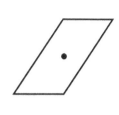

③ $17\frac{1}{2}\%$ ④ $24.70

⑤ $233\frac{1}{3}\%$ ⑥ $30

⑦ $24   ⑧ $277 200

⑨ the son $(7\frac{1}{2}\%)$ (mother's profit = 7%)

⑩ 12%, $26.10

⑪ $18.75, $24   ⑫ 20%

Exercise 22b (page 199)

① (a) $387   (b) $245   (c) $993.60
   (d) $69   (e) $23.96

② $10 500   ③ $1271.60

④ $1.20   ⑤ $4

⑥ $115, $35 $(23\frac{1}{3}\%)$

Exercise 22c (page 200)

① $4140

② (a) $10 500   (b) $31 500   (c) $2625

③ (a) $644   (b) $81.50

④ $1001

⑤ $21 529.80

⑥ $1336

Exercise 22d (page 201)

① (a) $92   (b) $276   (c) $333.50
   (d) $218.50   (e) $253

② (a) $90   (b) $135   (c) $157.50
   (d) $302.03   (e) $105.25

Exercise 22e (page 201)

① $11 980   ② $5238

③ $570   ④ $25.38

⑤ $56 800   ⑥ $3792.60

⑦ $316.32   ⑧ $3066

Exercise 22f (page 203)

① (a) (i) $110   (ii) $369.60
   (b) (i) $572   (ii) $1042.80
   (c) (i) $319   (ii) $682
   (d) (i) $1782   (ii) $2604.80
   (e) (i) $88   (ii) $330
   (f) (i) $862.84   (ii) $1431.89

② $820.60

③ (a) $5500   (b) $456   (c) $501.60
   (d) $8598.40

④ (a) $5400   (b) $257.40   (c) $603.55

⑤ $4540.25

Exercise 21c (page 196)

① (a), (b), (e), (f), (h)

② (a) uncountable number
   (b) uncountable number
   (c) 4
   (d) 1

⑤ 9

Practice exercise P21.1 (page 197)

① (a) A, B, H, M, W, X
   (b) H, X, Z
   (c) H, X

② (a) (i) twice   (ii) order 2
   (b) (i) four times   (ii) order 4
   (c) (i) three times   (ii) order 3

③ (a) (i) none   (ii) one
      (iii) two   (iv) four
   (b) (i) none   (ii) none
      (iii) order 3   (iv) order 2

④ (a) H, I, O, X   (b) H, I, N, S, X, Z

Exercise 22a (page 198)

① (a) (i) 80c   (ii) $4.80
   (b) (i) $1.50   (ii) $11.50
   (c) (i) 30c   (ii) $3.45
   (d) (i) 18c   (ii) $1.26
   (e) (i) 45c   (ii) $1.20

② (a) (i) 45c profit   (ii) 15%
   (b) (i) 45c profit   (ii) 25%
   (c) (i) $36 loss   (ii) 10%
   (d) (i) $36 profit   (ii) $37\frac{1}{2}\%$
   (e) (i) 63c loss   (ii) 15%

Practice exercise P22.1 (page 203)

①

	Cost price	Selling price	Profit	Loss
(a)	$2100	**$2420**	$320	X
(b)	**$63.45**	$44.95	X	$18.50
(c)	$120.00	$99.95	X	**$20.05**
(d)	**$2601.80**	$3290.00	$688.20	X
(e)	$84	**$56**	X	$28

② (a) $625
(b) $3125
③ Profit is $116; percentage profit is 29%
④ Profit on sale is $6000; selling price is $21 000
⑤ Percentage loss is 75%
⑥ Cost price was $4000
⑦ $6.90 per dozen
⑧ (a) $31 000
(b) $30 000
(c) Loss of $1000; $3\frac{7}{13}$%
⑨ $75
⑩ (a) 50
(b) $1450
(c) Profit of $250; 20.8%
⑪ (a) 12.5% loss
(b) 200 kg
⑫ Lucinda's car cost £8800; Jon's car cost $8500

Practice exercise P22.2 (page 204)

① $1080
② (a) 27.3%
(b) $2329
③ (a) $38
(b) $23.75
④

	Marked price	Selling price	Discount	VAT (17.5%)
(a)	$175	$155.75	**11%**	**$30.63**
(b)	$1998	$1458.54	27%	**$349.65**
(c)	**$120.63**	$110.98	8%	**$21.11**
(d)	**$512**	**$440.32**	14%	$89.60

⑤ 37.5%

⑥

	Marked price	Per cent discount	Amount of discount	Sales tax/VAT	Amount of Tax	Selling price
(a)	$150	20	$30	7%	$10.50	$130.50
(b)	$200	15	**$30**	8%	$16	**$186**
(c)	$1500	**30**	$450	8%	**$120**	$1170
(d)	$400	**6.25**	£25	**6.25%**	$25	**$400**
(e)	**$240**	10	**$24**	15	$36	**$252**

⑦ $385
⑧ $590
⑨ $718.58
⑩ (a) $1620 (b) $1530 (c) $1683
(d) $7650 (e) $6885 (f) $7695
(g) $13 500 (h) $13 356 (i) $13 482
It is cheaper to buy 10 desks than 9 because of the effect of the discount; 11 desks only cost $63 more than 9 desks.
It is cheaper to buy 51 desks than 50 desks because of the higher discount; 57 desks only cost $45 more than 50 desks.

Practice exercise P22.3 (page 205)

① $552.50
② Scheme A: bank percentage profit = 7.7%; $161 600 paid in instalments
Scheme B: bank percentage profit = 19%; $178 500 paid in instalments
③ (a) $9800
(b) $1401
(c) 16.7%
④ (a) $4095 (b) $163.80 per month
⑤ (a) $30 000
(b) $270 000
(c) $673 920
(d) $703 920

Practice exercise P22.4 (page 206)

① (a) $1400
(b) $1837.50
(c) $1435
(d) $1653.75
② (a) $4539.60 (b) $16 334.50
③ Trinidad and Tobago; take home pay greater by TT$4500

Revision exercise 8 (page 207)

1. 36 cm, 80 cm²
2. 54 cm²
3. 14 cm²
4. 72°
5. $a = 60°$, $b = 44°$, $c = 136°$
6. $a = 62°$, $b = 62°$, $c = 35°$, $d = 35°$, $e = 118°$
7. (a) G   (b) E   (c) BC   (d) △GFE
8. (a) B and H, C and G or D and F
   (b) △BAE, △HAE
   (c) △EBH
   (d) EDAF, ECAG
9. (a) 64°   (b) 118°
10. 20.2 cm

Revision test 8 (page 207)

1. C
2. B
3. C
4. C
5. B
6. 36 cm², $h = 9$ cm
7. $a = 118°$, $b = 118°$, $c = 112°$
8. $a = 70°$, $b = 38°$, $c = 72°$, $d = 38°$
9. 98 cm²
10. $A\widehat{B}D = B\widehat{D}C = 48°$, $D\widehat{A}C = A\widehat{C}B = 18°$
    $A\widehat{C}D = C\widehat{A}B = 23°$, $A\widehat{D}B = D\widehat{B}C = 91°$
    $B\widehat{O}C = D\widehat{O}A = 71°$, $C\widehat{O}D = A\widehat{O}B = 109°$

Revision exercise 9 (page 208)

1. (a) 15 km, 10 km, 9 km, 8 km, 6 km, 5 km,
      4 km, 2 km, 1 km
   (b) 6 km

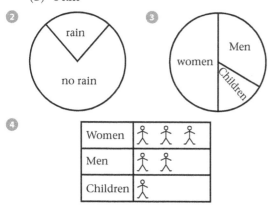

2.
3.
4.

5. (a) $\frac{3}{4}$ (b) 5%

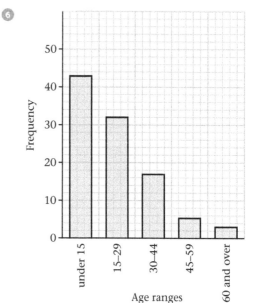

6.

7. Cislyn, Pearl, Rohan, Roland
   'is the sister of'

9. 1 → 1
   2 → 4
   3 → 9
   4 → 16
   5 → 25
   6 → 36
   7 → 49
   8 → 64
   9 → 81
   'is the square root of'

10. Robert, Andrew, Richard — tennis, football, cricket
    'likes playing'

Answers

Revision test 9 (page 209)

① C   ② B   ③ C   ④ B   ⑤ D

⑥

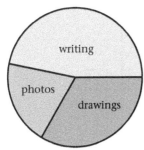

⑦

Drawings	□ □ □ □ □
Photographs	□ □ □
Writing	□ □ □ □ □ □ □

Revision exercise 10 (page 209)

① 12 litres
② 300
③ (a) $37\frac{1}{2}\%$ (b) 96   (c) 4 : 5
④ $37\frac{1}{2}$ refrigerators/hour
⑤ $65 637.60
⑥ $2134
⑦ allowances = $5500, tax = $950.40
⑧ (a) $13 832   (b) $17 472   (c) $3640
⑨ $1170
⑩ $192.48

Revision test 10 (page 210)

① B   ② C   ③ B   ④ D   ⑤ A
⑥ 2 h 24 min (2.4 h)
⑦ $200.80
⑧ 11.88 cm
⑨ (a) $2800   (b) $45 200   (c) $7560
⑩ (a) $255   (b) $1785   (c) $148.75

General revision test C (page 211)

① A   ② A   ③ C   ④ C   ⑤ D
⑥ B   ⑦ B   ⑧ C   ⑨ B   ⑩ C
⑪ $100
⑫ $a = 28°, b = 124°, c = 56°, d = 56°, e = 68°$
⑬ $a = 35, b = 110, c = 250$
⑭ $43.33

⑮ (a) 2
(b) 46%
(c)

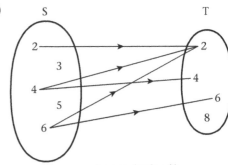

⑯

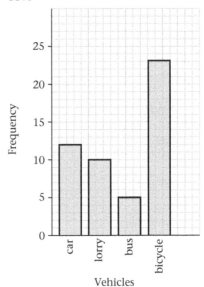

'is a multiple of '

⑱ (a) 'is three less than'
(b) 4 is not a member of set B

⑲ (a)

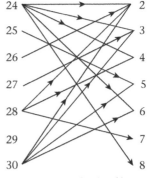

'is a multiple of '

(b)

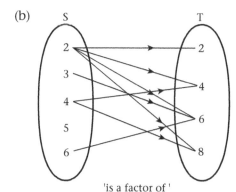

'is a factor of '

(c)

S   T

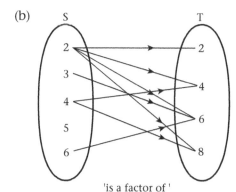

'is greater than'

Practice Examination
Paper 1 (page 213)

①	A	②	B	③	D	④	A	⑤	A
⑥	C	⑦	C	⑧	D	⑨	D	⑩	C
⑪	D	⑫	C	⑬	A	⑭	C	⑮	B
⑯	B	⑰	C	⑱	B	⑲	C	⑳	D
㉑	C	㉒	B	㉓	B	㉔	B	㉕	D
㉖	C	㉗	D	㉘	D	㉙	B	㉚	C

Paper 2 (page 215)

① (a) $2\frac{2}{3}$ (b) 19

② (a) (i)

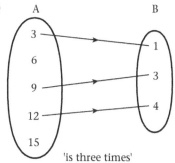

'is three times'

(ii) A is the domain; B is the range

(b) (i) 10:00 p.m.   (ii) 22:00 hrs

③ (a) (i) $3x + y + 10$ (ii) $3x - 2y + 17$
     (iii) $-\frac{7x}{12}$
  (b) (i) $7(a - 5)$ (ii) $3(x - 5y + 7)$
     (iii) $4x(x - 7y)$

④ (a) $296.40   (b) 20%

⑤ (a) (i), order 2   (iii) order 2
  (b) 9 m

⑥ $v = 45°$, $w = 128°$, $x = 37°$, $y = 98°$

⑦ (a) (i) $x = 2$ (ii) $x = -12$ (iii) $y = 1$
  (b) (i) $n + \frac{n}{3} = 24$ (ii) 18

⑧ 13.2 m

⑨

Rectangular faces	Triangular faces	Edges
3	2	9

⑩ (a) (i) 3   (ii) 15   (iii) 36
  (b) bus 80°
     motorcar 100°
     bicycle 150°
     walk 30°

(c)

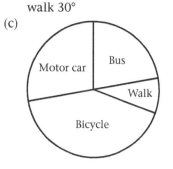

Index